Development with Identity

Community, Culture and Sustainability in the Andes

Development with Identity

Community, Culture and Sustainability in the Andes

Edited by

Robert E. Rhoades

University of Georgia

CABI Publishing

CABI Publishing is a division of CAB International

CABI Publishing
CAB International
Wallingford
Oxfordshire OX10 8DE
UK

Tel: +44 (0)1491 832111
Fax: +44 (0)1491 833508
E-mail: cabi@cabi.org
Website: www.cabi-publishing.org

CABI Publishing
875 Massachusetts Avenue
7th Floor
Cambridge, MA 02139
USA

Tel: +1 617 395 4056
Fax: +1 617 354 6875
E-mail: cabi-nao@cabi.org

A catalogue record for this book is available from the British Library, London, UK.

Library of Congress Cataloging-in-Publication Data

Development with identity : community, culture and sustainability in the Andes / edited by Robert E. Rhoades.
p. cm.
Includes bibliographical references and index.
ISBN-10: 0-85199-949-2 (alk. paper)
ISBN-13: 978-0-85199-949-4 (alk. paper)
1. Sustainable development--Ecuador--Cotacachi. 2. Sustainable agriculture--Ecuador--Cotacachi. 3. Land use--Ecuador--Cotacachi. I. Rhoades, Robert E. II. Title.

HC203.C68D48 2005
338.1′6′0986612--dc22

2005008120

Typeset by AMA DataSet Ltd, UK.
Printed and bound in the UK by Biddles Ltd, King's Lynn.

Contents

The colour plate section can be found following p. 102.

Contributors

Jenny Aragundy, *SANREM–Andes Project, Cuidadela Jardines del Pichincha, Pasaje B. N63-204, Quito, Ecuador; Tel: +593 9-781-4256; Email: JennyAragundy@web.de*

Mary García Bravo, *Heifer Project-Ecuador, Quito, Ecuador; Fax: +593 2-2501427 or +593 2-2556241; Email: marygarcia@heifer-ecuador.org*

Juana Camacho, *University of Georgia, Department of Anthropology, 250 Baldwin Hall, Athens, GA 30605, USA; Tel: +1 706-542-3922; Email: camachoj@uga.edu*

Florencia Campana, *Heifer Project-Ecuador, Quito, Ecuador; Fax: +593 2-2501427 or +593 2-2556241; Email: florenciacampana@heifer-ecuador.org*

B.C. Campbell, *University of Georgia, Department of Anthropology, 250 Baldwin Hall, Athens, GA 30605, USA; Tel: +1 706-542-3922; Email: eanthro@yahoo.com*

Ashley D. Carse, *University of North Carolina, Chapel Hill, Department of Anthropology; CB# 3115, 301 Alumni Building, Chapel Hill, NC 27599-3115, USA; Tel: +1 919-966-3160; Email: AshleyCarse@gmail.com*

William Deutsch, *Department of Fisheries and Allied Aquacultures, Auburn University, Auburn, AL 36849, USA; Tel: +1 334-844-4786; Fax: +1 334-844-9208*

Bryan L. Duncan, *Department of Fisheries and Allied Aquacultures, Auburn University, Auburn, AL 36849, USA; Tel: +1 334-844-4786; Fax: +1 334-844-9208; Email: bduncan@asesag.auburn.edu*

Edith Fernández-Baca, *Grupo Yanapai, Peru and Iowa State University, 107 Curtiss Hall, Ames, IA 50011, USA; Fax: +1 515-294-3180; Email: eferbaca@iastate.edu*

Cornelia B. Flora, *Iowa State University, 107 Curtiss Hall, Ames, IA 50011, USA; Fax: +1 515-294-3180; Email: cflora@iastate.edu*

Gabriela Flora, *American Friends Service Committee, Central Region Project Voice Organizer, 901 W. 14th Avenue, Suite #7, Denver, CO 80204, USA; Tel: +1 303-628-3464; Fax: +1 303-623-3492; Email: GFlora@afsc.org*

Jan L. Flora, *Iowa State University, 317 D. East Hall, Ames, IA 50011, USA; Fax: +1 515-294-0592; Email: floraj@iastate.edu*

Nicolás Gómez, SANREM-Andes, Cotacachi, Ecuador

Rafael Guitarra, *UNORCAC, Cotacachi, Ecuador; Tel: +593-06-916012; Email: unorcac@ecuanex.net.ec*

Auki Tituaña Males, *Municipio del Canton Cotacachi, Alcalde del Canton Cotacachi, Calle Pedro Moncayo entre Modesto Peñaherrera y García Moreno, Cotacachi, Ecuador; Email: alcalde@cotacachi.gov.ec*

William P. Miller, *University of Georgia, Department of Crop and Soil Science, 3107 Plant Science, Athens, GA 30602-7272, USA; Tel: +1 706-542-0896; Email: wmiller@uga.edu*

A. Shiloh Moates, *University of Georgia, Department of Anthropology, 250 Baldwin Hall, Athens, GA 30605, USA; Tel: +1 706-542-3922; Email: asmoates@uga.edu*

Virginia D. Nazarea, *University of Georgia, Department of Anthropology, Athens, GA 30602, USA; Tel: +1 706-542-3852; Email: vnazarea@uga.edu*

Lincoln Nolivos, *Universidad Central del Ecuador, Quito, Ecuador; Tel: +593 2-281-4048*

Marcia Peñafiel, *Alianza Jatun Sacha/CDC-Ecuador, Pasaje Eugenio de Santillán N 24-248 y Maurián, Quito, Ecuador; Tel: +593 2-243-2246; Email: mpenafiel@jatunsacha.org*

Maricel C. Piniero, *CATIE/NORAD, Casa No. 7, Avenida Libertad, Ciudad Flores, Peten, Guatemala; Email: mpiniero.catie.ac.cr*

Robert E. Rhoades, *University of Georgia, Department of Anthropology, 250 Baldwin Hall, Athens, GA 30605, USA; Tel: +1 706-542-1042; Email: rrhoades@uga.edu*

Fabián Rodríquez, *PO Box 17-10-7193, Quito, Ecuador; Tel: +593 2-330-0365; Email: fabian196@hotmail.com*

Sergio S. Ruiz-Córdova, *Department of Fisheries and Allied Aquacultures, Auburn University, Auburn, AL 36849, USA; Tel: +1 334-844-4786; Fax: +1 334-844-9208; Email: ruizcor@mail.auburn.edu*

Maria Claudia Segovia, *SEK International University, Department of Environmental Engineering, Campus Politécnico, Ecuador; Tel: +593 2-286-2427; Email: maclaudiasegovia@yahoo.com*

Kristine Skarbø, *Bygda, N-6200 Stranda, Norway; Tel: +47 97718299; Email: kristineskarbo@gmail.com*

Douglas Southgate, *Department of Agricultural, Environmental and Development Economics, The Ohio State University, 2120 Fyffe Road, Columbus, OH 43210, USA; Tel: +1 614-292-2432; Email: southgate.1@osu.edu*

Marco Tipán, *Direccion Nacional de Recursos Naturales, DINAREN, Av. Amazonas y Eloy Alfaro, Quito, Ecuador; Tel: +593 2-250-4753; Email: mepgiol@hotmail.com*

Karla Vásquez, *Universidad Central del Ecuador, Quito, Ecuador; Tel: +593 2-281-4048*

Xavier Zapata Ríos, *SANREM–Andes Project, PO Box 17-12-85, Quito, Ecuador; Tel: +593 9-781-4256 or +593 286-8578; Email: XavierZapata@web.de*

Franz Zehetner, *University of Georgia, Department of Crop and Soil Sciences, 3107 Plant Science, Athens, GA 30602-7272, USA; Tel: +1 706-542-0896; Email: FranzZehetner@web.de*

Foreword

When Dr Robert Rhoades visited the city of Cotacachi 7 years ago, we discussed at length his proposal for SANREM[1] to carry out a research project on agricultural sustainability and the management of natural resources in indigenous and peasant communities in the county's Andean area. The purpose of the study would be to contribute to the process of holistic development initiated on 10 August 1996, by residents of Cotacachi who belonged to the county's civil society organizations. We were extremely interested in Dr Rhoades' project, given that all proposed activities would be based on citizen participation, the central pillar of our Local Development and Democratic Participation model. In addition, the project would respect the process we had initiated, in its respect for the values of local cultures, as would the promise by SANREM's investigators to share with communities the results of their study in order to strengthen the development efforts in which we are engaged in Cotacachi.

In light of the research project's components, we decided to support Dr Rhoades' proposal and the institution he represents, the University of Georgia, because, unlike other research experiences the results of which are never shared with the communities involved, this experience guaranteed the involvement of social organizations and actors in the different stages of the SANREM programme's implementation.

Based on the kinds of cooperation described, professors and students from the USA, Ecuador and other countries came to Cotacachi to undertake their studies and to live with county residents. Through the years, we discussed a range of questions, developed friendships and implemented the research project.

In the year 2000, Cotacachi's multiethnic community declared itself an 'Ecological County' via a municipal ordinance intended to protect the environment and our cultures. This volume, with chapters by Dr Rhoades and his colleagues, represents the most complete and systematic synthesis available to date of our agricultural and natural resources, and will provide support to the local development process in which we are involved. We are pleased that Dr Rhoades has kept his promise to return to us the fruits of his research team's labours.

The title of the work, *Development with Identity: Community, Culture and Sustainability in the Andes*, provides guidance for the work carried out by men and women, today and in the

[1]SANREM CRSP is the Sustainable Agriculture and Natural Resource Management Collaborative Research Support Program.

future, in Ecuador and in other South American nations, tied to the conservation and preservation of mother nature, her biodiversity, land and water resources.

As the chapters demonstrate, Cotacachi County is located in the buffer zone of the Cotacachi-Cayapas Ecological Reserve, in the provinces of Imbabura and Esmeraldas. This protected area is one of the planet's most important treasures and must be preserved, not only for the citizens of our county, but for people throughout the world. Many decisions made in distant places affect the spectacular biodiversity of our county. One example of this is the phenomenon of global warming, the result of practices in industrialized countries, and another is the mining of minerals by rapacious, inhuman foreign companies.

Research by the SANREM team demonstrates the fragility of our natural environment and defines strategies for conserving the Earth's wealth. Our soils are rich in nutrients, but also vulnerable to erosion and depletion. By adopting the simple, low-cost solutions recommended by researchers, we can find ways to preserve these treasures. Water represents one of the most serious social problems in our county. Because of climate change and heightened demand, our communities, industries and services are competing with one another for access to this scarce resource. Cotacachi is looking for ways to conserve water, and the study on the monitoring, use and available sources of water will be a guide for current and future efforts in Cotacachi.

The research undertaken by SANREM demonstrates the usefulness and wisdom of our ancestral indigenous knowledge. While the studies included in these pages are based on the principles of Western science, researchers have not overlooked the importance of our millennial way of life and social organization.

Here in Cotacachi County, we believe that interaction among cultures and respect for cultural differences will result in true development. The SANREM study on the traditional stories of our elders, which includes the publication of a multilingual text that can be used in the classroom, will be very useful. The 'memory bank' and the 'Future Ancestors' Farm' have built a bridge between the past and the future, and will certainly contribute to the consolidation of our local development experiences.

The SANREM team has made an effort to learn about the importance of the *Pachamama* – our word for Mother Earth – and about our cosmovision, and thus, instead of relegating our ancient ways of relating to nature to the category of superstition, as have other foreign researchers, members of the team have made our customs and traditions and ways of being in nature a central part of their studies.

This work, which will be published in English and in Spanish, will be a body of knowledge that other scientists, technicians and local citizens and their leaders, will be able to consult in the coming years. The information contained in these pages, based on research in the field and in secondary resources, will contribute to the search for solutions to environmental and social problems. It will provide significant guidance in decision making and planning by social organizations and governmental institutions, including municipal government, in fields such as agriculture, ecotourism and the management of natural resources.

We thank the SANREM team for their efforts to return to us all the information shared with them by the citizens of Cotacachi and, in particular, by the Kichwa people, living on the flanks of Mount Mama Cotacachi. We would like to emphasize the importance of the county atlas developed by the technical team as part of this book, as information found in that volume will be incorporated into the County Natural Resources Management Plan.

We value highly the friendships we have made with Dr Robert Rhoades and his technical team here in Cotacachi, City of Peace. Our dream is that Cotacachi becomes a light that will shine on communities throughout the world in their struggle for a world of peace and solidarity, and in their efforts to care for our mother earth, the home of all peoples and cultures.

Auki Tituaña Males
Mayor of Cotacachi County

Acknowledgements

As is so often the case in science, serendipity decides how and where research will be carried out. Our work in Cotacachi was no exception. Quite by chance, in late June 1996, my wife, Virginia Nazarea, and I took a break from our research in Nanegal, where our first SANREM project was undertaken, and drove north to visit the famous market at Otavalo. Out of curiosity, we decided to drive a bit further north and westward across the Rio Ambi into the town of Cotacachi. For both of us, Cotacachi was love at first sight. The incredible beauty of the mountain setting combined with the colours and hues of the quaint houses in the town attracted us as had no other place in Ecuador. After wandering the largely empty streets, we arrived in the main plaza only to encounter what has to be one of the richest and most colourful ritual celebrations in the Andes: the 'dance' of *San Juan* (Quichua: *Inti Raymi*) by indigenous people who now were streaming from the mountain into the town, one community after another. Although I had lived more than a decade in the Andes, I had never witnessed such an animated and symbolically powerful event. This indigenous ceremony, which has pre-hispanic roots, represents the annual ethnic re-charging for more than 18,000 *indigenas* who live in more than 40 communities around the base of the sacred mountain Mama Cotacachi. We went away from Cotacachi that day with the distinct feeling that this community was unique and that we must return to experience more.

Later, we learned that one of our SANREM researchers in the frontier area of Nanegal, Mr Segundo Andrango, is a native of Cotacachi. He arranged for us to return later that year and meet with the leaders of the indigenous organization, UNORCAC, and with the newly elected indigenous Mayor, Auki Tituaña Males. In these exploratory meetings, we explained the purposes of SANREM and they informed us of their interests and needs. We were invited to present research proposals to the various leaders and community assemblies. This was the beginning of a collaborative effort that has lasted until the present. By mid-1997, the first SANREM researchers arrived and began establishing ties and friendship with our future partners.

In the course of the past 7 years, we have acquired an enormous debt to many individuals and organizations in Ecuador and the USA. The kind and generous people of Cotacachi gave us countless uncompensated hours to help us to understand the complexity and dynamics of their mountain home and the importance of their cultures. I am sure they wished many times that the *preguntónes* (big question-askers) would go away and give them some peace and quiet. We can never repay them for their time and kindness in letting us visit their homes, fields and social activities. This book is dedicated to the Cotacacheños as a small token of great thanks.

We also wish to express our gratitude to all the US and Ecuadorian institutions which have supported this research. It is never easy to cut through the bureaucratic 'red tape', budget uncertainties, language differences and diverse needs to make a large project like SANREM succeed. Research today in large interdisciplinary teams from multiple institutions working with diverse partners and stakeholders is very different from that of an earlier age when only a few disciplines and stakeholders were involved. Our research was supported by the Office of Agriculture, Bureau for Economic Growth, Agriculture and Trade, USAID through the Sustainable Agriculture and Natural Resource Support Program (SANREM CRSP) under the terms of Cooperative Agreement Number PCE-A-00-98-00019-00. We wish to thank Christine Bergmark and Robert Hedlund of USAID-Washington for their guidance. The grant was administered by the Office of International Agriculture, University of Georgia. Edward Kanemasu, Director, always encouraged us even when times were difficult. From the SANREM Management Entity, we are grateful to Directors Carlos Perez and his predecessor, the late Bob Hart, for their leadership and vision. Constance Neely and Carla Roncoli who served as Deputy Director at different times were always supportive and the best of friends and colleagues. Rex Forehand, Steven Beach, Diana Shelnutt, Sandy Gary and Natalie Gude of the UGA's Institute for Behavioral Research provided administrative support and accounting, always with a sense of humour. In Ecuador, we were supported through the USAID-Ecuador office by Jill Kelly, Natural Resource Officer, and her colleague, Monica Sukilanda. At the Catholic University-Quito, we are appreciative of the support of Nelson Gómez, Juan Hidalgo, Olga Mayorca and Monsarrath Mejía of the Department of Geography. Hernán Velásquez of the Ministry of Agriculture and Galo Rosales of the Ministry of Environment gave us guidance and backing from their respective agencies. Special thanks goes to Susana Cabeza de Vaca, Executive Director, and her staff of the Ecuador Fulbright Commission for their warm welcome during the period when I was also a Fulbright Fellow. Fernando Larrea of Heifer Project-Ecuador provided logistics to our research on social capital and institutions.

Many indigenous community and UNORCAC leaders were instrumental in making our research happen. They opened doors and gave us insights as no one else could. Without their support, SANREM–Andes would have never succeeded. Rafael Guitarra, President of UNORCAC, and Magdelena Fueres of Jambi Mascaric were especially critical for linking with the communities. Cornelio Orbes, former President of UNORCAC, always took up our cause with good humour and kind words. Others who always stood behind us and helped guide us through uncharted social waters were Alfonso and Segundo Morales and Inéz Rodriguez. We would be amiss if we did not also give thanks to the parents of the 'memory banking students' for their extra commitment to the recovery of traditional Andean crops of Cotacachi.

Special appreciation must be given to our field assistants who kept our research going even when the principal investigators were absent for months at a time. Carlos Guitarra, Rosita Ramos and Nicolás Gómez, all of UNORCAC, served as links between our projects and local communities. Carlos and Rosita also spent many hours translating from Spanish to Quichua. Nicolás was always the loyal assistant ready to drop anything he was doing to take us to the far ends of the canton. A number of SANREM field coordinators provided support to the project over the years: Eric Jones, Maricel Piniero, Natalia Parra, Shiloh Moates and Xavier Zapata. Each served admirably under less than ideal working conditions. Others who assisted with field research and activities for special projects were Rocio Alarcón, Mika Cohen and Stella Lima.

This book and other SANREM–Andes publications would not have been possible without the dedication of our editors. Tim Hardwick of CAB International was patient and helpful even as we missed several deadlines. Anabel Castillo, Executive Editor of Abya Yala

Press in Quito, has gone beyond the call of duty to see our various monographs to publication. Mary Ellen Fieweger edited and translated many of the manuscripts in this volume. Robbie Mixon and Milan Shrestha helped in processing early drafts of this volume. I owe a deep gratitude to Danila Rhoades who joined the editing process in the final critical stages. Without her hard work, keen eyes and organization skills, I doubt that this volume would have seen the light of day.

The institutions and individuals recognized in this acknowledgement are not responsible for any erroneous conclusions, misrepresentation of facts or inappropriate judgements found in this book. The authors are responsible for their own chapters and validity of the findings and conclusions. We welcome criticism and re-interpretations of our research. If we are wrong, please let us know where, how and why. Cotacachi is a complex and dynamic corner of the Andean world, and sustainable development is still an idea in progress. Only through open and honest debate and questioning of basic assumptions will sustainability scientists – and society – make progress towards the universal human dream of a sustainable future.

Robert E. Rhoades on behalf of the SANREM–Andes Team
Professor of Anthropology
University of Georgia
Athens, Georgia

1 Linking Sustainability Science, Community and Culture: a Research Partnership in Cotacachi, Ecuador

Robert E. Rhoades
University of Georgia, Department of Anthropology, Athens, GA 30605, USA

Introduction

This volume is a contribution to understanding the intersection of two emerging concerns in international development: sustainability and self-determination of indigenous communities. The common goal of combining these themes is to achieve sustainable development's challenge of 'meeting fundamental human needs while preserving life-support systems of planet earth' (Kates *et al.*, 2001, p. 641; http://sustainabilityscience.org). Sustainability science seeks a place-based understanding of nature–society interactions through an interdisciplinary research framework that integrates global and local perspectives (Obasi, 2002, p. 10). It also aims to link rigorous scientific method and knowledge with social learning and action by concerned decision makers. The corollary, therefore, of the science is an emphasis on full participation of multiple stakeholders in both the research and developmental outcomes, especially participation by local communities which will live by the consequences of programmatic or policy decisions. This means that a sustainability science project's research questions, design and implementation should respect open, democratic involvement of relevant stakeholders from problem diagnosis to action.

Participatory approaches reverse the values and methods of conventional science wherein researchers and development practitioners comfortably pursue management of agriculture and natural resources with limited or no input from local communities (Chambers, 1997). The Green Revolution is the most famous case of the former science-driven transfer of technology approach. In this previous paradigm, agricultural researchers largely defined the problem from afar, farm households became passive recipients of scientific products and social scientists were assigned to do mop-up *ex post facto* analysis on how farmers reacted to introduced technologies. Under the new sustainability paradigm, however, de-contextualized, top-down planning for people was no longer acceptable. A sustainable future, and how to get there, must account for local values, perceptions and capabilities, and not just what outsiders or distant policy makers assume would be desirable. Ultimately, the pursuit of sustainability is a local undertaking not only because each community is ecologically and culturally unique but also because its citizens have specific place-based needs and requirements.

Socially conscious development practitioners and field scientists working directly with communities, especially indigenous groups, had realized well before the 1992 Rio Earth Summit and the publication of *Agenda 21: Program of Action for Sustainable Development* (United Nations Conference on Environment and Development, 1992) that a new way of doing development work was in order (World Commission on Environment and Development (1987)). Since the mid-1970s, rural communities around the world have grown weary of development experts, whether national or foreign, who arrive in their towns and villages with both the problem and the solution already defined before consulting with local inhabitants. Among the early proposals for reversing development approaches were the Farmer-Back-to-Farmer model for technology generation in which research must begin and end with user needs (Rhoades and Booth, 1982), and subsequently the more general and popular Farmer-First philosophy (Chambers and Ghildyal, 1987). In these early formulations, as well as the whole participatory movement that followed, science and development could no longer proceed according to their own principles but had to take local values, beliefs and needs into consideration.

Complementing these new participatory trends has been the global indigenous rights movement in which ethnic minorities and tribal populations have organized socially and politically to demand greater rights, access and sovereignty over their ancestral land, knowledge and resources (Gray, 1997; Warren and Jackson, 2002). As a result of community resistance to outside agendas, scientists and development practitioners increasingly found themselves rebuked, ignored, shut out and sometimes physically ejected by the very people who were to benefit from their efforts. Conversely, those outsiders who explicitly recognized local values and knowledge and worked collaboratively with indigenous leaders and communities found that local doors opened and created space for even better science (Rhoades, 2001). The difference between the former way of doing development research and the new way has been labelled by Waters-Bayer (1994) as 'extractive research' versus 'enriching research' (see also Nigh, 2002). Extractive research aims to provide support and information to development agencies, non-governmental organizations (NGOs) or the academic community, while enriching research operates within a collaborative framework wherein local people's values and priorities are addressed and the research helps them deal with the outside world to achieve their own culturally defined goals.

Achieving enriching research with indigenous people, however, is not a readily understood process by the new genre of sustainability scientists (Rhoades, Chapter 21, this volume). Agenda 21-inspired research is far more complex and difficult than the previous transfer of technology model. It is no longer possible for scientists to design a 'solution' or technologies in the laboratory or on the experiment station and send them down to the extension agent who in turn delivers them to the farmer. The new paradigm of sustainable development requires long-term research in interdisciplinary teams of biological and social scientists examining multiple scale environments (landscapes, watersheds, catchments and ecoregions) inhabited by diverse stakeholders with different values. Additionally, many scientists are uncomfortable with the notion that their research needs to be locally relevant and approved by community leaders or assemblies.

This volume presents findings and synthesis from the Sustainable Agriculture and Natural Resource Management (SANREM) Andean research partnership conducted jointly with the Quichua-speaking people of Cotacachi, Ecuador (see Fig. 1.1, map of the area).[1]

Since 1997, the SANREM–Andes team has been working with the indigenous organization, Union of Campesino and Indigenous Organizations of Cotacachi (UNORCAC) and the Cotacachi cantonal government to provide research findings to help make better informed decisions about the management of natural resources and agriculture. The study area is located in

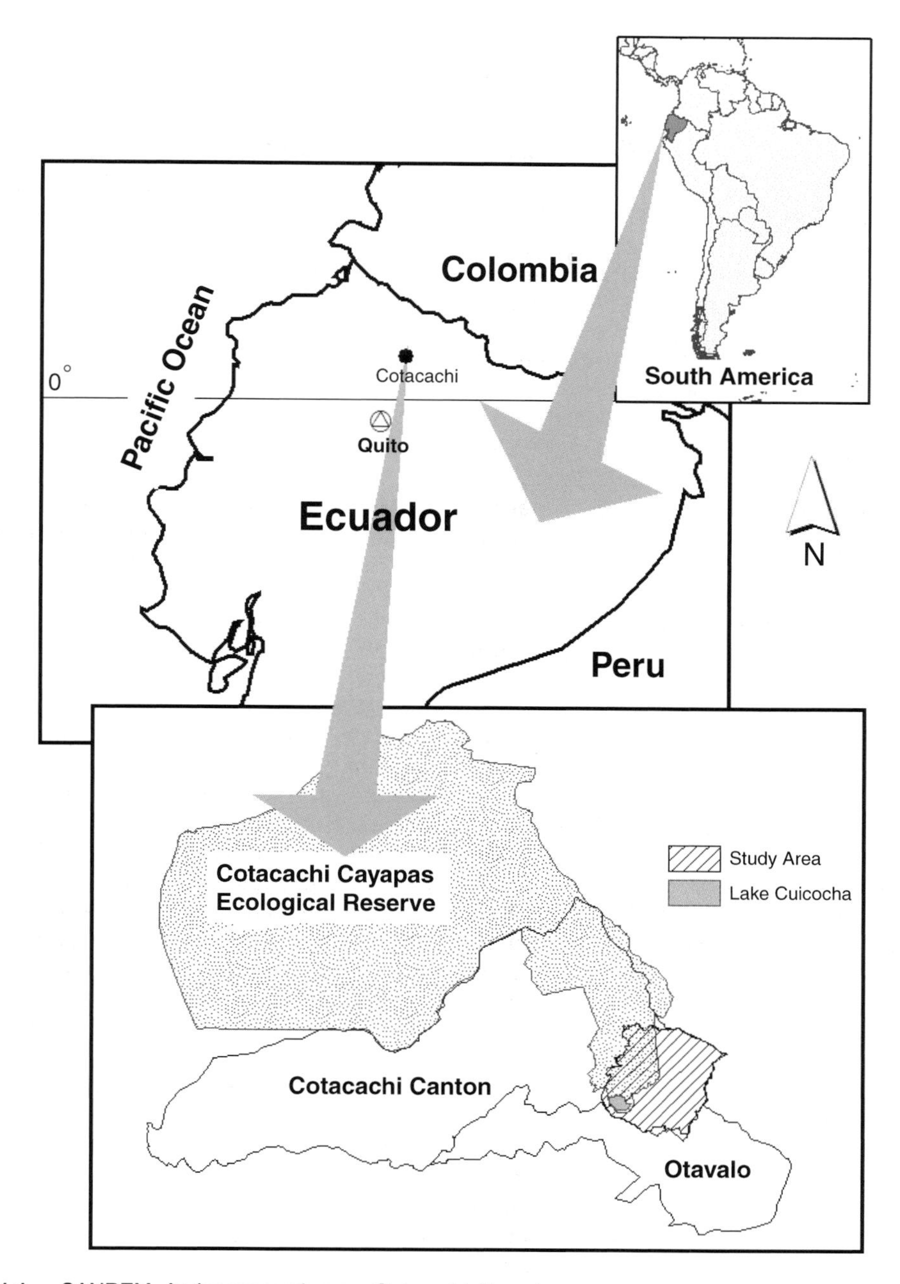

Fig. 1.1. SANREM–Andes research area: Cotacachi, Ecuador.

the highland area just north of the equator in the eastern part of Cotacachi canton where approximately 18,000 indigenous people live in 40 communities (*comunas*) distributed around the 'skirt' (*falda*) of *Mama Cotacachi*, a 4993 masl (metres above sea level) volcano that dominates the landscape and local cosmology (Fig. 1.2).

The indigenous communities are sandwiched between a growing urban zone around the town of Santa Ana de Cotacachi (population: 7500) and the Cotacachi-Cayapas Ecological Reserve. This national ecological

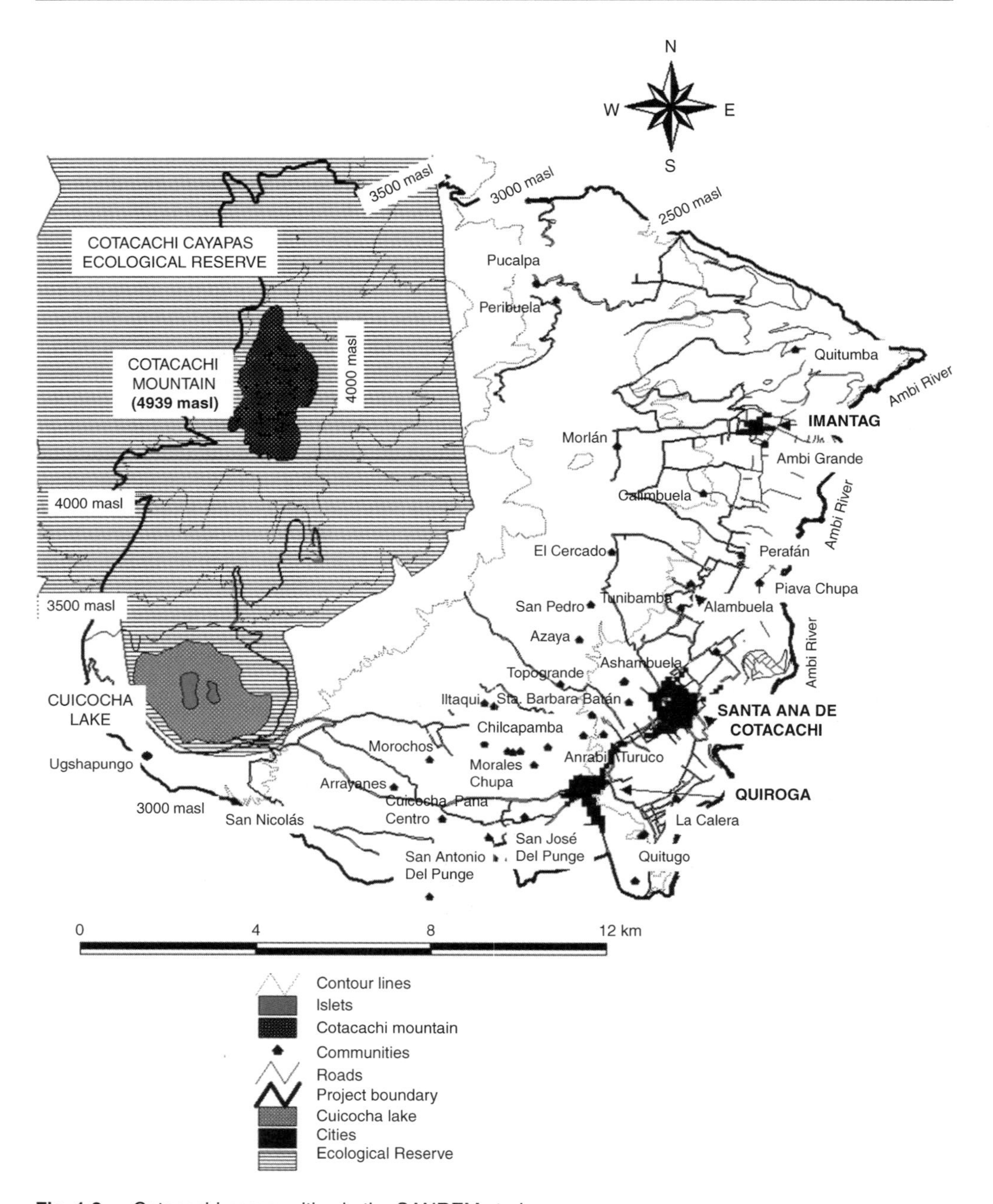

Fig. 1.2. Cotacachi communities in the SANREM study area.

reserve was established by executive order in 1968, covers 204,420 ha and extends downward from alpine ecosystems in the western Andean cordillera to the western humid tropical lowland forests not far from the Pacific coast (< 500 masl). This region is considered one of the world's 'hotspots' characterized by an extraordinarily high number of species per unit area (Alarcón, 2001). The reserve contains vast extensions of contiguous forested areas, while in the adjacent buffer zones the primary forest and associated species are rapidly disappearing. Inside the reserve are hundreds of critical watersheds supporting dozens of endangered species of mammals and birds, including the spectacle bear, jaguar, ocelot, mountain tapirs, monkeys, plate-billed mountain toucan and the endangered Andean condor (Rhoades, 2001).

As a globally defined research activity within the spirit of Agenda 21, SANREM faced the dual challenge of addressing broader questions related to society–environment interactions and impact while at the same time making sure that our efforts were relevant to the Cotacachi communities serving as our hosts during the project's existence. To help the reader appreciate this shifting of scale in our concepts and activities, this introductory chapter will provide a basic grasp of how global processes and interests connect with the local. First, I will place our work within the broader context of the Andean ecoregion and the mountain challenges of sustainable development as outlined by Chapter 13, ('Managing Fragile Mountain Ecosystems') of Agenda 21. Second, I will provide a brief overview of the place, people and development philosophy of highland Cotacachi, Ecuador. Finally, the goals and methods of the SANREM partnership programme will be outlined along with a brief discussion of the overall organization of this book.

The Andes and Sustainable Development: an Agenda 21 Challenge

The rationale for the SANREM–Andes research resides in an interest in discovering and promoting reliable information and decision-support tools for sustainable development in mountain regions of the world. As a case study of sustainability science in a region of the northern Andes, the authors make a direct contribution to the global mountain initiative which emerged from the 1992 Earth Summit in Rio (Messerli and Ives, 1997). Mountains and uplands represent the world's most diverse and fragile ecosystems, cover about 20% of the world's terrestrial surface, and are distributed across all continents and major ecoregions (Price, 1998). While about 10% of the world's population lives in these environments, mountains provide important economic resources (e.g. food, wood, water and minerals) for more than half of the world's population residing in adjacent lowlands. Mountains support the world's 'water towers' situated at the upper ends of the earth's river catchments, providing water, nutrients and energy to communities living downstream. In addition, mountains are crucial for global ecosystem functioning due to their important 'biodiversity' reserves of wild and domesticated plants and animals. Due to their historical isolation and difficult terrain, mountains also harbour many of the remaining indigenous peoples of the world who, at present, are undergoing rapid economic and social change. Mountains are also important for their great spiritual, aesthetic and tourism resources (Denniston, 1995).

The Andes, with its concentrations of biodiversity, major watersheds of global importance and critical levels of land degradation and rural poverty, is a significant world region where environment–society interactions need to be understood and addressed. Along with the Hindu Kush Himalaya of Asia, the Andes of South America can claim the largest, most diverse, and – by most measures – the most economically and ecologically important mountain setting in the world (Rhoades, 1997). Traversing this stunning 2000 km long landscape of glacial peaks, gorges, forests and human settlements are deep gorge river watersheds which feed the great Amazon Basin to the east and the coastal littorals and lowlands of the western Pacific coast. No other landscape on earth is characterized by so much biotic and geomorphological diversity in such a short distance as the Andean 'highland–lowland' interaction system. As in all mountain regions, the Andes are characterized by a 'three dimensionality' of latitudinality, horizontality and altitudinality which has the effect of producing contrasting environments at different elevations (Troll, 1968). Superimposed on this altitudinal zonation, moreover, are natural variations and human adaptive strategies that derive from aspect, slope and topography of the region. Despite economic progress in urban areas, some of the highest malnutrition and poverty rates in the world are found in the rural Andes. Out of 178 ecoregions in Latin America identified by the World Wildlife Fund and the World Bank (1995), 137 are listed as 'critical', 'endangered' or 'vulnerable'.

The sustainability of this Andean landscape and lifescape must be understood against the backdrop of these unparalleled environmental contrasts between the lowlands and the high zones. Carl Troll (1968), the great mountain geoecologist, noted: 'Nowhere in the world have I seen a more striking example of climato-ecological differentiation than in these Andean valleys.' The Andes rise from an arid coast (Chile, Bolivia and Peru) and tropical montane coastlines (Ecuador) on the west to volcanic and glaciated massifs well over 5500 m high and then abruptly drop to less than 100 m into the tropical Amazon Basin (Ecuador, Peru and Bolivia). The distance traversed in the narrowest distance is approximately 200 km in the Ecuadorian Andes while across the *altiplano* of Peru and Bolivia the distance is 500 km. Due to tectonic processes over millions of years, an unstable landscape has evolved which is unsurpassed in variety and complexity (Zehetner and Miller, Chapter 2, this volume). The climates range from the driest deserts on earth to the wettest rainforest jungles.

The hydrological resources of the Andes are virtually unique in the world in that they can be harnessed along most points of their rapid 4000–5000 masl descent. Traversing both the Pacific and Amazonia slopes of the Andes are > 100 major river systems which ecologically link the highlands and the lowlands. The rains, which are irregularly distributed over the year, are the main source of water, although rapidly disappearing mountain glaciers remain important to local communities. The steepness of gradient, the shortness of most rivers (between 100 and 60 km in length) and an abrupt drop of > 4000 masl make it difficult to impound water into lakes. Winter rains often bring rapid runoff, thereby creating torrential rivers which frequently cause devastation in the valleys below. Deforestation, road building, erosion from open pit mining and other human activities along the rivers have contributed to problems downstream. Due to locally and globally driven climate change and diversion of water for urban uses, the shortage of water and rainfall is becoming acute in many regions of the Andes (Rhoades *et al.*, Chapter 5, this volume). If the hydrological cycle is allowed further disruption, biodiversity and other natural resource conservation projects may be waging losing battles (Stadel, 1991).

The natural complexity of the Andes has given rise to equally complex and layered lifescapes of human settlements, cultures and economic systems which are interdependent on each other due to the need to exchange labour, food, goods and other resources between zones. Humans have lived in this landscape for at least 15,000 years, first as hunters and gatherers, and as agro-pastoralists over the past 4000 years. Human adaptation to the diverse Andean agroecological zones has resulted in vertical arrangements of production regimes, population movements and human settlements (Moates and Campbell, Chapter 3, this volume). The highlands are mainly populated by indigenous Quechua- (Quichua in northern Ecuador) speaking populations, the lower slopes and the coast are typically inhabited mainly by mestizo and, to a lesser degree, African-descended populations, while the more isolated tropical jungle areas on both the east and west are inhabited by other Amerindian groups who practice horticulture, fishing and hunting. Ethnohistorically, the landscape–lifescape is a direct outcome of the application of an indigenous set of subsistence technologies to the vertical landscape (Rhoades and Thompson, 1975). These technologies involve locally derived plants and animals adapted to different climate and biotic belts, agricultural techniques, settlement patterns and exchange between areas of diverse production and dispersed communities (Brush, 1982).

The rapid entrance of the Andean region into the global economy over the past 50 years has dramatically altered the landscape–lifescape of both eastern- and western-facing watersheds. Today, a mosaic of agricultural systems is found, involving combinations of subsistence strategies and production for national and, increasingly, international export. At all elevations, a mixture of large-, medium- and small-scale farming operations are embedded in an increasingly regionalized and internationalized market economy often based on 'bust and boom' plantation crops. The integration of rural communities into centralized national systems and towards increasingly urban lifestyles has been brought about by the development of improved communication, education and transportation. Andean populations have become even more mobile in response to new employment, a process made easier by the expanding road networks. In addition to seasonal migration for work and land, there is now a more permanent migration towards the urban centres and the coastal lowlands (Flora, Chapter 18, this volume). The rural area supplies growing urban areas with food, raw materials, labour and even rural capital from property rents, commercial ventures and purchase of urban consumer goods. In general, the rural areas are becoming more impoverished and dependent on urban regions while continuing to overexploit the declining resources of their rural base. Increasingly, urban pollution, mining discharges, deforestation and agroindustrial ventures in the highland areas are beginning to have negative impacts along the mountain river systems (Dollfus, 1982).

The challenge of sustainable development in the Andes is as complex as the mountain environment itself. In addition to the tremendous variability of human cultures across the region, Andean peoples have been at the forefront of indigenous self-representation and political mobilization for autonomous development. Throughout the region, from Chile to Colombia, indigenous groups have organized economic blockades, helped topple national governments, pressed for legal rights in national and international courts, and taken the reins of development in their own locales. Andean communities are in many respects the leaders of the global indigenous movement through the creation of self-help organizational structures (such as UNORCAC in Cotacachi). They have been quick to organize resistance to outside interventions that do not consider their needs, and have taken steps to revitalize traditional Andean culture. While the native communities of the Andean region as a whole are mobilizing, the Quichua-speaking people of northern Ecuador – especially in Cotacachi – are spearheading many of these changes. In the case of the SANREM research project, it was not only a matter of scientists deciding that participatory research was the right thing to do, but it simply was the only option if we wanted to work in the region (see Rhoades, Chapter 21, this volume). The Cotacacheños have taken their destinies in their own hands without turning insular. This is the essence of their belief in 'development with identity', a theme we will explore in depth throughout this volume.

Sustainable Development in Cotacachi, Ecuador

The welcoming sign shown in Fig. 1.3 sits at the entrance of Cotacachi and is both a political and cultural statement to the outside world. Revelling in the area's cultural richness, the Spanish sign declares (my translation):

> Cotacachi
> Land of the Sun
> Living Cultures
> for Development

Cotacachi's Andean territory is a setting in which indigenous people are increasingly defining the local 'rules of the game' for globally initiated development agendas. Seemingly indifferent to or unaware of the North–South academic debates over whether development carries negative impacts for

Fig. 1.3. Welcome sign at the entrance of Cotacachi (photo: Robert E. Rhoades).

indigenous peoples (Escobar, 1995; Bebbington, 2000), the Quichua communities of Cotacachi have during the past decade overwhelmingly embraced their own conception of development as a worthwhile goal. Today, Cotacachi is a 'hotbed' of development activity involving UNORCAC, the cantonal government, dozens of NGOs, and international and national government projects, including some multimillion dollar initiatives, engaged in various forms of directed change or programme assistance targeted specifically at indigenous communities. The focus is on education, health and nutrition, rural infrastructure, income generation projects, tourism, agriculture or natural resource management.

This international attention, however, stands in stark contrast to a legacy of brutal social neglect and disenfranchisement from Ecuadorian society which has characterized the conditions of indigenous people in Cotacachi over most of the past five centuries. As late as the 1960s and 1970s, Cotacacheños existed as little more than 20th-century serfs in the feudalistic hacienda legacy of colonial and national Ecuador. Largely disenfranchised and economically dependent on the haciendas, Indians survived under a system called *huasipungo* or debt in servitude to the white mestizo landlords or *hacendados.* In exchange for labour on the hacienda they were paid no wages but were given access to small parcels of land (*huasipungo* means doorgardens in Quichua) and other amenities such as firewood and water on the hacienda. Typically, the peasants ended up in debt to the hacendado with little chance to escape their lot in life. Although a subjugated people, Cotacacheños maintained their traditional culture while resisting assimilation through their customs, communal society and a rich oral history (see Nazarea *et al.*, Chapter 6, this volume). When the *huasipungo* system was abandoned as a result of broader economic structural changes and a partial land reform in 1963 and again in 1974, the Quichuas began to revitalize their traditional communities (*comunas*) while fighting for land titles, communal houses and bilingual schools.

Following the 1977 assassination of a charismatic indigenous leader, Rafael Perugachi, by local police, local indigenous intellectuals organized most of the rural communities into a 'second degree' organization called UNORCAC (Unión de Organizaciones Campesinas Indigénas de Cotacachi or Union of Peasant and Indigenous Organizations of Cotacachi). In Ecuador, 'first degree' refers to the community level (*comunas*), 'second degree' to a voluntary grouping of local communities and 'third degree' to national federations of indigenous organizations. Initially, UNORCAC was established to fight for civil rights against widespread racism in Ecuador, but by the 1980s the interest of the organization had turned to fighting illiteracy and building infrastructure and services. In 1978, after being frustrated by rejections for civil infrastructure support, indigenous people physically occupied the municipal offices of the cantonal seat of government and thus ushered in a new social era for Cotacachi. By the early 1990s, the indigenous organization's agenda had come to focus on development and gaining their own access to and control over projects and funds. Increasingly, development has blended with political interests and the indigenous movement at the local, national and international levels (see Flora *et al.*, Chapter 19, this volume). Today, virtually the entire budget of UNORCAC is derived from funds of international development projects, often mediated by Ecuadorian NGOs who are linked with international donors and agencies.

The success of the indigenous movement has been so marked that today the canton is led by a three-time elected (1996, 2000 and 2004) indigenous Mayor – Auki Tituaña Males – who has implemented a number of reforms in government which draw strongly on explicitly indigenous forms of social organization and values (Fig. 1.4). Mayor Tituaña is a member of the political party *Pachakutik* which means 're-awakening' in Quichua. Pachakutik has enjoyed considerable success in mobilizing indigenous people and mestizo sympathizers throughout Andean Ecuador. The Mayor has replicated at the cantonal level the idea of an assembly (*asemblea*) which involves all citizens of Cotacachi in developing, voting on and executing collective action projects in health, education and environment (Ortiz Crespo, 2004). The success of this effort, at least in outside image, has earned the Mayor a national reputation and two international awards (Dubai Award for democratization and UNESCO 'Peace City' Award). Recently, the canton has declared itself an 'Ecological Canton' indicating the creation of policies and actions to protect the region's environment. Whether or not Cotacachi can become the 'model' for alternative development in the developing world, it has unabashedly embraced development and conservation as its own, but with a strong indigenous flavour.

The SANREM Project: Global Science Meets Local Participation

The Sustainable Agriculture and Natural Resource Management Project (SANREM) was established by the US Congress in 1992 as a commitment to fulfilling Agenda 21 requirements of signatory nations. As a

Fig. 1.4. Auki Tituaña Males, indigenous Mayor of Cotacachi, receiving a tribute from communities during his 2000 inauguration (Photo: Robert E. Rhoades).

Collaborative Research Support Program (CRSP), the programme engaged US and host country university researchers with NGO partners and local communities 'to advance the principles, methods and research, and collaborative breakthroughs for a new paradigm of sustainable development' (National Research Council, 1991). SANREM was mandated to 'think globally' and 'act locally' through basic and applied field research in representative agroecological sites involving the full participation of local people and regional decision makers. The lessons learned in the individual sites would be 'extrapolated' or 'scaled up' through sharing information, methodologies, technologies and especially 'decision support tools' for making better agricultural and environmental decisions in the future. Our home institution, the University of Georgia, was selected to lead this 'global partnership' of US, national and international agencies, and local communities. Three representative landscapes were selected for long-term research: a tropical watershed in Mindanao, Philippines (Coxhead and Shively, 2005); semi-arid landscapes in Mali (Moore, 2005); and two microregions (Nanegal and Cotacachi) within the Andean buffer zone of the Cotacachi-Cayapas Ecological Reserve of northwestern Ecuador (see Rhoades, 2001). The purpose of the SANREM–Andes partnership, in keeping with the core questions of sustainability science, is to research the dynamic interactions between nature and society so as to provide insights, planning tools and information for better natural resource management for mountain agroecosystems (Kates *et al.*, 2001).

Landscape–lifescape research framework: SANREM–Andes

In line with Agenda 21, the global goal of SANREM was to develop principles and methodologies for long-term sustainable ecosystem management of multizonal and multiscale units such as landscapes, watersheds, catchments, river basins or ecoregions where multiple stakeholders utilized and often competed for the same natural resources. Figure 1.5 depicts the general

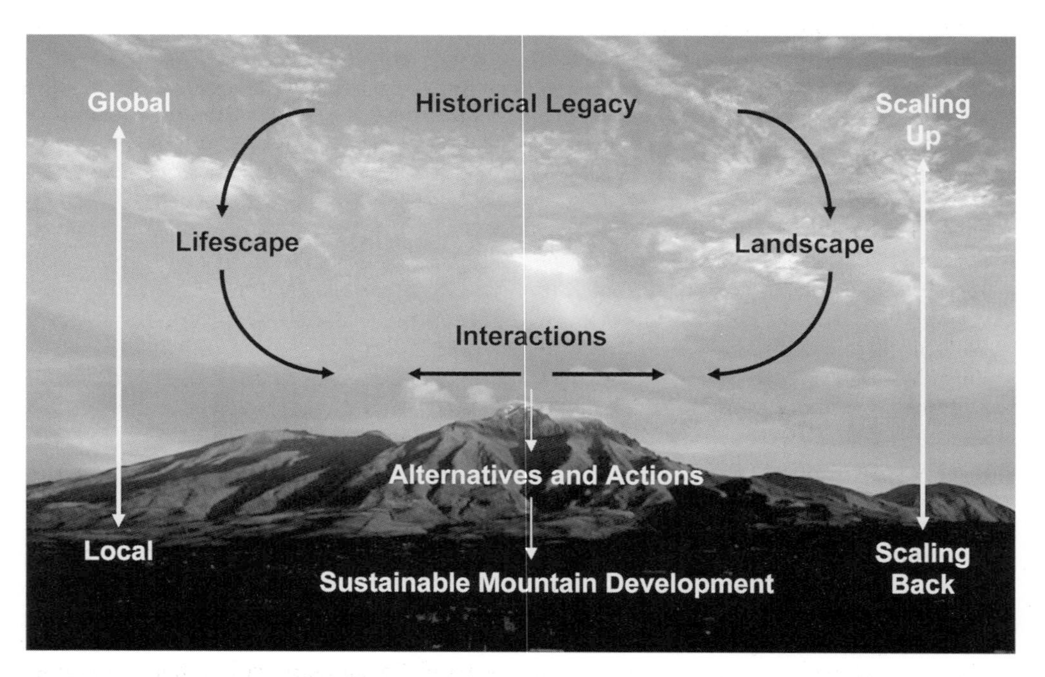

Fig. 1.5. The SANREM–Andes research framework.

SANREM–Andes research framework with an emphasis on linkages between social (lifescape) and environmental (landscape) dimensions. The framework begins with an appreciation and understanding of the historical legacy of the region (see various authors, Chapters 2–5, this volume). By focusing beyond the individual farm household and field plot, SANREM adopted as its research framework the ecological concept of landscape to describe and understand the complex, interactive processes within and between individual ecosystems of a toposequence transecting two or more agricultural zones (Rhoades, 2001). Landscape was, however, understood as more than just topography across which animals, plants, soils, water and other materials moved, but also as the dynamic, interconnected spatial patterns of biological and physical processes. Coupled with and complementing the natural science concept of landscape, social scientists added the notion of lifescape which included the economic, political, cultural and social aspects. Lifescape means the human dimension relative to the spatial template and involves how the natural world is perceived and acted upon by local people. Thus, a lifescape can be visualized as the superimposition of human intentions, purposes and viewpoint over the landscape. The landscape, or what is out there before us, is processed through human perception, cognition and decision making before a plan or strategy is formulated and an individual or collective action is executed (Nazarea, 1999, p. 1). As the research framework shows, however, our research did not focus narrowly on social–biophysical interactions within the Cotacachi research site but on how global processes impacted local activities and vice versa. Similarly, we looked for ways in which our research findings could be scaled-up (e.g. principles or lessons learned for sustainability in other mountain regions) as well as how experiences elsewhere might be applied or scaled-back to Cotacachi. Finally, through our analysis of lifescape–landscape system interactions, we actively sought with our local partners practical alternatives and action to solve real environmental and livelihood problems.

Andean integrated research methodology

The SANREM–Andes team was interdisciplinary and international, involving US and Ecuadorian scientists, NGO practitioners, governmental officers and local partners. Our SANREM team was made up of a consortium of four US universities (University of Georgia, Iowa State University, Ohio State University and Auburn

team from Ohio State University and the Antisana Foundation (PIs: Doug Southgate and Fabian Rodriguez) researched issues surrounding human behaviour and the allocation of water, an increasingly scarce resource in Cotacachi (Rodriguez and Southgate, Chapter 15, this volume). Anthropologists Robert E. Rhoades and Virginia Nazarea from the University of Georgia and their graduate students conducted numerous studies of history, society and culture related to natural resources (Moates and Campbell, Chapter 3; Skarbø, Chapter 9; Piniero, Chapter 10; Camacho, Chapter 11; Campbell, Chapter 17; Carse, Chapter 8; Flora *et al.*, Chapter 19, this volume).

Scientists did not exist in isolation, but worked in tandem with other disciplines and local partners to examine system

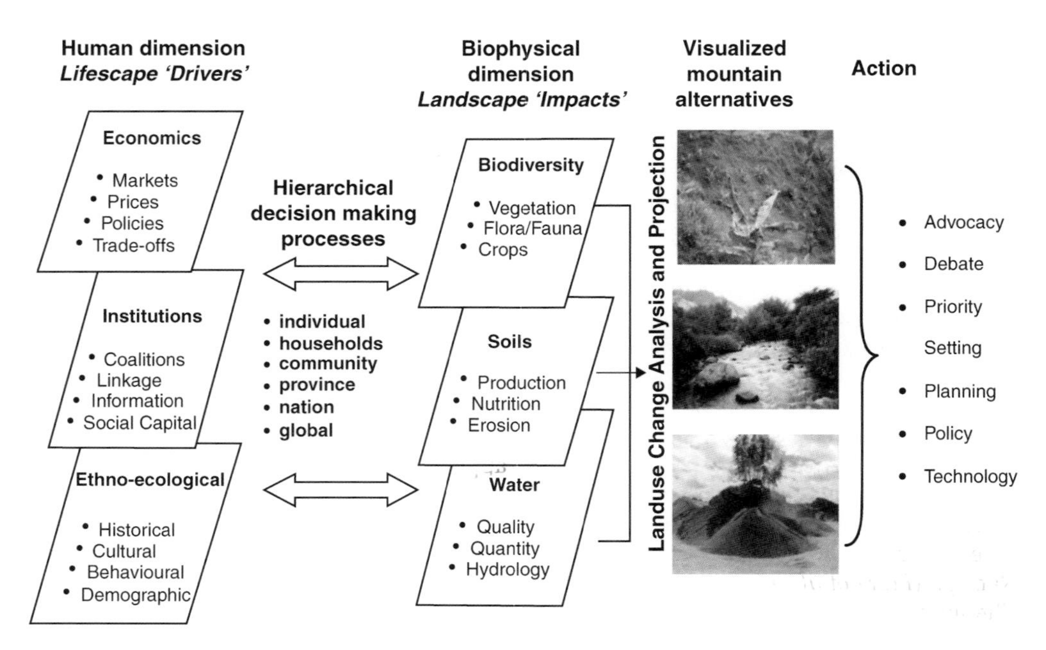

Fig. 1.6. Integrated research methodology: sustainable mountain development (Photos: Robert E. Rhoades).

interactions and multiple environmental trajectories and outcomes. Such collaboration in the field was the only means to achieve sustainability science's requirements of: (i) comprehending multiple scales and hierarchical decision making from local to global (e.g. globalization's effects on local agriculture); (ii) involving multiple stakeholders who often utilized or claimed the same resource (e.g. floriculturists' versus farmers' conflicts over water); (iii) understanding complex interactions leading to natural resource degradation (e.g. climate change and loss of agrobiodiversity); and (iv) incorporating local perceptions and knowledge in an immediate search for solutions.

The main natural resource themes for research were biodiversity, soils and water. Soil scientists from the University of Georgia conducted research and modelling on Cotacachi's soils, crop production and irrigation (see Zehetner and Miller, Chapter 12 and Zehetner *et al.*, Chapter 13, this volume). Biologists and botanists from the Ecuadorian NGO Fundación Jatun Sacha with students from Central University-Quito conducted a flora and fauna inventory of the highland Cotacachi area (see Peñafiel *et al.*, Chapter 7, this volume). Major attention was paid in the project to water resources, Cotacachi's main resource issue today in terms of scarcity and conflict. The Ecuador SANREM field coordinator, Xavier Zapata Ríos, with colleague Jenny Aragundy conducted a study of water quality and human needs (see Aragundy and Zapata Ríos, Chapter 14, this volume). The community-based water monitoring programme of Auburn University was one of the more visible applications of SANREM research (see Ruiz-Córdova *et al.*, Chapter 16, this volume).

While each team, whether social or biological science oriented, utilized their own methods, every effort was made to communicate research findings to each other. We pursued several integrating activities throughout the life of the project. A central research activity was the land use change analysis and projection project which analysed land use over >40 years from 1963 until 2000 (Zapata Ríos *et al.*, Chapter 4, this volume). A future projection until the year 2030 was utilized as part of an effort to assist Cotacacheños envisage their own futures and mountain alternatives (Rhoades and Zapata Ríos, Chapter 20, this volume). The ultimate purpose of the research is to provide decision support, information and alternatives which underpin real-world action, including a national resource plan for the canton.

Research Partnership Approach for Sustainability Science

Since the purpose of this book is to present research results, our participatory methodology utilized in the project will not be discussed in detail here. This methodology has been presented in previous publications based mainly on the research site of Nanegal (Flora *et al.*, 1997; Rhoades, 2001). The authors of individual chapters in this volume outline the specific methods used in their research. The methods used were highly varied and ranged from fairly narrow scientific techniques to open fora with indigenous people. The human hours invested in this project were immense. We maintained a permanent presence in the community over 5 years, renting an apartment in front of the bus station on '10 de Agosto' street near *Jambi Mascaric*, the women's health centre of UNORCAC. The apartment served as a social hub and meeting place for our scientists, visitors and especially indigenous collaborators. Most of the daylight hours, however, we were in the villages or up in the páramo. Our results, therefore, represent thousands of hours of listening, interviewing, recording and being guests in the communities of Cotacachi. In addition, we spent endless hours in front of computers with stacks of data trying to understand what we had seen, heard, experienced and measured. The end result is a mass of basic data far more voluminous than we can present here. A data and knowledge base involving GIS and development of a Natural Resource Atlas for the Cotacachi canton was headquartered at Catholic University-Quito with similar database nodes in Cotacachi and at the University of Georgia. All of this information is stored in a readily accessible CD format which has been returned to our collaborators and deposited in the municipal library in Cotacachi (Rhoades, Chapter 21, this volume).

In addition to this book, numerous other publications – many with indigenous collaborators – have also been published (e.g. Nazarea and Guitarra, 2004). The challenge of this volume has been to boil down the rich materials we have gathered and the insights gleaned to key themes which will make sense not only to other sustainability scientists but most of all to the people of Cotacachi. Given the pluralism of approaches to sustainability within our own Andes team, prior to each major section's chapters, I will provide a general integrative overview on how the chapters fit together as a unified theme.

Although the purpose of individual chapters is to present our final scientific research outputs, we also want to convey something of the modifications that we had to make as scientists working in the new paradigm of participatory sustainable development. In any major research undertaking of this magnitude and with so many voices and perspectives being presented, it is not an easy task to connect the many components which make up the dynamic and intricate landscape–lifescape of Cotacachi. Our research, on the one hand, was conducted within the basic scientific methods of problem formulation, data gathering and hypothesis testing. On the other, our research proposals were reviewed by the indigenous councils of each Cotacachi community where we worked and often by larger assemblies to make sure that they were sufficiently interested to give their time and resources to us (see Rhoades, Chapter 21, this volume). Often the research design and products were changed to accommodate the people's needs and wishes. By making the research a collaborative enterprise, we believe we achieved not only better basic science but science which will have local meaning and application.

A primary motivation to complete this end-of-project book was to finalize our promise to Cotacachi that our research would be 'enriching,' not just 'extracting' (Waters-Bayer, 1994). The reciprocal agreement between SANREM and UNORCAC stated that while our researchers would have open access to people, their homes, their fields and even their memories, we – in turn – would return project data, findings, publications and other products to the people (see Rhoades, Chapter 21, this volume). Although never a simple undertaking, we have tried our best to adhere to the spirit that research must be useful and delivered back to those who supported the effort and provided time and information.

Although this book alone will not in itself save the Cotacachi environment or dramatically improve its agricultural or natural resources, we hope the information contained therein will become an inspiration for and a starting basis for a participatory canton-wide strategic natural resource plan. We hope this book makes Cotacacheños proud and enriches their indigenous vision of 'development with identity' that draws strength from an ancestral past.

Note

[1] The place-name 'Cotacachi' is variously used throughout the larger area to designate different political units or geographical features within our study area (see Fig. 1.1). Cotacachi properly designates the name of the canton, an Ecuadorian administrative unit equivalent to the notion of county in the USA, located in the Province of Imbabura. This canton stretches from the eastern high Andean region (above 2500 masl) where our research took place westward and downward through subtropical hilly landscapes of the Intaq river watershed (750–1000 masl) finally narrowing along the near sea level banks of Rio Guayabamba as it enters Esmaraldas province. Cotacachi is typically used to refer to the canton capital, Santa Ana de Cotacachi, and its parish. In this book, however, when we use the term Cotacachi, we are referring mainly to our study area in the eastern Andean part of the canton between 2500 and 4000 masl inhabited principally by indigenous Quichua-speaking people. The term Cotacacheños, while correctly referring to all citizens of the canton regardless of ethnicity, is used in this book to refer to indigenous people of Cotacachi's highland zone.

References

Alarcón, R. (2001) Biological monitoring: a key tool in integrated conservation and development projects. In: Rhoades, R. and Stallings, J. (eds) *Integrated Conservation and Development in Tropical America*. SANREM CRSP and CARE-SUBIR, Athens, Georgia.

Bebbington, A. (2000) Reencountering development: livelihood transitions and place transformations in the Andes. *Annals of the Association of American Geographers* 90, 495–520.

Brush, S. (1982) The natural and human environment of the Central Andes. *Mountain Research and Development* 2, 19–38.

Chambers, R. (1997) *Whose Reality Counts? Putting the First Last*. Intermediate Technology Publications, London.

Chambers, R. and Ghildyal, B.P. (1985) Agricultural research for resource-poor farmers: the Farmer First-and-Last Model. *Agricultural Administration and Extension* 20, 1–30.

Coxhead, I. and Shively, G.E. (eds) (2005) *Land Use Changes in Tropical Watersheds: Evidence, Causes and Remedies*. CAB International, Wallingford, UK.

Denniston, D. (1995) *High Priorities: Conserving Mountain Ecosystems and Cultures. World Watch Paper 123*. World Watch Institute, Washington, DC.

Dollfus, O. (1982) Development of land-use patterns in the central Andes. *Mountain Research and Development* 2, 39–48.

Escobar, A. (1995) *Encountering Development: the Making and Unmaking of the Third World*. Princeton University Press, Princeton, New Jersey.

Flora, C., Larrea, F., Ehrhart, C., Ordonez, M., Baez, S., Guerrero, F., Chancay, S. and Flora, J. (1997) Negotiation participatory action research in an Andean Ecuadorian sustainable agriculture and natural resource management program. *Practicing Anthropology* 19, 20–25.

Gray, A. (1997) *Indigenous Rights and Development*. Berghahn Books, New York.

Kates, R., Clark, W., Corell, R., Hall, J., Jaeger, C., Lowe, I., McCarthy, J., Schellnhuber, H., Bolin, B., Dickson, N., Faucheux, S., Gallopin, G., Grubler, A., Huntley, B., Jager, J., Jodha, N., Kasperson, R., Mabogunje, A., Matson, P., Mooney, H., Moore, B. III, O'Riordan, T. and Svedin, U. (2001) Sustainability science. *Science* 292, 641–642.
Messerli, B. and Ives, J.D. (eds) (1997) *Mountains of the World: a Global Priority.* Parthenon Publishing Group, New York.
Moore, K. (ed.) (2005) *Conflict, Social Capital and Managing Natural Resources.* CAB International, Wallingford, UK.
National Research Council (1991) *Toward Sustainability: a Plan for Collaborative Research on Agriculture and Natural Resource Management.* National Research Council, Washington, DC.
Nazarea, V. (1999) Lenses and latitudes in landscapes and lifescapes. In: Nazarea, V. (ed.) *Ethnoecology: Situated Knowledge/Located Lives.* University of Arizona Press, Tucson, Arizona, pp. 91–106.
Nazarea, V. and Guitarra, R. (2004) *Cuentos de la Creacion y Resistencia.* Abya Yala, Quito, Ecuador.
Nigh, R. (2002) Maya medicine in the biological gaze: bioprospecting research as herbal fetishism. *Current Anthropology* 43, 451–477.
Obasi, G.O.P. (2002) Embracing sustainability science, the challenge for Africa. *Environment* 44, 8–19.
Ortiz Crespo, S. (2004) *Una Apuesta por la Democracia Participativa.* FLACSO, Quito, Ecuador.
Price, M. (1998) *Why Are Mountains Important?* The International Mountaineering and Climbing Federation, Bern, Switzerland.
Rhoades, R.E. (1997) *Pathways Towards a Sustainable Mountain Agriculture for the 21st Century: the Hindu-Kush Himalayan Experience.* International Centre for Integrated Mountain Development, Kathmandu.
Rhoades, R.E. (ed.) (2001) *Bridging Human and Ecological Landscapes: Participatory Research and Sustainable Development in an Andean Agricultural Frontier.* Kendall/Hunt Publishers, Dubuque, Iowa.
Rhoades, R.E. and Booth, R.H. (1982) Farmer-Back-to-Farmer: a model for generating acceptable agricultural technology. *Agricultural Administration* 11, 127–137.
Rhoades, R. and Stallings, J. (2001) *Integrated Conservation and Development in Tropical America.* SANREM CRSP and CARE-SUBIR, Athens, Georgia.
Rhoades, R. and Thompson, S. (1975) Adaptive strategies in alpine environments. *American Ethnologist* 2, 58–71.
Stadel, C. (1991) Environmental stress and sustainable development in the tropical Andes. *Mountain Research and Development* 11, 213–223.
Troll, C. (ed.) (1968) *Geo-ecology of the Mountainous Regions of the Tropical Americas.* Ferd. Dummlers Verlag, Bonn, Germany.
United Nations Conference on Environment and Development (1992) *Agenda 21: Programme of Action for Sustainable Development. Rio Declaration on Environment and Development.* United Nations Publication, Rio de Janeiro.
Warren K.B. and Jackson, J.E. (2002) *Indigenous Movements, Self-representation, and the State in Latin America.* University of Texas Press, Austin, Texas.
Waters-Bayer, A. (1994) The ethics of documenting rural people's knowledge: investigating milk marketing among Fulani women in Nigeria. In: Scoones, I. and Thompson, J. (eds) *Beyond Farmer First: Rural People's Knowledge, Agricultural Research, and Extension Practice.* Intermediate Technology Publications, London, pp. 144–150.
World Commission on Environment and Development (1987) *Our Common Future.* Oxford University Press, Oxford, UK.
World Wildlife Fund and World Bank. (1995) *A Conservation Assessment of the Terrestrial Ecoregions of Latin America and the Caribbean.* The World Bank, Washington, DC.

Part I

Time and Landscape in Cotacachi

An *anciano* of Cotacachi (photo: Natalia Parra).

Sustainability science differs from previous approaches to agriculture and environment by seeking a greater time depth analysis of society–nature interaction. In the past, researchers rarely dealt with time horizons beyond the annual cropping cycle or, at most, 3 or 4 years of a research project's funded life. Sustainability, however, must consider intergenerational decisions and impacts reaching across decades, centuries and even millennia. The modern Cotacachi landscape mosaic is the result of long-term interaction and feedback between natural and human processes. To understand how this landscape evolved, we must begin in pre-historic epochs tens of thousands of years before the first human ever set foot on the base of Cotacachi mountain. In Chapter 2, soil scientists Franz Zehetner and William Miller trace contemporary geoecological conditions to the ongoing collision of the Nazca and South American plates, the uplifting of the Andes and resultant volcanism. These processes shaped the area's soils, hydrology and topography which, in turn, determine potentialities for agriculture. Both a blessing and a danger, Cotacachi's volcano has been a source of rich soils and glacial water while at the same time inspiring fear among Cotacacheños as earthquakes and eruptions have time and time again destroyed fields, villages and lives. Despite our tendency to emphasize human actions over nature's power, Zehetner and Miller leave little doubt that age-old Andean geoecological forces still set limitations on what is possible for human survival.

Anthropologists Shiloh Moates and Brian Campbell introduce the historical ecology of the Cotacachi area in Chapter 3, beginning with the earliest archaeological evidence of villages 2400 years ago and ending with the present agricultural system. They demonstrate how the contemporary landscape represents a culmination of past events, including local adaptations in the pre-Inca period and the influences of domination from the Inca, the Spanish, the Catholic Church, the Ecuadorian nation-state and the present global economy. The traditional vertical system of complementary production and exchange has demonstrated resilience and continuity even in the face of dramatic change over long periods. Despite the historical transformations of the landscape, the traditional principles used by local inhabitants to exploit the mountain hold lessons for modern efforts in sustainable development.

In Chapter 4, our land use change (LUC) analysis team (Xavier Zapata Ríos, Robert Rhoades, Maria Segovia and Franz Zehetner) document dramatic changes in land use and land cover since 1963, just before agrarian land reform, up to 2000. This analysis was based on interpretation of an aerial photographic time series and subsequently verified through community workshops and field ground-truthing. This analysis over four decades demonstrates a growth of the urban zone, intensification of agriculture, reduction of field size through break up of the haciendas, reduction of native primary and secondary forests, and expansion of tree plantations and agroindustries such as floriculture. Indigenous landholdings after land reform are further fragmenting through inheritance, while new systems of land use, such as plantation forests and floriculture industries, take root in the landscape. Our LUC analysis over > 40 years became the primary window through which other research projects measured changes in biodiversity, soils and water, as well the basis for future projection of land use planning under different decision scenarios.

The 19th and 20th centuries provide an abundance of historical documents and captured images of the Cotacachi landscape. Thanks to the beauty of the region (and the mountain's lure to European and North American alpinists and chroniclers), accounts, paintings and photographs are available stretching back to Humbolt's observations of the Cotacachi volcano in 1802. In Chapter 5, Robert Rhoades, Xavier Zapata and Jenny Aragundy draw on this archival evidence, complemented by oral histories by elders in community workshops on climate change, to document the demise of Cotacachi mountain's glacial zone. The glacier disappeared virtually before our eyes during our project residency in Cotacachi between 1997 and 2005.

Finally, in Chapter 6, anthropologist Virginia Nazarea with Rafael Guitarra and Robert Rhoades discuss Cotacachi's folk tales and legends as a window to indigenous Andean cosmology. Playing on local memory and resistance as embodied in the imaginary and lively characters such as the rabbit and the wolf who inhabit the cultural landscape, the oral traditions tell the story of how Cotacacheños have actively engaged the lifescape–landscape to create their own unique understanding of human–nature interactions. The legends not only document how the small but crafty animal (rabbit: *indígena*) outwitted the large, clumsy animal (wolf: *hacendado*), but also stories of creation of life and orgins of culturally significant landscape features such as the sacred Cuicocha lake. These stories remind us that local peoples' view of their homeland is distinct from the tunnel vision of scientists who look at components such as soils, water, crops and fields. Cotacacheños see a place inhabited by many spirits and magical forces that provide moral and practical guidance.

2 Shaping an Andean Landscape: Processes Affecting Topography, Soils and Hydrology in Cotacachi

Franz Zehetner and William P. Miller
University of Georgia, Department of Crop and Soil Sciences, 3107 Plant Science, Athens, GA 30602-7272, USA

Volcanism and Topography

The uplifting of the Andes and the associated volcanism are caused by the collision of two tectonic plates: the continental South American plate moving east to west and the oceanic Nazca plate moving west to east. In the Ecuadorian Andes, two parallel chains of stratovolcanoes result stretching north–south and enclosing the 50 km wide *inter-Andean valley*. The study area is located about 35 km north of the equator, on the inner slopes of volcano Cotacachi oriented towards this temperate *inter-Andean valley*. The topography of the region is dominated by high volcanic peaks, including Cotacachi (4939 metres above sea level (masl)), Imbabura (4630 masl) and Cayambe (5790 masl), as well as by the enormous calderas of Cuicocha (3064 masl) and Mojanda (3716 masl). The pronounced verticality of the Cotacachi area, stretching from 2080 to 4939 masl, is shown in Fig. 2.1. Landscape development in the area has been heavily influenced by volcanic phenomena, such as lava and pyroclastic flows, pumice and ash falls, and subsequent mudslides induced by heavy rainfall events and earthquakes. Streams have deeply carved into the land forming ravines and dissecting the landscape into plateau-like upland areas stretching parallel to streams.

The volcanic complex of Cotacachi has a long history of volcanic activity involving several different eruption centres, of which only the side vent Cuicocha has been active in the Holocene. The other centres have not erupted in the past 40,000 years (Hall and Mothes, 1994), and the long extinct main vent of Cotacachi, the oldest of the eruption centres, has undergone heavy erosion in periods of glaciation. Volcano Cuicocha has had three phases of activity that occurred over a period of a few hundred years, ending about 3000 years ago (Mothes and Hall, 1991; Hall and Mothes, 1994; Athens, 1999). The present caldera of Cuicocha was formed by explosive eruptions that resulted in massive pyroclastic flows and tephra falls. These relatively young deposits have shaped the southern part of the study area, whereas the northeastern part is covered with older deposits originating from other eruption centres.

Climate and Hydrology

The climate in the area is that of an equatorial high-altitude environment, with temperatures almost constant throughout the year, but showing pronounced diurnal oscillations. Variations of climatic parameters over the landscape are largely a function of elevation. The mean annual

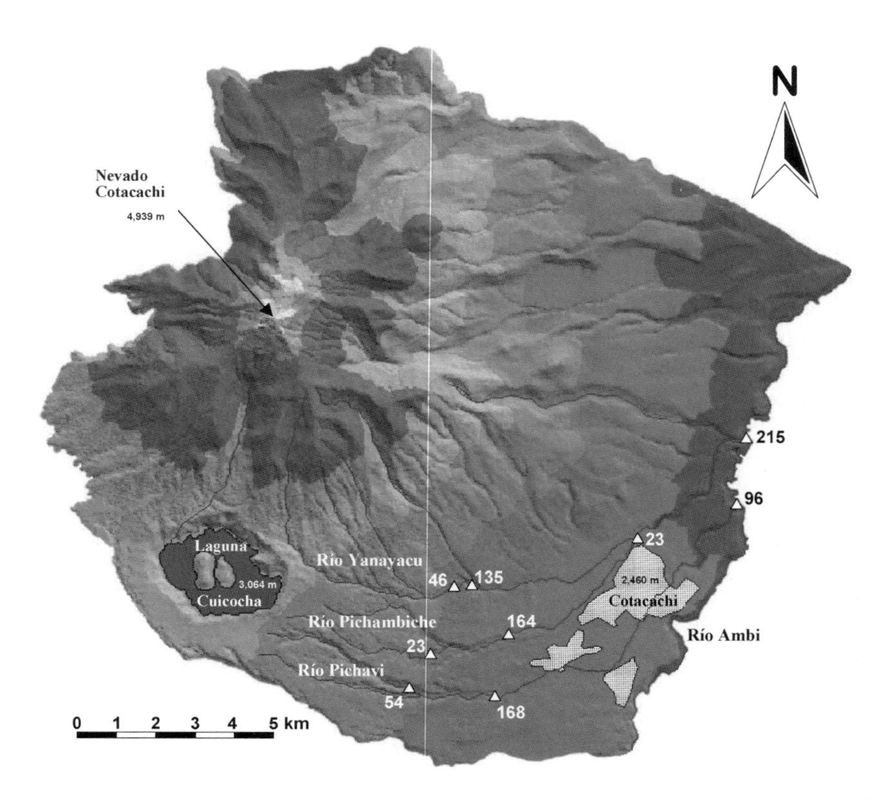

Fig. 2.1. TIN (triangulated irregular network) of the study area; sites of dry season base flow measurements are marked with triangles (values in l/s).

temperature is about 15°C at 2500 masl, and drops by about 0.6°C per 100 m of elevation increase. Rainfall in the area is generally dominated by low-intensity events. The mean annual precipitation is about 900 mm at 2500 masl and increases with elevation to about 1500 mm at 4000 masl (Nouvelot *et al.*, 1995). Mean annual PET (potential evapotranspiration) amounts to about 900 mm at 2500 masl and decreases with increasing elevation due to lower temperatures and higher humidity. The annual distribution of rainfall and PET for the nearby town of Otavalo (2550 masl) is shown in Fig. 2.2. The climate is characterized by an expressed seasonality, with a dry season of pronounced water deficit from June to September. With increasing elevation on the volcano, the climate becomes more humid, the dry season shorter and the summer water deficit less pronounced.

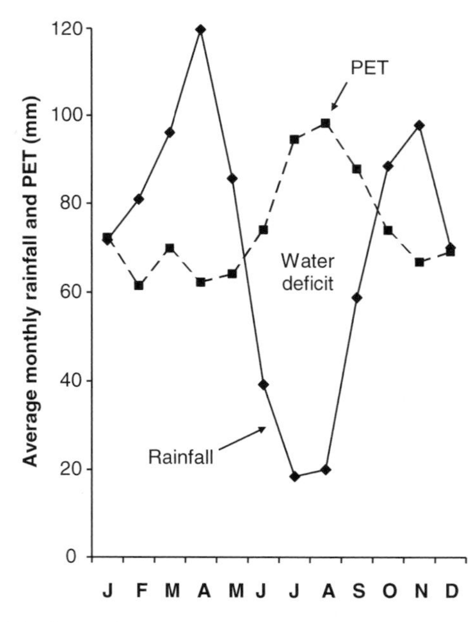

Fig. 2.2. Annual distribution of rainfall and potential evapotranspiration (PET) for Otavalo.

In her master's thesis on the volcanic complex of Cotacachi, von Hillebrandt (1989) reports that 'the valleys that stretch from the peak [of Cotacachi] are [. . .] heavily glaciated'. The icecap von Hillebrandt encountered 15 years ago has entirely disappeared now, which may be evidence of climate change in the region (Rhoades *et al.*, Chapter 5, this volume).

The Cotacachi study area is drained by several small tributaries of Río Ambi (Fig. 2.1) and is located at the headwaters of the Mira watershed, which drains Ecuador's northwestern corner into the Pacific Ocean. Apart from the Cuicocha crater lake, which does not have an outflow, water availability is very limited at elevations above 2700 masl, especially during the dry summer months. Figure 2.1 shows dry season base flow in the three principal streams draining the southern part of the study area (Yanayacu, Pichambiche and Pichaví). The discharge measurement taken furthest upstream represents the origin of each stream in the dry season, which is located below 2700 masl for all three streams. Discharge increases from the origin downstream; however, below about 2600 masl, much of the stream flow is diverted into irrigation canals, sometimes leaving very little water in the streambeds. The second discharge measurement was taken just upstream of the first irrigation canal, and the third just upstream of the stream outflow. Due to its location at the headwaters, the study area is drained by small streams with low discharge, and the glacial retreat over the past years may have lowered the stream base flow further in the area. As a consequence, water is a scarcely available resource for competing agricultural, industrial and domestic demands in the Cotacachi area.

Soils and Agriculture

Soil types

The volcanic soil parent materials in the area are generally pumiceous and have andesitic to dacitic composition. The soils in the southern part of the study area have formed on the 3000-year-old Cuicocha deposits and are in their early stages of development, whereas the soils in the northeastern part have formed on deposits older than 40,000 years and are thus more advanced in their development.

Apart from differences in age and composition of parent materials, soil formation in the area is heavily influenced by climatic differences with elevation along the volcanic slopes. The soils' organic matter contents (Fig. 2.3), water storage capacity, structural stability and phosphate retention increase with altitude. At high elevations, the soils' clay mineralogy is dominated by active amorphous constituents, whereas at low elevations the clay mineral halloysite predominates (Zehetner *et al.*, 2003). Andic soil properties increase with elevation and, according to US Soil Taxonomy (Soil Survey Staff, 1998), the high-elevation soils are classified as Andisols and the low-elevation soils as Inceptisols and Entisols (Zehetner *et al.*, 2003).

The recent volcanic deposits overlie an older, more developed surface formed on volcanic parent materials of preceding eruption episodes. A typical soil profile is presented in Fig. 2.4 and shows recent soil development on a series of Cuicocha tephra overlying a buried soil (paleosol) formed on

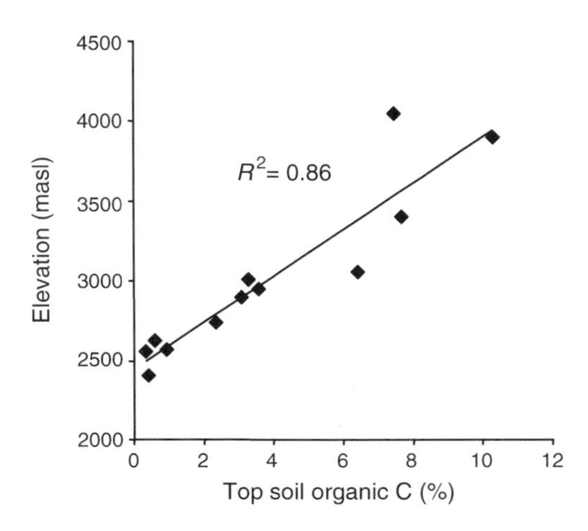

Fig. 2.3. Altitudinal variation of soil organic matter contents.

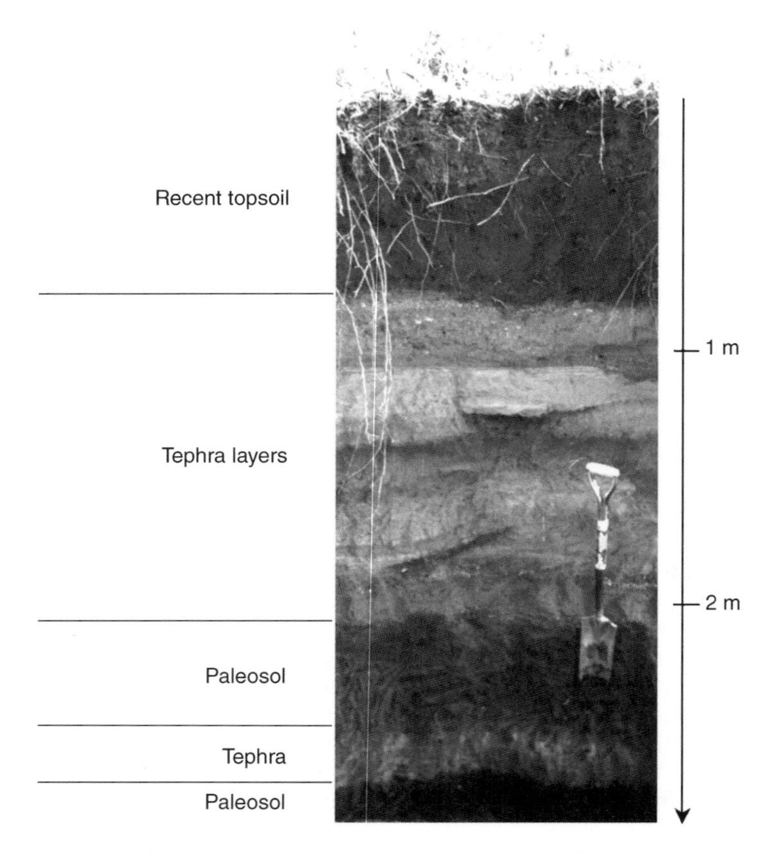

Fig. 2.4. Typical profile of a volcanic ash soil in the area (photo: Franz Zehetner).

older tephra that in turn overlies an even older paleosol. The parent material of this deepest paleosol is volcanic ash that has been cemented and indurated over time and is locally referred to as *cangagua*. In areas where the recent soils have been eroded, the older paleosol strata can reach close to the surface or crop out entirely and thus play an agronomic role once again.

Land use and land management

The majority of the region's vegetation cover has been altered from its natural state by human activity. In high-elevation zones (above 3000 masl), which are frequently burned during the dry summer months, *matorral*-scrubland and *páramo*-grassland are predominant, and only remnants of native forest are left (see Zapata *et al.*, Chapter 4, this volume). At elevations below 3000 masl, the landscape is dominated by agricultural land use and introduced eucalyptus forests (*Eucalyptus globulus* Labill.). The major portion of agricultural lands lies on upland plateaus exhibiting slope gradients between 0 and 20%. The steep sides of ravines offer refuge for native brushland vegetation; however, occasionally, even these areas are planted with eucalyptus or cultivated with agricultural crops. The bottomland areas on the narrow floodplains inside ravines are typically used as pastures. The major agricultural crops in the area are maize (*Zea mays* L.), bean (*Phaseolus vulgaris* L.) and potato (*Solanum tuberosum* L.).

Indigenous peoples have inhabited the region for thousands of years (Athens, 1999) and have employed farming practices well suited to the climate and topography of the area (see Moates and Campbell, Chapter 3, this volume). Ancient bench terraces, probably of pre-Colombian origin, are a proof of an early awareness of natural resource conservation. Over the past 500 years, the social and agricultural structures have undergone dramatic changes involving the centuries-long enslavement of the indigenous population within the *hacienda* system, a half-hearted land reform in the 1960s, the advent of the Green Revolution in the 1970s and lately a newly awakened consciousness among the indigenous population to preserve their own heritage.

At present, agriculture shows marked differences between *hacienda*-type operations on the one hand and smallholder farms in the mostly indigenous peasant communities on the other. The large-scale *hacienda* agriculture is characterized by intensive management with high inputs and a high degree of mechanization. The situation in the indigenous communities is different. Many farmers own < 2 ha of arable land and very little livestock. Due to limited resources and the desire to produce organically, the use of chemical fertilizers and pesticides is uncommon. Manure application rates are generally low, and many farmers do not fertilize their land at all. The limited amount of available land forces many farmers to crop continuously and avoid prolonged fallow periods. Land management operations, such as tillage and cultivation, are generally done by hand or with the use of oxen, and irrigation is only available in some low-elevation communities.

Soil erosion

Soil erosion is a complex process that involves a variety of factors including rainfall intensity, soil stability, topography, ground cover and land management. Erosional land degradation is a widespread phenomenon in Andean South America, where many regions are covered with soils derived from volcanic parent materials.

The runoff–erosion characteristics of the volcanic ash soils in the Cotacachi area are strongly altitude dependent (Fig. 2.3). At high elevations, accumulation of organic matter and formation of active amorphous constituents have led to the development of soils with stable aggregate structure, high infiltration capacity and consequently low potential for runoff generation and soil erosion. At low elevations, low organic matter contents and absence of active amorphous constituents have led to the formation of weakly aggregated soils with lower infiltration capacity and higher susceptibility to runoff generation and soil erosion. However, in comparison with other soils of different origin and composition from the USA (Kinnell, 1993), Australia (Sheridan *et al.*, 2000) and Spain (Duiker *et al.*, 2001), the erodibility indices determined for these more erodible low-elevation soils are classified as low. This and the comparatively low rainfall intensities in the region lead to the conclusion that soil erosion is not a major threat to sustainability in the Cotacachi area, which is generally corroborated by field observations.

In the steeply sloping high-elevation zones of the area, the soils are very permeable and stable, and the soil surface is well protected from raindrop impact by dense scrub and grassland vegetation. However, burning of this protective vegetation cover may result in the formation of a water-repellent surface layer that promotes runoff and soil erosion. In the lower zones, where the soils are more susceptible to runoff and erosion, lower slope gradients and the presence of structural barriers such as bench terraces, earth walls and border hedgerows effectively decrease soil loss and sediment export. However, the removal of such barriers in large-scale agricultural operations may lead to increased sediment export and related adverse effects on water quality. Presently, most of the sediment in streams is probably from unpaved roads and trails, which in some places are deeply sunken below the surrounding fields, and from the sides of ravines, which are prone to mass wasting due to very steep slopes and underlying indurate ash strata.

References

Athens, J.S. (1999) Volcanism and archaeology in the northern highlands of Ecuador. In: Mothes, P. (ed.) *Actividad Volcánica y Pueblos Precolombinos en el Ecuador.* Ediciones Abya-Yala, Quito, Ecuador, pp. 157–189.

Duiker, S.W., Flanagan, D.C. and Lal, R. (2001) Erodibility and infiltration characteristics of five major soils of southwest Spain. *Catena* 45, 103–121.

Hall, M.L. and Mothes, P.A. (1994) Tefroestratigrafía holocénica de los volcanes principales del valle interandino, Ecuador. In: Marocco, R. (ed.) *El Contexto Geológico del Espacio Físico Ecuatoriano: Neotectónica, Geodinámica, Volcanismo, Cuencas Sedimentarias, Riesgo Sísmico.* Estudios de Geografía, Vol. 6. Colegio de Geógrafos del Ecuador, Corporación Editora Nacional, Quito, Ecuador, pp. 47–67.

Kinnell, P.I.A. (1993) Interrill erodibilities based on the rainfall intensity – flow discharge erosivity factor. *Australian Journal of Soil Research* 31, 319–332.

Mothes, P. and Hall, M.L. (1991) El paisaje interandino y su formación por eventos volcánicos de gran magnitud. In: Mothes, P. (ed.) *El Paisaje Volcánico de la Sierra Ecuatoriana: Geomorfología, Fenómenos Volcánicos y Recursos Asociados.* Estudios de Geografía, Vol. 4. Colegio de Geógrafos del Ecuador, Corporación Editora Nacional, Quito, Ecuador, pp. 19–38.

Nouvelot, J.-F., Le Goulven, P., Alemán, M. and Pourrut, P. (1995) Análisis estadístico y regionalización de las precipitaciones en el Ecuador. In: Pourrut, P. (ed.) *El Agua en el Ecuador: Clima, Precipitaciones, Escorrentía.* Estudios de Geografía, Vol. 7. ORSTOM, Colegio de Geógrafos del Ecuador, Corporación Editora Nacional, Quito, Ecuador, pp. 27–66.

Sheridan, G.J., So, H.B., Loch, R.J., Pocknee, C. and Walker, C.M. (2000) Use of laboratory-scale rill and interill erodibility measurements for the prediction of hillslope-scale erosion on rehabilitated coal mine soils and overburdens. *Australian Journal of Soil Research* 38, 285–297.

Soil Survey Staff (1998) *Keys to Soil Taxonomy,* 8th edn. USDA-NRCS, Washington, DC.

von Hillebrandt, C.G. (1989) Estudio Geovulcanológico del Complejo Volcánico Cuicocha-Cotacachi y sus Aplicaciones, Provincia de Imbabura. Thesis. Escuela Politécnica Nacional, Quito, Ecuador.

Zehetner, F., Miller, W.P. and West, L.T. (2003) Pedogenesis of volcanic ash soils in Andean Ecuador. *Soil Science Society of America Journal* 67, 1797–1809.

3 Incursion, Fragmentation and Tradition: Historical Ecology of Andean Cotacachi

A. Shiloh Moates and B.C. Campbell
University of Georgia, Department of Anthropology, 250 Baldwin Hall, Athens, GA 30605, USA

Introduction

The Andes provides an extraordinary landscape for historical ecology research because of the diversity of vertically layered agroecological production zones and the concomitant sociopolitical developments that emerge as adaptations to this diverse environment (Rhoades and Thompson, 1975). Andean experts propose the concept of 'verticality' to explain the pre-historic attempts to exploit and 'control a maximum of floors and ecological niches' (Murra, 1985). Yet, northern Ecuador remains distinct from the rest of the Andes because of its proximity to the equator and the comparatively lower altitude of its mountains. In effect, the agroecological zones are condensed; the distance between distinct zones is much shorter than in the central and southern Andes. Therefore, while verticality is strongly marked in the Peruvian and Bolivian Andes, where ecological zones span larger distances, the concept must be adjusted to represent agroecology accurately in the Ecuadorian highlands by emphasizing that agroecological zones are exploited differently because of their compactness (Salomon, 1986).

This chapter discusses the historical ecology of the Cotacachi region of northern Ecuador, exploring traditional agroecological strategies and the changes that have occurred as a result of pre-historical, historical and contemporary sociopolitical events and processes. We employ the concept of ecological complementarity as a contemporary formulation of verticality to demonstrate the traditional exploitation of proximate agroecological zones (Oberem, 1978; Salomon, 1986; Knapp, 1991). While focusing on the agricultural practices and cultivars of the region and how they have changed through time, we also demonstrate that traditional approaches have been restricted, modified and discontinued based on sociopolitical intervention.

Changes in the human ecology of the region must be understood as a culmination of past events. For this reason, we use a diachronic analysis of land use changes, beginning with pre-Incaic land management, followed by an investigation of Inca, Spanish and contemporary influences on aboriginal land use. The presentation of pre-Incaic agricultural, cosmological and land use practices focuses on characteristics strictly representative of the northern Andes. Some of these characteristics, however, are Pan-Andean in nature. Spanish chroniclers and archaeological excavations have shed light on the impacts of the 'Inca incursion' into the Ecuadorian Andes, allowing for a better understanding of the transformative effects of the Inca presence on local land management (Plaza Schuller, 1976). Shortly after the Inca conquest of Ecuador, in the late 15th century, the Spaniards arrived, taking over where the Inca left off in the subjugation process. The Inca and Spanish impacts on local land practices were numerous and profound, as have been the more recent introductions by national and international institutions and organizations. We explore these most recent agricultural and ecological shifts through interviews with local indigenous peoples, surveys, archive research, participant observation, literature reviews and field plotting.

Traditional Andean agroecological and sociocultural practices constituted adaptations to a unique vertical landscape. The landscape is characterized by inconsistent and fluctuating climatic conditions that force farmers to innovate and diversify agriculturally to avoid malnutrition and famine (Zapata Ríos *et al.*, Chapter 4, and Rhoades *et al.*, Chapter 5, this volume). Ecological complementarity and communal socioeconomic relationships provided protection against the inconsistency of the environment through risk aversion. Ecological complementarity minimized risk because of the diversity of crops and staggered planting and harvest seasons in diverse ecozones (D'Altroy, 2000). Traditional Andean sociocultural institutions emphasized communal solidarity and redistribution over material accumulation by individuals. These traditional strategies ensured that families had sufficient labour when necessary. Contemporary farmers, however, are no longer able to employ the myriad traditional methods developed by their ancestors. Historical incursions and more recent exogenous introductions have left the contemporary indigenous populations of the region with mere fragments of these traditional adaptations. Through historical ecology, we elucidate this process of landscape fragmentation by demonstrating how communities have systematically been denied the ability to practice ecological complementarity because of intervention (Crumley, 1993).

The Cotacachi highlands span four major agroecological zones, including a variety of ecosystems and an ecological reserve, la Reserva Ecologica Cotacachi-Cayapas. Figure 3.1 illustrates the four agroecological zones, which can be divided by altitude into the páramo (> 3000 masl) and inter-Andean cultivated lands (2300–3000 masl), which may be divided further into a cereal zone (2700–3000 masl), maize zone (2500–2700 masl) and intercropped short-cycle crop zone (2300–2500 masl). These zones differ significantly in rainfall, soils and, therefore, vegetation and crops. The páramo has a native vegetative cover of bushes and a high capacity for water retention that exceeds 200% of its dry weight. The highest population density in the highlands is found in the inter-Andean zone between 2300 and 2700 masl, where land tenure conflicts and hacienda presence have scarred the once-forested landscape (Fig. 3.1).

The Pre-Inca North Ecuadorian Sierra

The richness of Ecuadorian pre-history has been largely overlooked because of the impressive pyramids and highly developed pre-Colombian civilizations to both the north in Mesoamerica and to the south in the central Andes. Archaeological efforts in the Andes traditionally focused on Peru, which was considered the 'nuclear area' of civilization in South America. Yet, recent findings date ceramic remains from the lowlands in Ecuador to 1000 years before the earliest pottery in Peru, and plant domestication was taking place in Ecuador by approximately 6000 BC (Marcos, 2003; Raymond, 2003). Around 1400 BC, the semi-nomadic, incipient agriculturalists in the highlands began to establish permanent villages, intensify agriculture and participate in trade with other groups. The floral remains show that the people were taking full advantage of an array of crops that we now know as typically Andean: potatoes, achira, oca, chochos, beans, quinua and maize.

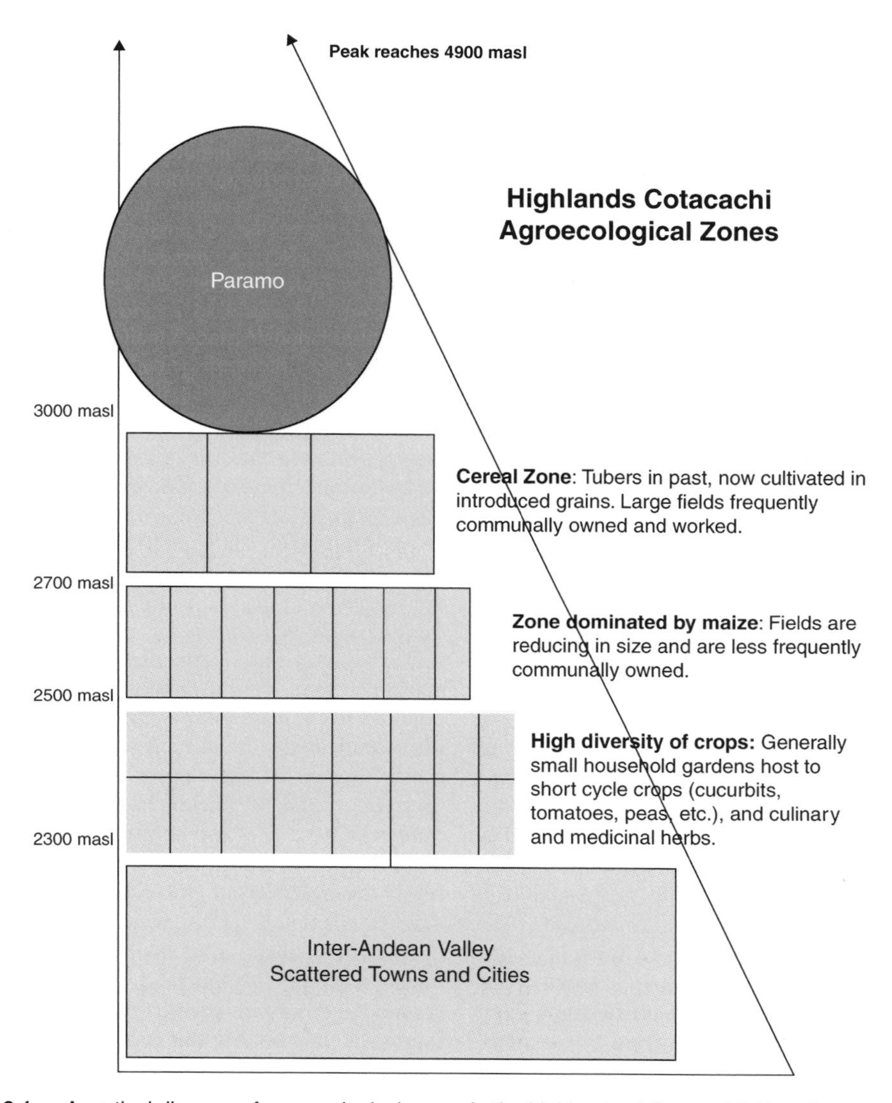

Fig. 3.1. A vertical diagram of agroecological zones in the highlands of Cotacachi, Ecuador.

Faunal remains show high frequencies of deer and rabbit consumption, as well as Andean domesticates, camelids and guinea pigs (Bruhns, 2003).

By 1000 BC, the highlands had developed a sophisticated trade network with coastal populations. Extensive remains of marine species such as *Spondylus*, *Strombus*, *Conus*, *Anadara*, *Tyropecten* and *Pinctad* as well as obsidian and artefacts related to coca use clearly illustrate this point. It was during the period between 1400 BC and European contact that these early highland chiefdoms of Ecuador developed very specialized agricultural systems to exploit the distinct ecological zones of the Andes (Bruhns, 2003).

Agroecological complementarity

The strong ideological, religious and cultural connections between the contemporary indigenous peoples of Cotacachi and their natural environment gives a mere glimpse of how tightly interconnected they were prior to the subjugations by the Incas and Spaniards. Ethnographic data revealed that the indigenous peoples of Cotacachi conceptualize their world in a highly dualistic sense: the communities at a higher altitude represent masculinity, the sun, the head (on a body), white, heat and day, whereas the communities of lower altitudes represent femininity, the earth and moon, the body, darkness, cold and night. Indigenous Cotacacheños perceive these cosmological elements as necessary and complementary. They surface in all facets of life, including religion, agriculture and land use, and have existed since pre-Hispanic times.

The pre-Hispanic communities of the northern Andes exploited the proximity of multiple agroecological zones through several approaches. Many *llajtakuna* (villages) were situated near or within ecotones in order to exploit the natural resources more thoroughly on both sides and within the ecotone (Mayer, 1985; Knapp, 1991; Crumley, 1993). The ecological variation in a steep environmental gradient, such as in the Andes, is characterized by more discontinuous plant communities, which leads to a wider variety of resources (Odum, 1997). Besides the location of communities on ecotones, the pre-Hispanic groups of the region also created *anexos*, or satellite villages, to be situated strategically in lowland zones (Mayer, 1985). These *anexos* consisted of either a group of individuals from the *ayllu*, referred to in Quichua as *kamayuj*, that resided temporarily or permanently in lowland villages, or lowland communities that were coerced through force or attractive trading opportunities. Both the *kamayuj* and the trading partner villages served to supply the 'mother' community with highly desired or needed goods (Mayer, 1985; Salomon, 1986; Alchon, 1991). Specific examples of this abound in the ethnohistorical literature, with Spanish chroniclers mentioning lowland communities of the humid tropics being dependent on or somehow subservient to the sierran chiefdoms (Plaza Schuller, 1976; Barros, 1980). In Cotacachi, the Caranqui chiefdom of the sierra obligated western lowland communities (Intag) to economic dependence and complemented their highland goods with cotton (*Gossypium* spp. Cav.), hot peppers (*Capsicum annuum* L.), coca (*Erythroxilon coca* Lam.) and exotic goods for status (Barros, 1980; Salomon, 1986; Knapp, 1991).

In addition to linking themselves with lowland goods through *anexos*, *curacas* (chiefs) also used *mindalaes* (long-distance traders) to achieve ecological complementarity (Pease, 1985). These merchant traders, who tended to inherit their position, usually had higher status than other members of the *curacazgo* (chiefdom) and specialized in the acquisition of specific lowland or distant products of markedly exotic origin such as gold, silver, coca, fish, cotton and beads (Salomon, 1986, p. 106). Besides the *mindalaes*, there were also more local traders employed by the *curacas*. These traders dealt more strictly with non-sumptuary import goods and were also more limited spatially than the *mindalaes*. They only enacted trade with nearby lowland communities and would have typically established trades such as the exchange of local surpluses of maize (*Zea mays* L.) or beans (*Phaseolus vulgaris* L.) for lowland common goods such as red capsicum peppers, cotton or salt. At least three different interzonal forms of exchange existed during pre-Incaic times in northern Ecuador; *anexos*, *mindalaes* and local markets (*tiangueces*), all of which served to channel exotic goods from the lowlands to the highlands (Barros, 1980; Pease, 1985; Alchon, 1991).

The wide variety of agricultural and wild resources from the varying regions was available at markets (*tiangueces*) that typically were situated on the borders between zones. Salomon (1986) explains that *tiangueces* existed in the north of Ecuador, including the Cotacachi region, and represented 'a far-flung general web of relationships.' He situates and describes the phenomenon in detail:

> The upper layer of the system would have consisted of a web of 'tiangueces' located at strategic intersections of transport routes in both highland and lowland zones, frequently near major ecological frontiers (marine/tropical forest; montana/sierra;

xerophytic/humid), each sponsored by the elite specialist organization(s) of a given area. Their function, from the sponsors' point of view, would have been primarily to afford access to prestigious goods from remote, politically and ecologically foreign zones, and their mode of operation would have included some relatively 'naked' or 'single-stranded' exchange although not necessarily a fully monetarized or price-making market.

The lower tier of the system would have consisted of a web of relationships passing in part through 'tiangueces' but fundamentally organized on a llajta-to-llajta basis, its links consisting of relatively enduring, perhaps ceremonious, relationships of reciprocity between households or other small units belonging to communities in neighboring but ecologically complementary zones. From the viewpoint of participants, the goal would have been a relatively secure supply of ecologically foreign, but culturally indispensable, consumption goods.

(Salomon, 1986, p. 115)

The cotton, peppers, coca and other crops of the lowlands were exchanged for the tubers, maize, quinua and other products of the highland communities at these regional markets. These regional markets represent another link in the chain of the ecological complementarity of the region. To the chagrin of the *curacazgos* of the northern Andes, these markets first attracted the attention of the Inca (Salomon, 1986).

A number of small chiefdoms, each with their own language, customs, and political and economic institutions, persisted in the northern highlands until the end of the 15th century AD. It is difficult to know exactly how many polities there were before the Incan expansion, but the Otavalos, Caranquis, Pastos, Caras and Panzaleos were a few of the larger groups in the north of Ecuador, and the Puruha and Cañaris were two that resided in the south. There is evidence for exchange and violence between these groups, but there is no evidence for any one of them having political dominion over another (Alchon, 1991).

Agricultural crops and animals

The Indians . . . nourish themselves with toasted maize, made into dumplings and gruel; they eat potatoes and some worms that grow in the ground, fat ones, which they call *cusos*, and another kind of little fish which grows in the rivers, called *choncho*, and *ocas*, *ollocos*, *maxuas*, *arracachas*, *zapallos*, *jiquimas*, and *avincas*, roots which grow under the ground, like potatoes, which are called truffles; and likewise red pepper, together with another food which is called *chiche* and which tastes and smells like shrimp; and as well with some herbs which they call *yucas* . . .

(Rodriguez Docampo (1650) in Salomon, 1986, p. 73)

Despite the lack of sources on pre-Incaic agricultural practices in northern Ecuador, the ethnohistorical record proves highly useful. Due to the timing of the Spanish arrival, shortly after the Inca conquest, the observations of Spanish chroniclers tend to refer to the practices of pre-Incaic peoples (Plaza Schuller, 1976; Knapp, 1991). The pre-Inca diet mainly revolved around maize. It was consumed in a variety of forms, as it still is today, including: *chicha* (fermented drink), toasted kernels, flour, popcorn, *choclo* (green on the cob) and mote (hominy). The other major seed crops grown in the highlands included beans (*Phaseolus vulgaris* L.), *chocho* (*Lupinus tricolor* Sod.), cucumber (*Solanum muricatum* Ait.) and *quinua* (*Chenopodium quinoa* Wild.); the tubers grown in the highlands were the *oca* (*Oxalis tuberosa* Sav.), *mashua* (*Tropaeolum tuberosum* R. y P.), *melloco* (*Ullucus tuberosus* H.B.K.) and potato (*Solanum tuberosum* L.). Fruit trees were also grown in the highlands, including avocados (*Persea americana* Mill.), *lucuma* (*Lucuma obovata*), *grenadilla* (*Passiflora* sp.), highland papayas (*chihualcan, babaco*), *chirimoya* (*Annona muricata*), tree tomato (*Cyphomandrea betacea*), *guaba* (*Inga* spp.) and *guayaba* (*Psidium guajaba*). In the lower ecological zones, the pre-Incaic crops most probably included: manioc (*Manihot esculenta*), groundnuts (*Arachis hypogaea* L.), coca (*Erythroxilon coca* Lam.), cotton (*Gossypium* spp.), sweet potato (*Batata edulis*

Choisi), Jicama (*Pachyrhizus* sp.) and red capsicum pepper (*Capsicum annuum*).

The pre-Incaic domesticated animal economy consisted mostly of guinea pigs, camelids and dogs. The camelids that preceded the Inca into northern Ecuador were scarce and their use was limited to the chiefs, whereas guinea pigs and dogs were extremely common (Salomon, 1986; Knapp, 1991). A 1575 Spanish source noted that 'In the second or back room of the house they have their pantry, full of big pots and small ones . . . In this place they have great crowds of cuyes, which we call guinea pigs, and this is where they feed them with quantities of grass . . .' (quoted from Salomon 1986, p. 75–76). Due to the scarcity of domesticated animals, meat was typically only eaten on festival days, and the majority typically went to the chiefs. Other sources of protein for common people, besides the maize and beans, were local game, edible insects, fish and arthropods. As detailed in the earlier quote, the people of the region ate thick grub worms and continue the practice today, actually referring to them as 'shrimp of the soil'. Deer, rabbits and partridges, the principal game animals, were hunted in the páramo (Cieza de Leon, 1553). Domesticated dogs generally were not eaten. Rather, they were used in hunting endeavours, and good hunting dogs were highly prized (Salomon, 1986, p. 82).

Agricultural practices

The paucity of domesticated animals precluded the application of large amounts of organic fertilizer, but the soils of the region – due to the volcanic ashes of the mountains Cotacachi and Imbabura – were extremely fertile and therefore required little fertilization. The household manure sources were limited to that of humans, dogs and guinea pigs, but colonial documents suggest that fertilization with manure was practised at the time of the Spanish arrival (Knapp, 1991). The ethnohistorical record on specific cultivation practices is scarce, but one anonymous source (1573) states:

> They cultivate maize in ridges (*camellones*), measuring little more than one foot from one to the other. When they wish to sow, they make a hole with their finger and throw in it two grains of maize and a bean; the latter winds around the maize stalks' as it grows . . . The Indians sow potatoes in their maize fields, although separate from the maize. They always keep their fields weeded and clean and the soil well cultivated, which helps increase production . . .
>
> (quoted in Knapp, 1991, p. 271)

The interplanting strategy with beans and maize most probably included *zapallo* (*Cucurbita maxima*) or *zambo* (*Cucurbita pepo* L.; *ficifolia*) as is practised today in the region. The interplanting of maize, beans and *zapallo* is referred to as the 'three sisters'; implying that they always grow up together. Knapp (1991) interprets some ethnohistorical literature as alluding to potato–maize rotation, but concludes that due to the small amount of cultivars grown in the region at the time and the limits of the ethnohistorical data, the crop rotations and fallow length are difficult to determine.

Evidence of irrigation in northern Ecuador varies; while most sources claim that irrigation systems have only been constructed within the last 150 years, some archaeologists argue that they were there before the Inca (Myers, 1974; Knapp, 1991). Besides facilitating irrigation for agricultural intensification, raised field systems also increased production. Raised fields are virtually non-existent in present-day agriculture, yet many agronomists and archaeologists have been investigating their existence and role in pre-historic Ecuadorian agriculture (Knapp and Denevan, 1985; Knapp, 1991). Raised fields or *camellones* have been discovered throughout northern Ecuador and have been confused with *wachakuna*, which are modern potato and maize ridges employed in the region, but which most probably have been practised since pre-Inca times (Knapp and Denevan, 1985; Sherwood, 1999). Knapp (1991) explains that they are distinct in that 'raised fields always have standing water in their intervening ditches, and consequently have special soil fertility, subirrigation and frost management functions which *wachakuna* in Ecuador do not possess' (p. 147). These same characteristics were attributed to the *camellones* by an indigenous linguist/anthropologist from Cotacachi, although he claimed that these raised fields were actually an Inca invention brought to Ecuador. Either way, evidence definitely indicates that the most recently discovered 'raised fields' are pre-Colombian, either Incaic or pre-Incaic (Borchart de Moreno, 1995).

Another agricultural practice employed during pre-Incaic times and still in use today is reliance on lunar phases as indicator guides for myriad agricultural activities. Indigenous farmers planted aboveground crops as the moon waxed, tubers as the moon waned, and similarly applied manure, weeded and harvested according to the same logic. As the moon waxes, the energy in any organic entity is pulled upward, thereby increasing growth in aboveground crops. As the moon wanes, the energy is pulled into the earth, thereby increasing growth in tubers, and similarly drawing the energy of manure into the ground. According to local indigenous agriculturists today, one should never plant near the time of a full moon. One of our male informants explains: 'I see the lunar phases: [on a] full moon one doesn't plant because it will grow with fungi, like ash, dust, or mushrooms.' The older farmers recommend planting tubers 5 or 6 days after the full moon and aboveground crops approximately 5 days after a new moon.

The INCA Incursion

It was not until the early 16th century with the Incan expansion into the northern Andes from their political centres in Peru that the numerous polities in northern Ecuador were consolidated under one political force. The end of the 15th century and the beginning of the 16th was a time of great turmoil and political change in Ecuador. During the reign of the Incan ruler Pachacuti Inca (1438–1471), trade materials from the Incas were beginning to make their way north into Ecuador. The materials were luxury items such as fine ceramics, llama-wool textiles, and silver goods that eventually made their way into the hands of the local ruling chiefs. After a short period of this type of exchange with their northern neighbours, the Incan trade transformed into a military invasion headed by Topa Inca, Pachacuti Inca's heir (Alchon, 1991).

The Incan expansion was a relatively short military campaign in the northern territories, including present-day Cotacachi, though it was not done without resistance. Many of the chiefdoms resisted even after multiple military campaigns. Archaeological work shows that forts were erected by local chiefdoms throughout central and northern Ecuador to fight off the Inca (Plaza Schuller, 1976). As the Inca made their way further north, they continued to face stiff resistance. The Cayambes and Caranquis, two of the most northerly groups, withstood four military campaigns and 20 years of attempted conquest until Huayna Capac, Topa Inca's successor, finally defeated them at Lake Yaguarcocha (blood lake) in around 1500. According to writings by Pedro Cieza de Leon, more than 20,000 men from the northern region died in that battle. This was the decisive battle for the Inca and marks the end of their imperial expansion.

Immediately following the Inca conquest over the Caranqui, they enacted the *mitmaqkuna*, a requisite labour tribute, and transported the remaining men of the region to northern Peru to work in the coca plantations. The fact that the labour tribute requirement was so quickly enforced has been attributed to Inca fear of the inhabitants of the northern chiefdoms (Barros, 1980; Farga and Almeida, 1981). The effects of the battles and the *mitmaqkuna* on the local population were devastating. Spanish chroniclers noted that the fields no longer had workers, and accounts from locals told of the destruction of a once populous chiefdom (Barros, 1980; Farga and Almeida, 1981; Borchart de Moreno, 1985).

In 1525, only a few years after the lands had been conquered, upon the death of Huayna Capac, a civil war broke out between his sons, both being natural heirs to the throne. This threw the newly formed 'northern Incan Empire' into a new wave of violence. The war that ensued between Atahualpa and Huascar, the heirs to the northern empire, once again devastated the populations of the north Ecuadorian chiefdoms. The Inca rulers forced them to show their allegiance by fighting on their behalf, frequently against their own kind. This severely disrupted the economic and political stability of the northern Inca empire, making their dominion incomplete and short lived (Farga and Almeida, 1981, p. 178; Alchon, 1991).

Agricultural impacts of the Inca

The Inca impact on the agricultural practices of northern Ecuador appears somewhat insignificant, having only introduced squash and camelids; however, their impact on the culture is inestimable (Knapp, 1991). The major change was the introduction of more varieties and cultivars of existing crops. Inca technological advancements were minimal, but Inca tools for working the land were probably adopted. The pre-Inca tools for preparing the land, *chacllas*, were described as 'something similar to a wooden spade that they had in order to mash down and smooth out the land'. The Inca employed the use of a *chaquitaclla*, which a Spanish chronicler (1573) depicts in this account:

> The tools with which they work the soil are spades of hard wood, five to six hands long and about one hand wide. In the middle they have a notch which makes a handle for applying force and giving a stronger push. With these they work the soil more easily than with the azadones . . . because they crumble the soil better.
>
> (Anonymous: p. 217 in Knapp, 1991)

The Inca tool is definitely distinct from those used in the Cotacachi region prior to the Inca incursion, and it remains debatable whether or not the Inca introduced the *chaquitaclla* in northern Ecuador.

The Inca demanded tribute similar to the local *curacas*, but in much larger amounts. They expected high quality products to ship back to royalty in Peru. Therefore, they implanted textile manufacturing workshops in northern Ecuador where local artists were taught new techniques and also to use the wool of the llama and alpaca, which previously were rare in the region. The tribute demands of the Inca did not destroy the ecological complementarity of the region, rather they held stronger control of the trade links that previously existed (Barros, 1980; Farga and Almeida, 1981; Salomon, 1986). The Inca also enhanced the long-distance trading routes, and *mindala* function, of pre-Inca times. The connections between zones were strengthened rather than severed under the Inca because they were familiar with the utility and power inherent in the control of vertical systems.

The Spanish Conquest

In the year 1534, only a few years after the bloody wars between the two Incan heirs and approximately 40 years after the Inca had initially penetrated the northern territories of what is now Ecuador, the Spanish arrived in the area. The Inca culture had not taken a strong hold. In fact, many of the pre-Incan chiefdoms' languages and customs still persisted. Yet, the physical presence of storehouses, fortifications, Incan officials and conspicuously vacant villages throughout the region, in addition to the Inca language, Quechua, symbolized Inca political reign (Alchon, 1991). The Spaniards utilized the remnants of the Inca subjugation, their language, to extend their dominance. By the 1530s, the Spaniards began establishing Quechua as the *lingua franca* and language of administration in and around the Quito region of Ecuador. The *lingua franca* became known in Ecuador as Quichua. The Spanish reinforced the transition to a unified language for their own administrative ends and contributed to the fact that today, 500 years later, either Quechua or Quichua is spoken from Chile to Ecuador, despite the short-lived reign of the Inca (Salomon, 1986).

Demographic impacts

To acquire an estimate of the population in the northern Andes before 1492 is a complicated task. Cieza de Leon, a Spanish chronicler in the Andes, stated that, 'In the past the area must have been much more populate . . . One cannot travel anywhere (except for the most broken and difficult [terrain]) without seeing that the land had been populated and worked' (Cieza de Leon, 1553). By the time Cieza de Leon wrote this account in the 1540s, the populations had been subjected to the Incan conquest, the civil war between the heirs to the northern empire, the Spanish conquest and, arguably the most destructive force of all, Old World diseases. In fact, the worst epidemics in northern Ecuador preceded the Spanish, occurring between 1524 and 1533. The diseases had migrated north with Incas already in contact with Spaniards in Peru. It is possible that smallpox and measles could have worked their way south into Ecuador from central Mexico, where they were introduced by the Spanish in 1520, but the timing is such that it is more likely they were carried north by the Inca who were also greatly affected by the diseases (Alchon, 1991). The Old World diseases had such a debilitating effect on both the Inca and the chiefdoms of the north that the Spanish conquest was not one of great length or difficulty. Diseases such as smallpox, measles and plague cleared the way for the advancing Spanish and created an environment of severe virulence that raged in waves of epidemics throughout the mid-16th century. Besides disease, earthquakes, floods, frosts, unpredictable climatic cycles, volcanic eruptions and droughts also displaced large numbers of people and destroyed agricultural yields in every decade of the 1600s (Alchon, 1991; D'Altroy, 2000).

Impacts on agroecological complementarity

The situation when the Spanish arrived was explained previously; the exportation of men for tribute labour, disease and devastating battles had liquidated the once dense population of the north Ecuadorian landscape. The inhabitants who worked the fields only a generation or two before had disappeared. Due to the apparent absence of landowners or occupants, the conquistadors quickly assumed control of entire watersheds, selecting the choice valleys with water access for themselves. The women and children and few men that were still there could do little if anything to stop the appropriation of their traditional lands.

The Spaniards appropriated and employed the Inca and pre-Inca sociopolitical structures that remained in place (Oberem, 1978; Barros, 1980; Farga and Almeida, 1981). The Spaniards were familiar with the *mita*, the Inca tribute system, and quickly re-employed it to their own ends. They put an *encomienda* system into effect, which required that able-bodied indigenous people work on the haciendas (Oberem, 1978). They also demanded tributes of local goods, such as cotton articles and agricultural products (Cisneros, 1990). The ecological complementarity of the past was quickly dismantled as haciendas blocked traditional trade routes and denied local peoples access to the land where they had previously foraged or farmed (Wachtel, 1976; Farga and Almeida, 1981). Some of the traditional products grown, foraged and traded within the system of ecological complementarity began to be employed less, as substitutable products brought over from the Old World replaced them. The Spaniards strove to maintain independence within their haciendas; this in turn precluded the perpetuation of trading between ecological zones (Oberem, 1978; Farga and Almeida, 1981; Cisneros, 1990). The traditional uses of cotton and *Cabuya Blanca* (*Fourcroya andina* Trel.) (see Campbell, Chapter 17, this volume) in textile production quickly began to lose out to the sheep wool introduced by the Spaniards (Cisneros, 1990).

Sheep and other domesticated animals brought by the Spaniards had serious impacts on the landscape of the region. Because haciendas were located in the lush inter-Andean valleys rather than near the páramo, where animals were traditionally pastured, hacienda owners began to clear forest patches or appropriate indigenous maize fields for their flocks of sheep and cattle (Barros, 1980; Farga and Almeida, 1981). The result of hacienda appropriation of ideal agricultural land in the valleys, with primary access to water and firewood, was the movement of indigenous peoples further up the mountainside. An indigenous man from present-day Cotacachi describes the process:

> Before they didn't speak of private properties. Here they were communal . . . Due to the conquest our indigenous compañeros searched out refuges very high so that the Spaniards wouldn't arrive there . . . It was in the Spanish epoch when they searched out refuge so no one would bother them.

Compared with the indigenous inhabitants of the region, the arriving Spaniards had extremely different perceptions of the landscape. Not only did they see the natural surroundings strictly as a resource to be capitalized on and owned privately, which differed significantly from the native conception of the land, they also included the indigenous inhabitants as part of the land to be exploited. When the Spaniards usurped a valley or large plot of land, they expected and forced the indigenous peoples living there to remain and work for them. When the Spaniards first arrived in Ecuador, they began to force indigenous peoples, perhaps *mindalaes*, to return to their original communities so that they could force them to work for the *hacendado* who had stolen their land (Barros, 1980; Farga and Almeida, 1981). Records show that when a hacienda was sold, the indigenous inhabitants were included in the transaction. In one specific instance, in 1751, an *hacendado*: 'complained before the *Audiencia* against some caciques that he hadn't turned over the 24 tribute workers to whom he had rights. He said that without the right to the tribute workers his father never would have bought the hacienda' (translation – authors; Oberem, 1978, p. 53).

Colonial Period

While the origin of the *latifundio*, large estate system, can be located in the original distributions of indigenous lands to the conquistadors, the *huasipungo* system emerged in the 18th century. The *hacendados*, who were no longer content with the rotating, but free, work of the *mitayos* (*encomenderos*), began to devise ways to retain the constant services of the indigenous workers. The major mechanisms employed in order to tie the indigenous workers to their land (*huasipungo* system) were the monopolization of keystone resources such as water and firewood, and increased tribute demands, both of which increased the dependence of the inhabitants upon the *hacendados*. Within the *huasipungo* system, the Spaniards consolidated a process of usurpation, a more complex and thorough socioeconomic subjugation of masses of indigenous peoples, ensuring the monopolistic control of natural resources and a steady work supply. Eventually, the indigenous peoples were forcibly indebted to the *hacendado*. They were obligated to remain on the haciendas, working for access to resources that, only a generation or two previously, were shared communally among their families. As one indigenous farmer relates:

> . . . The poor campesino had to work for necessity and they forced him to work and indebted him, giving him credit for 10, then 15, then 20. Afterwards came months and from there years, then the poor man could never pay, right? Then that was turned into life on the hacienda, for one works for necessity not in order to pay. They became more and more in debt. Eventually they had to work for the hacendado forever.

Some indigenous peoples escaped this fate by running off to higher elevations, evading the Spaniards. Survival at higher elevations, however, was a difficult task because the majority of arable land and good water sources below were controlled by the *hacendados*. As Transito Guandinango, an indigenous woman from El Batan, explains:

> We couldn't even pick berries. If they found us, they hit us; like that they came to take my parents obligatorily to work and didn't pay them anything. They frequently took them far away for a week at a time or two

> weeks . . . Before it was really sad because the access to the hacienda was denied, one couldn't even enter, not even pick herbs nor firewood. If they saw you enter they were quickly behind to whip us.

Within her recounting, the elements of the *mita* appropriated by the Spaniards are evident when she speaks of her parents being taken elsewhere to do work without being paid. While some communities in Cotacachi have no recollection of haciendas on their land, most of the region was previously dominated by *hacendados* and many still exist. A former president of UNORCAC describes the situation: 'They came to colonize all the land . . . but here in Imbabura we see an experience that Cotacachi has lots of haciendas, each square from here to there, beginning from Cuicocha until the north there's a hacienda.' An agricultural census from 1943 indicates that at least 45 haciendas existed within canton Cotacachi at that time (Mendizabal, 1999).

Impacts on agriculture and land management

Spanish colonial rule broke up the very foundation of the natives' existence – vertical, multizonal, agroecological subsistence. The large estates damaged the subsistence systems that had developed over hundreds of years by restricting the movements of indigenous peoples and taking away access to their lands. Villages aligned themselves so that different resources and crops could be harvested from several agrozones simultaneously. In such a way, indigenous farmers minimized the risk of famine.

The land management practices of the haciendas, however, differed greatly from those of the indigenous peoples. Hacendados frequently told the indigenous peoples that in exchange for clearing more forest area, they would receive more access to cropland and pasture. The latter rarely occurred, but the former was usually realized. Salomon (1986) provides a synopsis of the commonality amongst several Ecuadorian altitudinal, ecological, agronomic and bioclimatic research report:

> All these authors take as their point of departure the fact that the inter-Andean environment has been radically transformed by human use. Agriculture in its lower parts and grazing in its upper have removed an original vegetation which once included extensive forests, and have rendered the soils susceptible to severe erosion. These processes had not advanced so far in pre-Columbian times and it must be borne in mind that the aboriginal chiefdoms inhabited countryside quite unfamiliar to modern Quiteños.
>
> (p. 35)

These processes occurred because of a drastic change in land management. Unlike the local practice of intercropping in small fields and exchange between zones, the hacienda employed large-scale, intensive agriculture in order to ship large surpluses back to Spain. The crops demanded in the Old World were not traditional Andean crops; therefore, the perpetuation of local cultivars and varieties occurred mostly on the small *huasipunguero* plots and in the páramo refuge communities. One must also assume that due to the destruction of the exchange system between zones, the traditional trading of landraces was disrupted.

The haciendas eventually adopted some local crops and, likewise, the indigenous peoples adopted many Spanish cultivars and tools. Specifically, according to the Ministry of Agriculture, in 1985, 90% of the cultivated land in three northern provinces, including Imbabura, was planted with eight crops, five of which were introduced by the Spaniards in the 16th century (Knapp, 1991). The three principal Native American crops that continue to be grown are maize, potatoes and beans; the five introduced Old World crops are barley, wheat, fava beans, peas and lentils. Wheat and barley appear to have been integrated into the diet quite rapidly, as evidenced by Cieza de Leon's remark from 1553: '. . . Knowing the benefit and utility of wheat and barley, many of the natives . . . sow them, consume them and make grogs of barley' (quoted from Knapp, 1991, p. 101). The local campesino organization, UNORCAC, however, downplays the significance of introduced crops even today, stating that in the local small-scale farms, the production of maize combined with beans and sambo is predominant and 'are the base of traditional alimentation of the indigenous population of Cotacachi. In certain cases, one rotates the cultivation of corn with peas, potatoes, wheat and barley' (UNORCAC, 1996). The main significance of this quote is that introduced crops have still not replaced native crops as the nutritional base.

The indigenous peoples also adopted the use of steel when available and utilized it for their hoes (*azadon*) for ploughing fields. They also quickly integrated the Old World domesticated animals into their agricultural practices. The most common example is the use of cattle in the *yunta*, or yoke, which was used to plough the fields and is currently a widespread practice. Besides the utility of introduced animals in agricultural work, they also served as an abundant and convenient fertilizer source. The Cotacachi autodiagnostic (1996) attests: 'Mechanical techniques in soil tillage are not applied, rather the *arado* with the yunta is used. Animal waste, crop residues, and organic waste are used as fertilizer' (UNORCAC, 1996).

The Catholic Church

Throughout the colonial period, the Catholic Church played a major role in the fragmentation of the landscape. The church held title to large tracts of land throughout the Cotacachi region; the same land that was traditionally worked by the indigenous people. Bishops and priests from surrounding cities and towns demanded tithes from the indigenous farmers that were paid to the church, usually 10% of the harvest. They also enacted *mita* demands on the indigenous peoples, expecting them to work on their haciendas in exchange for religious services, such as blessing their fields or seeds. Even today, some of the largest land holdings in Cotacachi, such as 'El Sagrado', 'La Compañia' and 'El Rosario', are haciendas that retain their original names from the era when the lands were controlled by the church. The impact of the Catholic ideology, forced upon indigenous peoples by both the *hacendados* and clergy members, is immeasurable, but definitely profound. Indigenous peoples came to believe in the power of Catholic saints and clergy and integrated the Catholic ritual and worldview into their indigenous cosmovision.

National Period

The majority of exploitative practices and injustices of the colonial period that deprived indigenous inhabitants of Cotacachi of the freedom to engage in traditional agriculture continued well after national independence. A local Quichua man recounts:

> Before it was terrible; when one wanted to enter the hacienda, or cross the land of the hacienda, if an administrator saw – they carried us to work for free . . . we worked and they didn't even pay us one real; then the things were painful, mistreatment of people. All the water, firewood, everything belonged to the *hacendado*.

Although independence from Spain was declared for Ecuador in 1822, the indigenous *huasipungueros* and *encomenderos* were not liberated. In fact, for the indigenous peoples of Cotacachi, freedom from colonial rule resulted in another set of 'official' oppressors. Provincial government leaders and employees forced the indigenous peoples to pay tribute to the state through obligatory labour on infrastructure projects. They were violently threatened and forced to participate in back-breaking work as if they were beasts of burden. The government administrators frequently abused their powers and committed acts of violence and rape in indigenous communities, and indigenous peoples had no recourse. Therefore, while the transition from colonial to national government may have transformed economic relations in Ecuador as a whole, it changed the exploitative social existence of indigenous peoples little. This local history comes from stories that we have collected from indigenous peoples throughout the Cotacachi region.

Through the latter half of the 20th century, these injustices continued and spurred the formation of UNORCAC, a union of organizations that originated to combat social injustice (Rhoades, Chapter 1, this volume). The primary difference between the national and colonial period is that indigenous peoples had a federal government to which they could decry their mistreatment and demand intervention. Frequently, indigenous peoples from Cotacachi walked for weeks from their homes to the capital of Quito in order to speak to officials about the oppression they experienced at the hands of *hacendados* and provincial government officials.

Agrarian reform

A top-down process that has had myriad impacts on the landscape of Cotacachi is the agrarian reform movement, which occurred in Ecuador between 1964 and 1973 (Valerezo, 1999). Although the explicit intent of the movement appears genuine and progressive, the implementation of the plan backfired in several ways. The intent was to break up the *latifundios*, or large estates and haciendas, in order to distribute the land more equitably; however, the major consequence of the restructuring was less access to key resources for the small farmer. Murra (1985) states:

> In the last decades, the agencies for agrarian reform of the various republics with Andean populations have continued the process of breaking up the archipelagoes and impoverishing the inhabitants, since they still have not realized the existence of Andean patterns of simultaneous exploitation of various ecological 'stories' by one and the same population.
>
> (p. 18)

Traditional ecological complementarity continued to be dismantled under the most recent attempts at equity. Oberem (1978) explains that any *huasipunguero* who had worked the land of an *hacendado* for at least 10 years should have received that property as his own (pp. 64–65). Examples from Cotacachi demonstrate that more often than not the local peoples were not aware of the new law. They either never received the land, or the more common scenario throughout Cotacachi was that the landowner gave a small parcel of degraded land without any water access to the *huasipunguero*, acting as if it was a kind gesture of his own volition. When government officials eventually came around, if they came around at all, the *huasipungueros* could no longer make any claims for more land. As the president of El Batan, a community surrounded by hacienda, illustrates:

> Regarding agrarian reform, I would say that we almost haven't practised it here in Cotacachi. We wouldn't have these haciendas here now if we would have practised it, already these lands would have belonged to our indigenous compañeros. But an application of the reform hasn't arrived here. Perhaps for no knowledge of the laws; the hacendados took advantage of that; they said: 'Look, we are very caring and we are going to give you this piece of land. Take it quick, make some boundaries. They took advantage in this manner and because of that we are a community within the hacienda. Because our ancestors didn't take advantage applying the good laws that were given as a priority to those that worked the land . . . Each kilometre one finds haciendas and the best lands are within them, while the bad lands are left to the indigenous communities . . . I have seen no change in land tenure for twenty years . . . I see that there haven't been changes, except our compañeros from Tunibamba who bought their hacienda.

It is evident from aerial photo interpretation and ground truthing that land reform was not carried out sufficiently, with as much as 80% of the population currently living on 20% of the land. Data collected during 2003 in the indigenous communities surrounding Cotacachi show that land reform merely granted the natives land that they had historically occupied on the large haciendas, but did not break up the large estates. Indigenous community lands are typically small and less desirable for cultivation compared to the haciendas. Similarly, the lands allocated to the indigenous populations are insufficient to withstand the pressures of generational population growth and result in out-migration, low agricultural productivity, erosion and social distress (see Zapata *et al.*, Chapter 4, this volume).

Green Revolution

Señor Flores, an indigenous man who resides in a community in the cereal zone in the Cotacachi highlands, illustrates the changes he has seen and felt because of the predominantly government-sponsored Green Revolution:

> Before we only planted with organic fertilizer. We never utilized chemicals. It wasn't necessary to put chemical fertilizers down, but some mestizos gave us a demonstration in small parcels and sure enough with the chemicals it produced some huge potatoes. After seeing that almost all of the community began to utilize chemicals. Now, I've utilized chemicals since I was young, because some people came to indicate how to use them. They came here with seeds to be planted together with the chemicals, but here sometimes we didn't have the seeds . . . Now we tend to fumigate up to 3, 4, 5 times during development, but when we fumigate various times we have the risk of contaminating the fruit, and then it's not worth eating. Before we had more or less 15 varieties of native potatoes, but now we have lost them; potatoes, varieties such as Pintadas and Navis blancos, white eyes. Now we put chemicals on everything: potatoes, mellocos, ocas, trigo, cebada. We didn't know the tractor before; it was only tilling by hand and with the yunta.

The issues that Señor Flores addresses are poignant and devastating. The Green Revolution technologies introduced into indigenous communities of northern Ecuador by the Food and Agriculture Organization (FAO), Ministry of Agriculture, Andean Mission, and other governmental and non-governmental organizations (NGOs) typically consisted of chemical fertilizers, herbicides, pesticides, fungicides, improved seed varieties dependent upon agrichemicals and access to tractor use (Brush, 1976; Field, 1991; Ponting, 1991; Frolich *et al.*, 2000). One group of agricultural researchers (Frolich *et al.*, 2000) in Ecuador explains the impact of the introduction of the Green Revolution:

> In the northern Andes of Ecuador, the recent and rapid spread of industrial technologies, in particular, tractors, synthetic fertilizers and pesticides, threaten the productivity and long-term sustainability of the farming system, in particular with regard to soils and pest management. These changes in agricultural practices have been accompanied by a dramatic and worrisome loss of genetic resources.

Improved varieties have instigated a significant loss of traditional cultivars. Some farmers are now reliant on chemicals and no longer self-sufficient. Larry Frolich, a professor and researcher at the Catholic University in Ibarra, Ecuador, explains that in Carchi, in Ecuador's north sierra, 'Only 50 years ago, dozens of potato varieties were consistently grown with no use of fungicides, while today, even the most resistant chauchas (traditional varieties) need several applications' (Frolich *et al.*, 2000). UNORCAC (1996) reports that while some communities and individuals have utilized agrichemicals, they are not frequently applied because of their prohibitive cost. While UNORCAC claims that there is no tractor use by small-scale farmers, our findings demonstrate an increasing implementation. This may be attributed to availability of tractors for rent by the hour and also the recent phenomenon of cattle theft. Numerous indigenous people lamented their inability to till their land with a *yunta* because their cattle had been stolen. Although the tractor costs more per hour, the land preparation may be done in an hour, whereas a *yunta* may take several days. These trade-offs have serious consequences. Local researchers explain the repercussions:

> The use of tractors on relatively moderate to steep inclines has resulted in the translocation of tremendous quantities of soil down below. There has been a consistent tendency throughout the Ecuadorian Andes; in Carchi the mechanized tillage on slopes has grown dramatically in the last

> decade, up to the point that the use of tractors has become the paramount cause of physical erosion and degradation of soils.
> (translation – authors; Sherwood, 1999)

Current Trends in Land Management and Agriculture

Land tenure

The contemporary indigenous communities of Cotacachi are products of the historic displacement and fragmentation of the landscape by haciendas and the Catholic Church. They are small fractions of haciendas that were granted either by the estate owners or the government during agrarian reform. In some cases, land was purchased from *hacendados* or the church by indigenous farmers in the early part of the 20th century.

Tunibamba, a village within Cotacachi, has been fighting in court over the rights to the hacienda where they and their ancestors worked as *huasipungueros* for centuries. Recently, they won the court case and the sign outside the land now reads: EX-HACIENDA TUNIBAMBA and TUNIBAMBA LUCHA POR LA TIERRA (Tunibamba fights for the land) (UNORCAC, 1996). Similarly, other communities, with the assistance of UNORCAC, are taking *hacendados* to court for control of the lands. Some of the most salient present-day problems of Cotacachi residents tie directly into the dismantling of the microverticality system by Spanish conquistadors. The inaccessibility of essential resources, such as water and firewood, because of hacienda monopolization continues: the 1996 census shows that 70% of water sources in Cotacachi are located on hacienda property (UNORCAC, 1996). To exacerbate this problem, the haciendas have utilized agrochemicals of all sorts for at least 30 years, which has led to serious contamination of drinking water. The president of a local Cotacachi community explained that his community remains under pressure from a Spanish NGO to use agrochemicals on communal land. They received credit from the NGO and have used chemicals on communal lands in the past, but consensus in the community now leans toward the abandonment of agrochemicals on communal land.

During the research period, this same community called a town meeting to discuss the state of agriculture whereby questions could be directed to the entire community. The communal discussion and decision making was not anomalous for the community. The main question relevant to this historical ecology research that was addressed by the community revolved around the *sabiduria de nuestros abuelos*, the knowledge of our ancestors. The community divided into 'focus groups', discussed the issue, came up with responses and then shared these with the entire community. The responses included the following: a long list of cultivars that the ancestors traditionally grew with an emphasis on a wide variety of crops; the tradition of planting by the phases of the moon; the exchange of cultivars within the community and family; and the loss of traditional crops and agricultural practices. The emphasis tended toward a loss of traditional practices. As one group leader explained, 'We have lost a significant amount of traditions, we now sell our traditional crops and buy foreign goods and foods in the market.' As this communal example demonstrates, things are changing in the indigenous communities and the indigenous people do not want to lose their traditions. They see their culture very closely connected to their subsistence strategies and land.

Agricultural practices

Figures 3.2 and 3.3, graphic representations of the contemporary agricultural calendars, illustrate this interconnection between sociocultural traditions and agricultural practices. As the figures demonstrate, ecological complementarity continues, yet the thorough exploitation of multiple zones has decreased significantly because of landscape fragmentation. Introduced wheat and barley dominate the fields in the high zone. Traditionally, this would be where the native tubers such as *mashua* and *oca* would be grown. According to interviews, mashua, a native Andean root, was an important staple crop only one or two generations ago, yet now it is regarded primarily as a medicine and rarely eaten as food. Maize, however, continues to be the culturally central crop of the indigenous people in Cotacachi. Even farmers in the high zones, who do not have access to fields in the lower zones, grow at least small plots of maize for ritual and nutritional purposes, despite the fact that it takes twice the time to mature than in the lower zones. The cultural significance of maize cultivation cannot be downplayed. The agricultural calendars, as seen in Figs 3.2 and 3.3, indicate the planting dates for various crops in relation to the religious festivals and holidays celebrated by the Quichua in Cotacachi. The various interventions throughout history have not only left an indelible mark on the landscape, they have also transformed agricultural perceptions and practices. The traditional Pan-Andean Inca holiday of the sun, Inti Raymi, corresponds to the Catholic saint's day, San Juan, and the maize harvest, which constitute an enormous celebration because of their cultural significance.

In addition to a fluctuating climate, various factors currently contribute to the degradation of traditions (see Campbell, Chapter 17, this volume). With the robbery of animals (lack of organic fertilizer) and growing dependency on agrochemicals, tractors and improved varieties, traditional practices become more and more difficult to employ. As an indigenous woman despairs:

> So, now, already, after 20 years, around here the people are asking themselves what happened to their traditional agricultural technologies. I think that also the pressure from large companies, they say, 'look, I am going to offer you this, but you have to do a promotion.' So, the product gets spread around and they get more money, and it

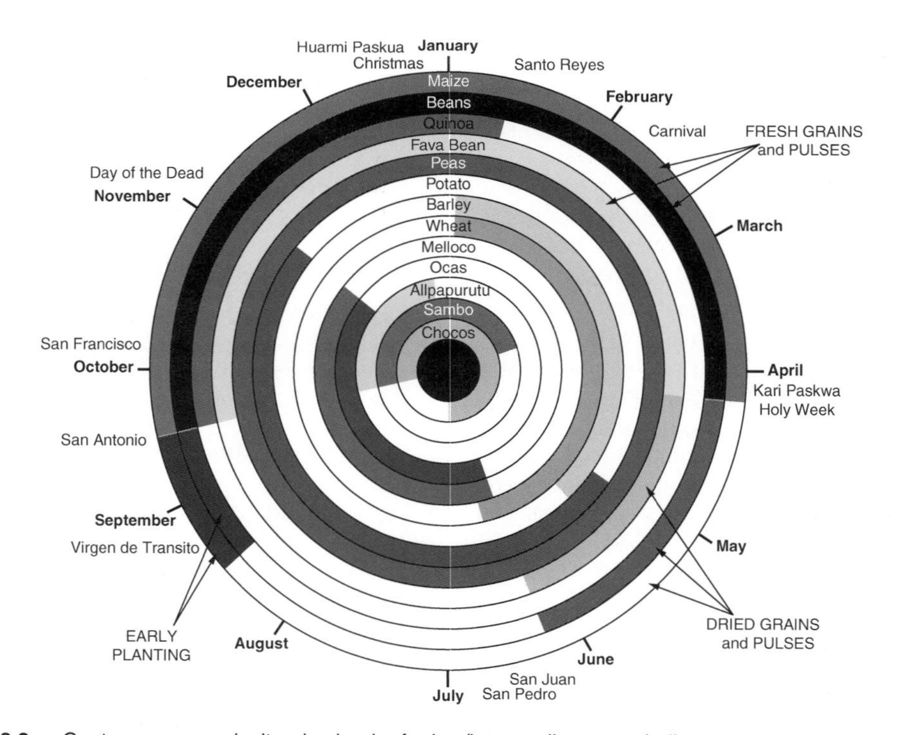

Fig. 3.2. Contemporary agricultural calendar for low/intermediate zone indigenous communities.

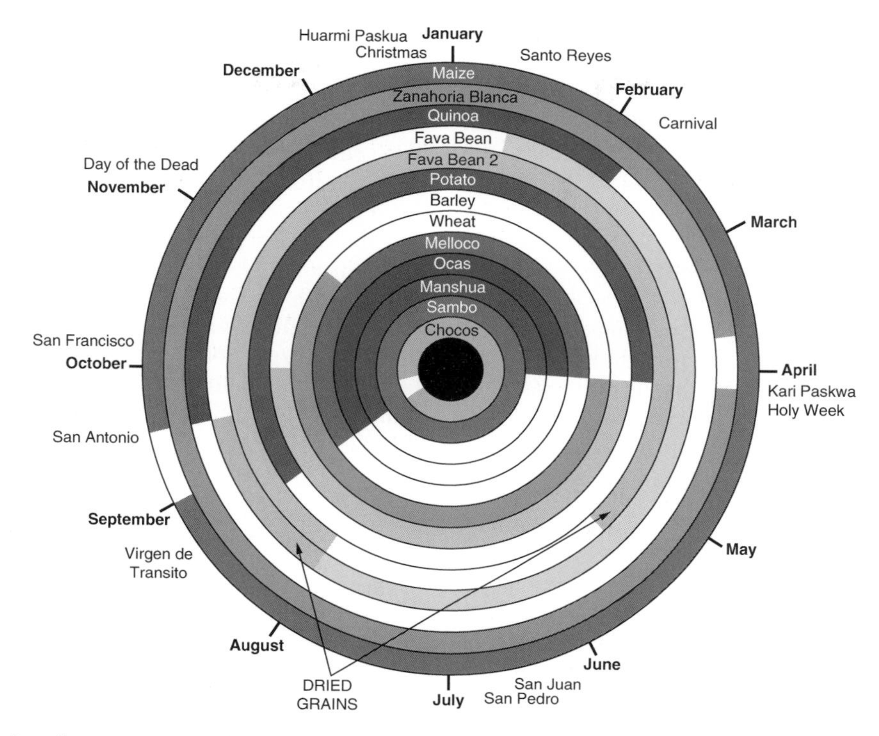

Fig. 3.3. Contemporary agricultural calendar for high zone indigenous communities.

> doesn't matter to them if they're damaging the environment, the land; but it's the other way around for us because we have a scarcity of food sources. What happened with our nutrients? Now, the government tells us that we have to return to our traditional agriculture; but how do we return if the lands are already destroyed?

This is definitely the question. Yet, one may also ask how the agricultural, cultural and biological diversity of the region has survived until now, especially with such impediments. Ecological complementarity proved an aboriginal solution to their vertical environment and has overcome exogenous attempts to dismantle it (Mayer, 1985). Murra (1985) professed:

> It is remarkable that, in spite of the pressures exercised against everything Andean and those who created them during the 450 years of colonial and republican regimes, we still encounter among highland peasants a preference for locating their fields in complementary fashion, on several different ecological tiers, sometimes located several days' walk from the center of population. There is well-documented, contemporary evidence of sizeable groups who have managed to maintain their ethnic self-awareness, along with access to their outliers in the lowlands . . . ecological complementarity was a major human achievement, forged by Andean civilizations to handle a multiple environment, vast populations, and hence high productivity. It helps us understand the unique place of the Andean achievement in the repertory of human histories; it may even point to future possibilities.
>
> (pp. 9–11)

Conclusion

While this historical ecology has demonstrated that the indigenous peoples of Cotacachi developed a very effective subsistence strategy based on the management of several agroecological zones, these practices have been continuously hindered. The underlying theme has been the obstruction or inability of the indigenous inhabitants to pursue their traditional adaptations. These obstructions are twofold: sociopolitical and environmental. In terms of sociopolitical disruptions, the Inca, Spanish and national interventions all denied the local inhabitants the ability to engage in their traditional subsistence strategies. The Inca violently subjugated and deported, the Spanish enslaved and fragmented the landscape, and the national governments forced tributes of labour and harvests, and these atrocities barely describe the impositions. The sociopolitical exploitation of the indigenous inhabitants cannot be underestimated and represents significant obstruction to traditional ways of life. The second obstruction to traditional subsistence strategies resides in the environmental inconsistencies and catastrophes that plague the region. Although Alchon (1991) describes the region as abundant, we must not forget that the area is plagued by volcanic eruptions, drought, heavy rains, frosts and El Niño, which 'has been a significant factor in human life [in the area] over the last 5,000 years' (D'Altroy, 2000, p. 359). These interventions, fragmentations and traditions demarcate the turbulent historical ecology of Cotacachi.

References

Alchon, S. (1991) *Native Society and Disease in Colonial Ecuador.* Cambridge University Press, Cambridge, UK.

Barros, H.L. (1980) *Demografia y Asentamientos Indigenas en la Sierra Norte del Ecuador en el Siglo XVI.* Vols 11 and 12. IOA, Otavalo, Ecuador.

Borchart de Moreno, C. (1995) Llamas y Ovejas: el desarrollo del ganado lanar en la Audiencia de Quito. In: Yala, A. (ed.) *Colonizacion Agricola y Ganadera en America Siglos XVI–XVIII.* Ediciones Abya Yala, Quito, Ecuador, pp. 153–190.

Bruhns, K.O. (2003) Social and cultural development in the Ecuadorian highlands and eastern lowlands during the formative. In: Quilter, J. (ed.) *Archaeology of Formative Ecuador.* Dumbarton Oaks Research Library and Collection, Washington, DC, pp. 7–32.

Brush, S. (1976) Man's use of an Andean ecosystem. *Human Ecology* 4, 2.

Cieza de Leon, P. (1553) *La Cronica del Peru.* Espasa-Calpe, Madrid, Spain.

Cisneros, H.J. (1990) Tecnicas textiles artesanales en Imbabura. *Sarance (IOA)* 14, 21–41.

Crumley, C. (1993) *Historical Ecology: Cultural Knowledge and Changing Landscapes.* School of American Research Press, Santa Fe, New Mexico.

D'Altroy, T. (2000) Andean land use at the cusp of history. In: Balee, W. and Crumley, C.L. (eds) *Imperfect Balance: Landscape Transformations in the Precolumbian Americas.* Columbia University Press, New York, pp. 357–390.

Farga, C. and Almeida, J. (1981) *Campesinos y Haciendas de la Sierra Norte.* Vol. 30. IOA, Otavalo, Ecuador.

Field, L. (1991) *Sistemas Agricolas Campesinos en la Sierra Norte.* Magenta, Centro Andino de Accion Popular (CAAP).

Frolich, L., Sherwood, S., Hemphill, A. and Guevara, E. (2000) *Eco-papas: Through Potato Conservation Towards Agroecology.* ILEA, Ibarra.

Knapp, G (1991) *Andean Ecology: Adaptive Dynamics in Ecuador.* Westview Press, Boulder, Colorado.

Knapp, G. and Denevan, W.M. (1985) The use of wetlands in the prehistoric economy of the North Ecuadorian highlands. In: Farrington, I.S. (ed.) *Prehistoric Intensive Agriculture in the Tropics.* BAR International Series, Oxford, UK, Vol. 1, pp. 185–208.

Marcos, J.G. (2003) A reassessment of the Ecuadorian formative. In: Quilter, J. (ed.) *Archaeology of Formative Ecuador.* Dumbarton Oaks Research Library and Collection, Washington, DC, pp. 7–32.

Mayer, E. (1985) Production zones. In: Masuda, S., Shimada, I. and Morris, C. (eds) *Andean Ecology and Civilization: an Interdisciplinary Perspective on Ecological Complementarity.* University of Tokyo Press, Tokyo.

Mendizabal, T. (1999) *Medicina Tradicional e Interaccion de Sistemas Medicos en Las Comunidades Andinas del Canton Cotacachi*. Medicos Sin Fronteras, Quito, Ecuador.
Murra, J. (1985) The limits and limitations of the 'vertical archipelago' in the Andes. In: Masuda, S., Shimada, I. and Morris, C. (eds) *Andean Ecology and Civilization: an Interdisciplinary Perspective on Ecological Complementarity*. University of Tokyo Press, Tokyo, pp. 3–15.
Myers, T. (1974) Evidence of prehistoric irrigation in northern Ecuador. *Journal of Field Archaeology* 1, 309–313.
Naranjo, M.F., Pereira V.J.L. and Whitten N.E. (1984) *Temas Sobre la Continuidad y Adaptaciâon Cultural Ecuatoriana*. Ediciones de la Universidad Catâolica, Quito, Ecuador.
Newson, L.A. (1995) *Life and Death in Early Colonial Ecuador*. University of Oklahoma Press, Norman, Oklahoma.
Oberem, U. (1978) Contribucion a la historia del trabajador rural de America Latina: 'Conciertos' y 'Huasipungueros' en Ecuador. *Sarance (IOA)* 6, 21–49.
Odum, E. (1997) *Ecology: a Bridge Between Science and Society*. Sinauer Associates, Inc., Sunderland, Massachusetts.
Pease, F. (1985) Cases and variations of verticality in the southern Andes. In: Masuda, S., Shimada, I. and Morris, C. (eds) *Andean Ecology and Civilization: an Interdisciplinary Perspective on Ecological Complementarity*. University of Tokyo Press, Tokyo, pp. 3–15.
Plaza Schuller, F. (1976) *La Incursion Inca en el Septentrion Andino Ecuatoriano*. Instituto Otavaleno de Antropología, Otavalo, Ecuador.
Ponting, C. (1991) *A Green History of the World*. Penguin Books, New York.
Raymond, J.S. (2003) Social formations in the western lowlands of Ecuador during the early formative. In: Quilter, J. (ed.) *Archaeology of Formative Ecuador*. Dumbarton Oaks Research Library and Collection, Washington, DC.
Rhoades, R. and Thompson, S.I. (1975) Adaptive strategies in alpine environments: beyond ecological particularism. *American Ethnologist* 2, 535–551.
Salomon, F. (1986) *Native Lords of Quito in the age of the Incas*. Cambridge University Press, New York.
Sherwood, S. (1999) *Reporte sobre Cultivos de Cobertura*. Centro Internacional de la Papa (CIP), Quito, Ecuador.
Temistocles, H.M. (1999) *La Revolucion Verde IndoAmericana: Tecnologias Agricolas Precolombinas para la Produccion de Alimentos en Armonia con la Naturaleza*. Desde el Surco, Quito, Ecuador.
UNORCAC (1996) *Memoria del Taller de Autodiagnostico en la UNORCAC*. UNORCAC, Cotacachi, Ecuador.
Valerezo, G.R. (1999) *A Social History From Aboriginal Times to the Present: People, Land, and Society in the Nanegal Parish*. SANREM.
Wachtel, N. (1976) *Los Vencidos*. Alianza Editorial, Madrid.

4 Four Decades of Land Use Change in the Cotacachi Andes: 1963–2000

Xavier Zapata Ríos,[1] Robert E. Rhoades,[2] Maria Claudia Segovia[3] and Franz Zehetner[4]

[1]*SANREM–Andes Project, PO Box 17-12-85, Quito, Ecuador;* [2]*University of Georgia, Department of Anthropology, 250 Baldwin Hall, Athens, GA 30605, USA;* [3]*Department of Environmental Engineering, SEK International University, Ecuador;* [4]*University of Georgia, Department of Crop and Soil Sciences, 3107 Plant Science, Athens, GA 30602-7272, USA*

Introduction

One of the key research areas of sustainability science and development focuses on land use/land cover change (LULCC). Land use refers to human exploitation of the land, while land cover denotes the biophysical nature of the land surface. Obviously, these two aspects of the landscape are interconnected: how the land is utilized by humans affects the land cover, and vice versa (Meyer and Turner, 1992). Due to the fundamental importance of LULCC for natural resource management, a central organizing core research activity involved understanding how and why the present landscape evolved over the past 50 years. By preparing land use maps of the area based on aerial photographs at different points in time, we have a deeper understanding of what was happening to both the natural systems and the human-derived use categories. This rather straightforward information, in turn, could be further integrated into insights from local people themselves to yield a much richer understanding of Cotacachi landscape change. Issues of soil erosion, land degradation, shift in agriculture and water uses, migration and inheritance patterns could be more accurately discussed and pinpointed. In fact, the LULCC study by SANREM researchers was considered to be the touchstone from which all other research emanated.

This chapter, therefore, will present the project's land use/land cover analysis from 1963 until 2000 for the Andean zone of Cotacachi canton. The methodology we used to create vegetation cover maps for 1962, 1978, 1993 and 2000 for the study area will be outlined. We will then identify the characteristics of different land use categories, quantify the area corresponding to each land use category by year, and determine the major factors affecting land use change over time.

Objectives

- Create a land use change map for 1963, 1978, 1993 and 2000 for the study area of the SANREM project (Sustainable Agriculture and Natural Resource Management).
- Identify the characteristics of different land use categories.

- Quantify the area corresponding to each land use category by year.
- Determine the major factors affecting changes in land use.

Study Area

The study area is located in the Andean zone of Cotacachi canton, which is in the province of Imbabura in Ecuador's northern sierra (Fig. 4.1). The canton has three general land use zones: urban, Andean and subtropical. Project activities took place primarily in the urban and Andean zones. The two zones mentioned have a surface area of approximately 21,902 ha in the parishes of Cotacachi, Quiroga and Imantag, with altitude ranging from 2080 to 4939 metres above sea level (masl). The study area includes territory within the Cotacachi Cayapas Ecological Reserve. The average temperature is 15°C and total annual precipitation is 1259 mm. Mount Cotacachi, rising to 4939 masl, is an important landmark in the project area. The nearly vertical flanks of the volcano are eroded due to the former action of glaciers. Cuicocha lake, produced during an eruption, is 200 m deep and lies on the southern side of Mount Cotacachi. The hydrography of the higher areas includes the Pichaví, Pichambiche and Yanayacu rivers. In the highland grasslands, known as *páramo*, there are a number of lagoons, including the Donoso de Piñán, Cristococha and Yanacocha. Since they are located near population centres, these lagoons serve as sources of water for irrigation, generation of electricity and domestic use.

The area's vegetation is classified as low montane dry forest (Cañadas, 1983) and humid montane heath (Sierra *et al.*, 1999a). In general, the majority of the study area's vegetation cover has been subject to human intervention, especially at altitudes below 3000 masl where introduced eucalyptus (*Eucalyptus globulus*) forests and croplands are found. The native vegetation forms brush lands and is generally located in ravines, on cliffs and on steep slopes that offer protection from agriculture, grazing and burning. Above 3000 masl, remnants of native forest are also present.

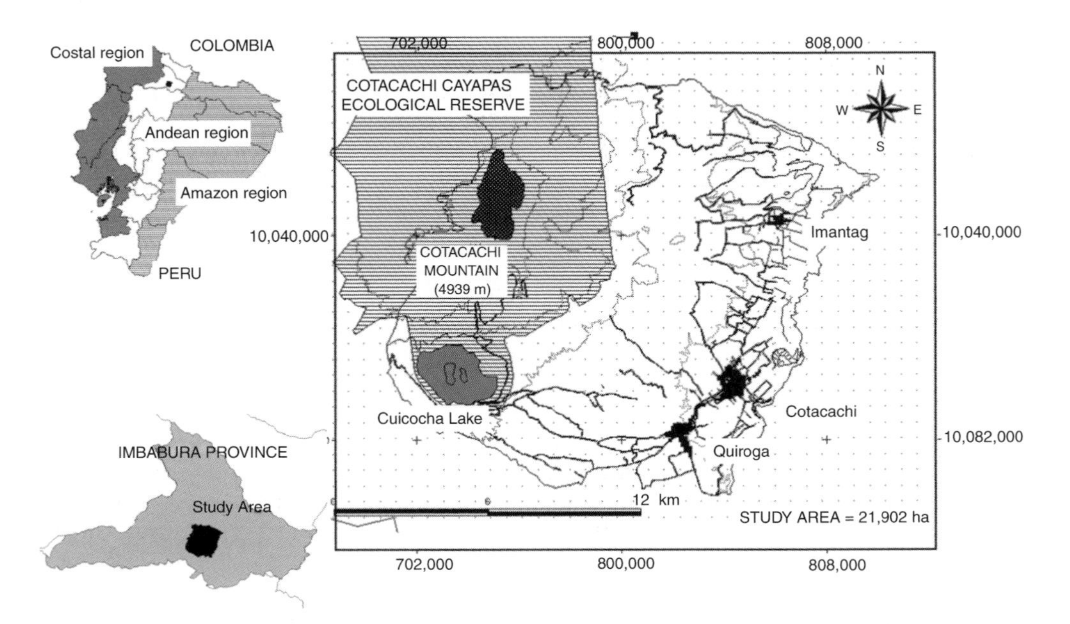

Fig. 4.1. Location of the land use change study area in canton Cotacachi, province of Imbabura, Ecuador.

Land Use Change Analysis Methodology

The methodology utilized to analyse land use change in Cotacachi follows standard techniques and tools of LULCC research. First, we have obtained cartographic information of the study area and prepared a base map at a scale of 1:50,000. In addition, we acquired aerial photos taken by the Military Geographical Institute (IGM, Spanish acronym) for the years 1963, 1978, 1993 and 2000. The available black and white photos are 23 × 23 cm with varying scales. With the base maps and aerial photos, we identified 12 land use categories, which will be discussed in detail here. These categories were confirmed during initial field trips to the study area. The IGM handled the ortho-correction of aerial photographs. This process involves a correction of scanned photographs using control points. Later, the ortho-corrected photographs were geo-referenced using control points obtained from topographic maps prepared by the IGM at a scale of 1:50,000. The control points used were intersections of paved roads, the edges of houses, the boundaries of parcels and the confluence of rivers. We obtained a high degree of precision by geo-referencing the aerial photos with control points obtained from topographic maps (Vanacker *et al.*, 2000). In this fashion, we created four mosaic photographs, one for each of the years under analysis (Fig. 4.2, as an example of 2000). Subsequently, using GIS, we created a grid with cells measuring 1 ha each, with sides of 100 × 100 m. We decided that 1 ha was the smallest mapping unit appropriate for a comparison of the various mosaics. We classified each of the cells on the basis of the 12 categories identified. This classification involved a photographic interpretation process and field verification. The photographic interpretation was undertaken by a single photo interpreter in order to ensure consistency and thus obtain a uniform classification.

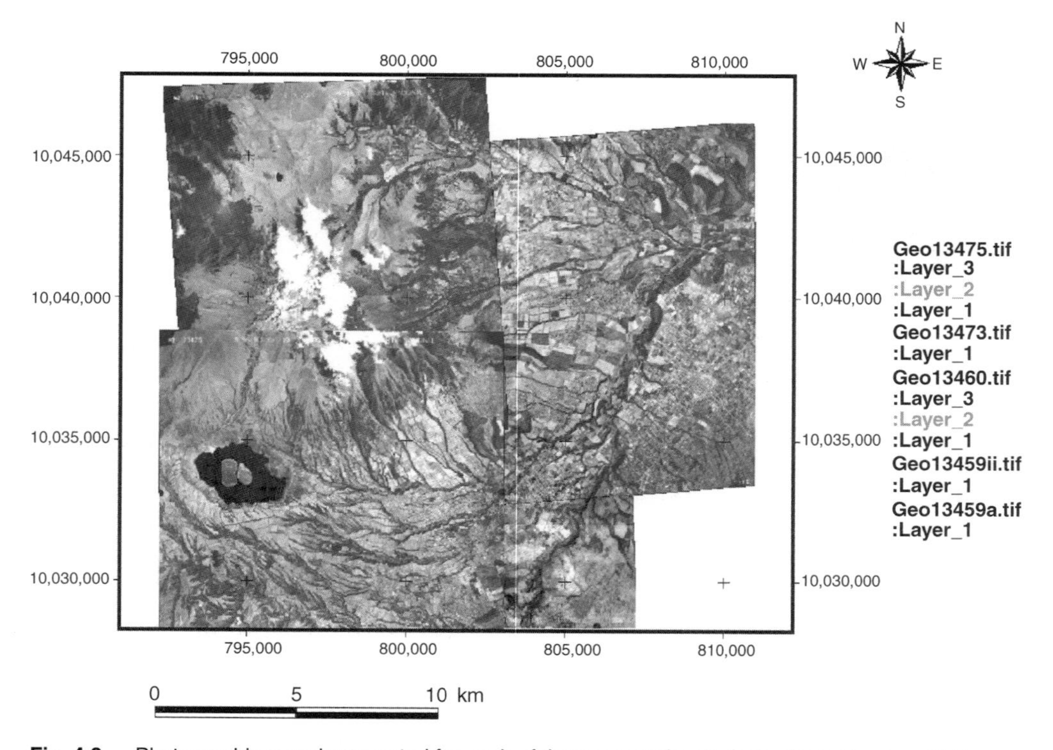

Fig. 4.2. Photographic mosaic corrected for each of the years under analysis. Example year 2000.

All photos were also enlarged up to 1 × 1 m at an approximate scale of 1:10,000. These photos were used as a basis for discussion with community members and other residents of the canton regarding land use. In addition, the photos provided for better interpretation in the classification process.

Results

Land use categories

Our analyses led to the identification of 12 land use categories: páramo grass, brush land, páramo–brush land mix, crops (three categories based on extent of croplands), pastures, native forest, introduced forest, urban areas, bodies of water, Cotacachi mountain and others. We were unable to classify croplands in any greater detail due to the scale of the aerial photos and the differences in the seasons in which they were taken.

Páramo (code 1)

Páramo is the name given to high altitude grassland ecosystems in South America. They are very diverse in flora and fauna. The general characteristics of these zones include: low night-time temperatures, high solar irradiation during the day, frequent fog and high humidity, among others. These unique ecosystems are very important especially in terms of hydroregulation. Ecuador has different kinds of páramos (Mena Vásconez and Medina, 2001). In the study area, we observed grassland páramos. The dominant grass genera in this type of páramo are *Calamagrostis*, *Festuca* and *Stipa* (Fig. 4.3).

The grasslands consist of clump grasses. The major representatives belong to the *Poaceae* and *Cyperaceae* families. These plants are characteristic of dry and arid soils. The long, narrow blades protect the young shoots found inside the clump (meristem). At the same time, the blade's morphology, with its reduced exposed surface area, protects it from the loss of water. In the study area, we observed how these plants came back in an unusual shade of green after a burning. The clumps or grasses are very resistant plants which fulfill a very important role within this ecosystem. The dead blades form part of the organic matter that enriches páramo soil and maintains nutrients on the surface of the soil (Hofstede *et al.*, 1998; Korner, 1999; Mena Vásconez and Medina, 2001).

The rosettes are another type of plant characteristic of the páramo. These are plants that have adapted to living close to the soil in order to avoid the cold. They have also demonstrated a high degree of resistance to being stepped on by animals and thus are often found in livestock grazing lands (Korner, 1999; Mena Vásconez and Medina, 2001).

(a)

(b)

Fig. 4.3. (a) View of a section of páramo (aerial photograph no. 31758 from 1993); (b) typical páramo landscape (photo: Xavier Zapata Ríos).

The páramo has been subject to human influence for thousands of years, and this fact has led to great controversy among scientists about the relationship between the original vegetation and its current situation. Many believe that grassland páramos began as bushy páramos. Others think that the typical páramo flora is better adapted to climate extremes than the trees of the Andean forest, and that this vegetation is produced when the forest disappears due to felling or burning, at which point páramo vegetation invades the area, and acts as an obstacle to regeneration (Hofstede, 2001).

In the study area, we observed a fairly extensive area that had been burned, a practice common in the Ecuadorian sierra. These burnings sometimes reach the edges of forests, leading to the occupation of these spaces by páramo species. The colonization of these plants may be rapid given that the *Poaceae* are dispersed by the wind and are successful colonizers that easily compete with pasture grasses. Once established, this páramo vegetation makes it difficult for the seeds of bushes and trees to take hold. Thus, in burned areas, there are few bush species. This factor inhibits the regeneration of the original vegetation, especially if burning is repeated. Due to these external factors, it is currently very difficult to establish boundaries between the Andean forest and the páramo (Ulloa and Jørgensen, 1995; Hofstede *et al.*, 1998).

When grassland is burned, the flames do not reach all the vegetation. After the burning, all dead matter disappears due to a rapid decomposition process. The nutrients released by the burning are quickly absorbed by the soil and thus cannot be used by new vegetation. Despite the belief that burning encourages the growth of vegetation and the improvement of the soil, it has been proven that there is no change in fertility and that the vegetation has a limited supply of nutrients, i.e. the vegetation that does sprout again does not grow more quickly than the original vegetation. All that is achieved is an ecosystem lower in productivity after the burning and a change in soil structure resulting in reduced water retention capacity. The consequences of burning are intensified by grazing (Laegaard, 1992; Hofstede *et al.*, 1998).

Another factor that affects the floristic composition of an area is the soil, especially the parent material. The soils of páramos included in this study are formed by recent volcanic ash, and are known as Andosols. The soils are young with little differentiation among horizons. Generally, they are francos to sandy, derived from pyroclastic material with < 30% clay in the first metre. The texture is medium, with andic, alofanic properties but also with a good deal of volcanic glass. The soil is black due to its high organic content. Base saturation is low. Water retention at 1500 kPa varies at about 100%. The soil temperature regime, from isomesics to isofrigids, is 10°C. The humidity regime varies from udic to perudic (Hofstede *et al.*, 1998; Medina and Mena Vásconez, 2001).

Brush (code 2)

The brush is composed of bushy vegetation formations, woody-stunted in appearance (Fig. 4.4). Generally, it is thought to be transition vegetation, and is found from 2300 to 2800 masl. Altitudinal variations have also been observed based on local ecological conditions. As transition vegetation, it also varies progressively in its floristic composition and physiognomy, depending on the altitude (Acosta-Solis, 1977; Ulloa and Jørgensen, 1995).

Among the major genera of arboreal and bushy plants in the brush land, we found the following: *Delostoma* (*Bignoniaceae*), *Laplacea* (*Theaceae*), *Saurauia* (*Actinidaceae*), *Weinmannia* (*Cunoniaceae*), *Brachyotum*, *Centronia* (*Melastomataceae*), *Plumbago* (*Plumbaginaceae*), *Caesalpinia* (*Caesalpinaceae*), *Sida*, *Pavonia* (*Malvaceae*) and *Vismia* (*Hypericaceae*), among others.

Páramo–brush land mix (code 21)

The area visited is a transition zone between vegetation of the Andean brow and the páramo proper (Fig. 4.5). This is due to the fact that the Andean brow is the lower boundary of the páramo. The area is also known as a bushy, grassy or low páramo (Jørgensen *et al.*, 1995; Sierra *et al.*, 1999a;

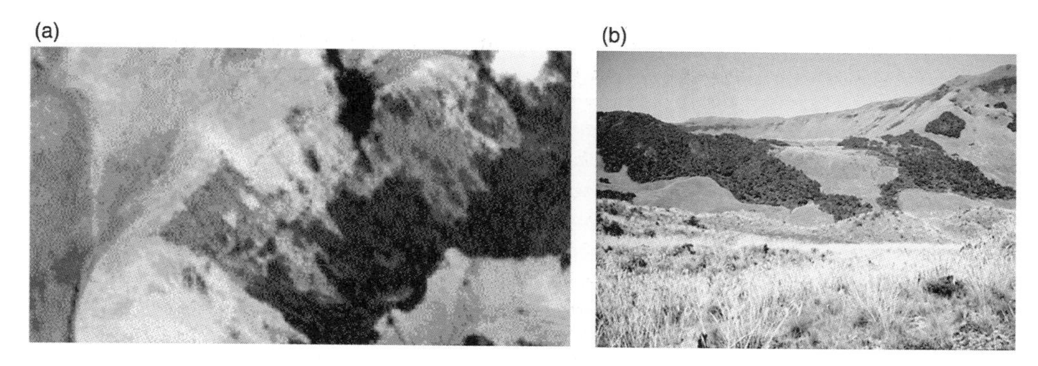

Fig. 4.4. (a) View of an area covered in brush (aerial photograph no. 13473 from 2000); (b) view of a ravine covered in brush (Photo: Xavier Zapata Ríos).

Fig. 4.5. (a) Páramo–brush land mix (aerial photograph no. 13475 from 2000); (b) view of an area of páramo–brush land mix (Photo: Xavier Zapata Ríos).

Valencia *et al.*, 1999; Mena Vásconez and Medina, 2001). This type of páramo is the zone with the greatest diversity of vascular plants because it includes elements of the Andean forest and the páramo (Hofstede *et al.*, 1998). In addition, it is a type of vegetation that is spreading due to the destruction of the Andean forests.

In this type of páramo, grassland is not dominant and the principal species are the cushion species (Mena Vásconez and Medina, 2001). The genera that form cushions are *Azorella*, *Werneria* and *Plantago*. At the same time, bushes and grasses from the *Lycopodium*, *Jamesonia*, *Gentiana*, *Gentianella*, *Satureja*, *Lachemilla* and *Hypericum* genera are found (Table 4.1), and these plants are also common in the high Andean forests (Ulloa and Jørgensen, 1995). All of these plants exhibit specific adaptations in order to resist the inclement climate of this ecosystem, particularly the wide daily temperature swings and the high level of solar irradiation. Among the most common adaptations are pubescence in leaves characterized by a thick cuticle, the small size of plants and coriaceous leaves.

Native forest (code 3)

The relicts of Andean forest are found in ravines, and on cliffs and steep slopes (Fig. 4.6). The vegetal composition of these relicts varies according to the exposure they have had to human activities, as well as to humidity and soil type. This can be seen especially in the size of plants and the presence or absence of woody plants. Many of the woody plants in these relicts have commercial value and have been harvested. As a result, new pioneer species move in and alter the vegetal composition of these relicts. Thus, in ravines near croplands and eucalyptus forests, plants are small bushes no higher than 2 m. The most representative plants are *Baccharis* sp. (*Asteraceae*),

Table 4.1. Representative páramo grasses.

Family	Scientific name	Common name
Apiaceae	*Azorella pedunculata*	
Asteraceae	*Culcitium* sp.	
Asteraceae	*Hypochaeris* sp.	Chicory
Asteraceae	*Werneria* sp.	
Asteraceae	*Hypochaeris sonchoides*	
Blechnaceae	*Blechnum* sp.	Fern
Caryophyllaceae	*Silene* sp.	
Caryophyllaceae	*Cerastium* sp.	
Caryophyllaceae	*Stellaria* sp.	
Coriariaceae	*Coraria ruscifolia*	
Cyperaceae	*Carex* sp.	
Cyperaceae	*Rhynchospora* sp.	
Ephedraceae	*Ephedra* sp.	
Ericaceae	*Pernettya* sp.	
Ericaceae	*Disterigma* sp.	
Ericaceae	*Gaultheria* sp.	
Ericaceae	*Vaccinium* sp.	Blueberry
Fabaceae	*Lupinus* sp.	Lupin
Gentianaceae	*Gentiana* sp.	
Gentianaceae	*Halagenia* sp.	
Gentianaceae	*Gentianella* sp.	
Geraniaceae	*Geranium* sp.	Geranium
Hypericaceae	*Hypericum laricifolium*	Romerillo
Juncaceae	*Distichia muscoides*	
Lamiaceae	*Satureja* sp.	
Lycopodiaceae	*Lycopodium* sp.	
Plantaginaceae	*Plantago* sp.	Llantén
Poaceae	*Calamagrostis* sp.	Grass
Poaceae	*Festuca* sp.	Grass
Poaceae	*Stipa* sp.	Grass
Pteridaceae	*Jamesonia* sp.	Fern
Rosaceae	*Lachemilla* sp.	
Scrophulariaceae	*Bartsia* sp.	
Scrophulariaceae	*Castilleja fissifolia*	Indian brush
Valerianaceae	*Valeriana* sp.	Valerian

Taxonomy according to Marbberley (1997).

Coriaria sp. (*Coriaceae*), *Calceolaria* sp. (*Scrophulariaceae*), *Miconia* sp. and *Blakea* sp. (*Melastomataceae*).

During visits, we observed that altitude is a factor correlated with the vegetal composition of the forest. While a more detailed study of this point is required, it is evident at a glance that the Andean forest relicts are located above 3000 masl in areas which have been subject to limited intervention. These relicts have species characteristic of mature forests, such as *Freziera canescens* (*Theaceae*), *Saurauia* sp. (*Actinidaceae*), *Weinmania* sp. (*Cunoniaceae*), *Oreopanax* sp. (*Araliaceae*) and *Prunus* sp. (*Rosaceae*). Among the bushes, we have *Acaena* sp. (*Rosaceae*), *Gaultheria* sp., *Disterigma* sp., *Pernettya* sp. (*Ericaceae*) and *Arcytophyllum*

(a)

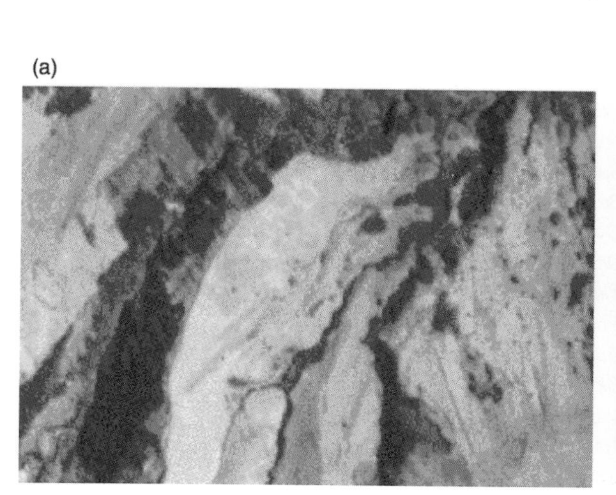

(b)

Fig. 4.6. (a) Native forest (aerial photograph no. 13473 from 2000); (b) view of native forest in the Sacha pasture at Hacienda El Hospital (Photo: Xavier Zapata Ríos).

sp. (*Rubiaceae*). The trees and bushes mentioned tend to form small, dense forests on the least disturbed sites. It is possible that this vegetation was dominant in the zone before the onset of human intervention (Table 4.2).

One of the factors that determine vegetation type is precipitation. Andean forests at sites up to 3500 masl are found in a condensation zone, and thus mist is more frequent. This provides for a high degree of humidity during most of the year, even though precipitation is not abundant. For these reasons, in this type of forest, we find low rates of evapotranspiration, slow growth and little decomposition of organic matter. The trees grow to a height of several metres. The high humidity in the Andean forest creates a microclimate that is ideal for epiphytes and bryophytes (Hofstede *et al.*, 1998; Sierra *et al.*, 1999b).

Temperature is another important factor. It will affect the vegetal composition of the Andean forest, with evident differences between relatively low forests (2400–3000 masl) and high forests (> 3000 masl). Above 3000 masl, low temperatures and poorer soils limit the growth and metabolism of trees, and thus the size of plants and their leaves diminishes.

Introduced forest (code 4)

Throughout the inter-Andean region, eucalyptus (*E. globulus*) forests are common. In the study area, we observed various patches of this type of forest on the edges of croplands. At high elevations, there are also pine forests, such as the forest that belongs to the community of Morochos (Fig. 4.7).

Pastures (code 5)

These are areas planted in pasture for livestock. In this category, we have included areas that have been used constantly as pastures during the last four decades. While it is true that farmers in the zone allow the soil to rest after periods of planting, such areas have been difficult to identify. For this reason, in the study area, there are short-term grasslands which are used later for crop production (Fig. 4.8).

Croplands (codes 61, 62 and 63)

A large percentage of the study area is given over to croplands (Fig. 4.9). A number of

Table 4.2. Representative plants of Andean forest relicts.

Family	Scientific name	Common name
Actinidaceae	*Saurauia* sp.	Moquillo
Aquifoliaceae	*Ilex andicola*	
Araliaceae	*Oreapanax ecuadoriensis*	
Asteraceae	*Loricaria* sp.	Palm
Asteraceae	*Barnadesia* sp.	Crown of thorns
Asteraceae	*Baccharis* sp.	Chilca
Asteraceae	*Gynoxys* sp.	Capote
Asteraceae	*Diplostephium* sp.	
Bromeliaceae	*Tillandsia* sp.	Huicundos
Bromeliacea	*Puya* sp.	Puya
Buddlejaceae	*Buddleja incana*	Quishuar
Campanulaceae	*Siphocampylus* sp.	
Caprifoliaceae	*Viburnum* sp.	
Chloranthaceae	*Hedyosmun* sp.	
Coriariaceae	*Coriaria ruscifolis*	Shanshi
Cunoniaceae	*Weinmania pinnata*	Cashco or oak
Equisetaceae	*Equisetum* sp.	
Ericaceae	*Pernettya* sp.	
Ericaceae	*Disterigma* sp.	
Ericaceae	*Lycopodium* sp.	
Lycopodiaceae	*Lycopodium* sp.	
Melastomataceae	*Miconia* sp.	
Melastomataceae	*Blakea* sp.	
Monimiaceae	*Siparuna* sp.	
Myrsinaceae	*Geissanthus* sp.	
Myrsinaceae	*Myrsine* sp.	
Myrtaceae	*Myrcianthes* sp.	Myrtle
Piperaceae	*Piper* sp.	
Poaceae	*Chusquea* sp.	Suro
Rosaceae	*Prunus* sp.	
Rosaceae	*Acaena* sp.	
Rosaceae	*Hesperomeles* sp.	Huagra manzana
Rubiaceae	*Palicournea* sp.	
Rubiaceae	*Psycotria* sp.	
Rubiaceae	*Galium* sp.	
Rubiaceae	*Arcytophyllum* sp.	
Rutaceae	*Zanthoxylum* sp.	
Sabiaceae	*Meliosma* sp.	
Smilacaceae	*Smilax* sp.	
Solanaceae	*Cestrum* sp.	
Theaceae	*Freziera canescens*	Quatze
Valerianaceae	*Valeriana* sp.	Valerian
Verbenaceae	*Durantha triacantha*	

Fig. 4.7. (a) Eucalyptus forest in the community of Quitugo (aerial photograph no. 13459 from 2000); (b) view of a eucalyptus forest in the community of Quitugo (Photo: Xavier Zapata Ríos).

Fig. 4.8. (a) Pastures (aerial photograph no. 13473 from 2000); (b) view of the Sacha pasture at Hacienda El Hospital (Photo: Xavier Zapata Ríos).

Fig. 4.9. (a) View of La María Hacienda and the community of Colimbuela (aerial photograph no. 13460 from 2000); (b) view of croplands (Photo: Xavier Zapata Ríos).

crops have been part of the diversity of high Andean regions for centuries. The most common of these include, principally, maize (*Zea mays*, *Poaceae*), potato (*Solanum tuberosum*, *Solanaceae*), quinoa (*Chenopodium quinoa*, *Chenopodiaceae*), broad bean (*Vicia faba*, *Fabaceae*) and lupin (*Lupinus mutabili*, *Fabaceae*). In addition, there are Old World crops that have adapted well to high Andean conditions, such as onion (*Allium cepa*, *Lilaceae*), cabbage (*Brassica oleracea*, *Brassicaceae*), barley (*Hordeum vulgare*, *Poaceae*) and wheat (*Triticum tritice*, *Poaceae*). It is possible to find areas left fallow, depending on harvest times for the various crops.

Adjacent to croplands is found vegetation characteristic of disturbed areas, where the dominant representatives are members of the *Asteraceae*, *Poaceae*, *Euphorbiaceae*, *Scrophulariaceae* and *Agavaceae* families. Members of these families, such as the tree known locally as the lechero (*Euphorbia laurifolia*, *Euphorbiaceae*), the century plant (*Agave* sp., *Agavaceae*), the sigse (*Cortaderia nitida*, *Poaceae*) and the alder (*Alnus acuminata*, *Betulaceae*), are used as living fences. These plants are typically found on roadsides.

We classified croplands into three categories in order to facilitate the analysis of landholding patterns. The first category, code 61, refers to parcels < 3 ha. In the second category, code 62, are those holdings between 3 and 5 ha. Finally, code 63 includes all properties > 5 ha.

Urban areas (code 7)

These are populated areas completely covered by infrastructure, including houses and roads. In the study area, there are three urban areas: Cotacachi, Quiroga and Imantag (Fig. 4.10).

Cotacachi mountain (code 8)

Mount Cotacachi, with its snow, rocks and sand pits, falls in this category (Fig. 4.11).

Bodies of water (code 9)

This category includes the lagoons of Cuicocha and Cristococha, which are easily identified in aerial photos (Fig. 4.12).

Greenhouses (code 10)

In the study area, we have observed the development in recent years of an area covered by greenhouses. These are structures of wood, metal or a mix of these materials, covered with plastic. The major crop produced in these structures is flowers. The principal purpose for the greenhouse is to protect crops from inclement weather, and to regulate temperature, humidity and airflow (Fig. 4.13).

(a)

(b)

Fig. 4.10. (a) View of the city of Cotacachi (aerial photograph no. 13459 from 2000); (b) streets of Cotacachi (Photo: Xavier Zapata Ríos).

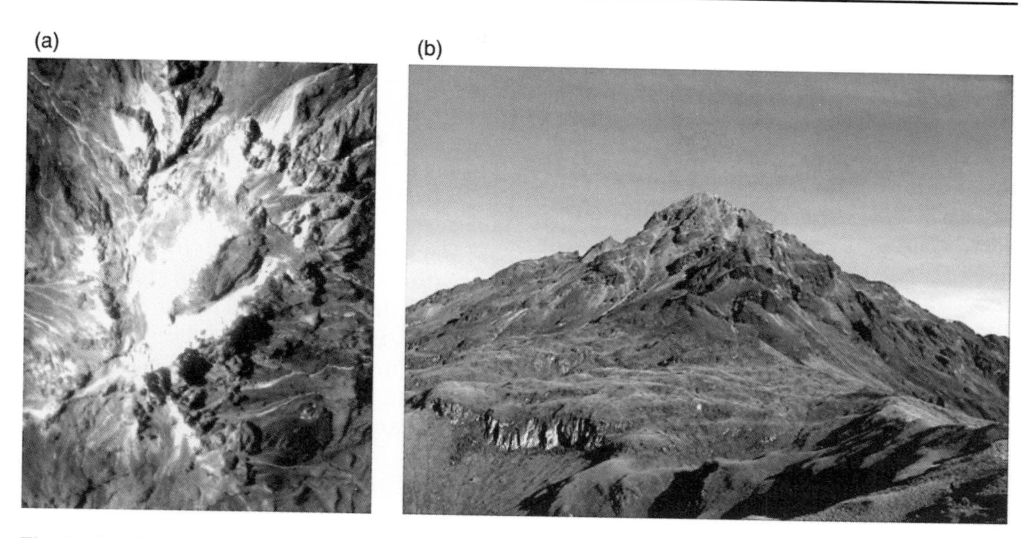

Fig. 4.11. (a) View of Cotacachi (aerial photograph no. 31758 from 1993); (b) southeastern face of Mount Cotacachi (Photo: Xavier Zapata Ríos).

Fig. 4.12. (a) Cuicocha lagoon (aerial photograph no. 13475 from 2000); (b) view of Cuicocha lagoon from the south (Photo: Xavier Zapata Ríos).

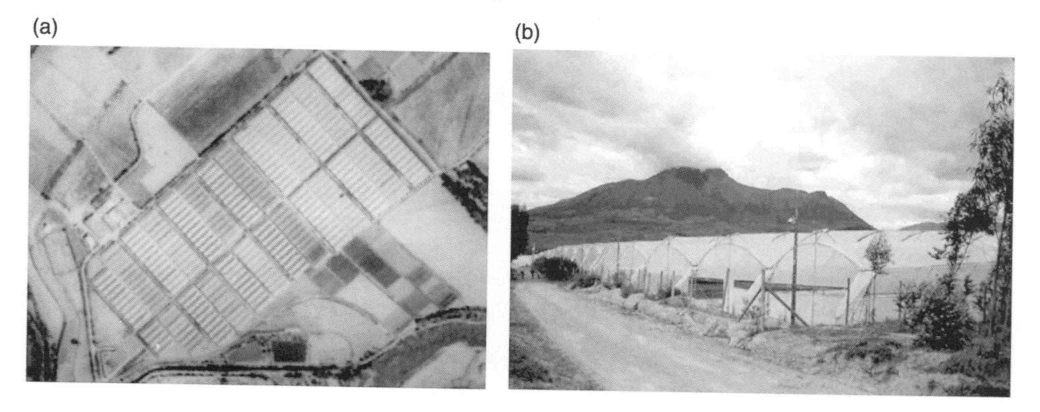

Fig. 4.13. (a) Greenhouses (aerial photograph no. 13459 from 2000); (b) crops grown in greenhouses on the San Martín Hacienda (Photo: Xavier Zapata Ríos).

Others (code 11)

These are difficult to classify zones, generally adjacent to rivers flowing through ravines. These zones bear signs of a high human impact. Generally they are a mixture of pastures, brush, trees and fallow lands. They are used to transport animals from low to high zones (Fig. 4.14).

Dynamics in Land Use Change

The analysis of photographic mosaics for the period from 1963 to 2000 indicates that there have been significant changes in land use (see an example Fig. 4.15 and land use maps Plate 1 and Plate 2).

Table 4.3 and Fig. 4.16 contain the area and percentages corresponding to each type of vegetation cover for the 4 years under analysis.

An analysis of changes in each type of vegetation indicates that for the páramo (code 1), there is no observable increase or decrease in area. Throughout the four decades under analysis, the páramo has covered 18.7% of the study area. The reason for this can be attributed to the creation of

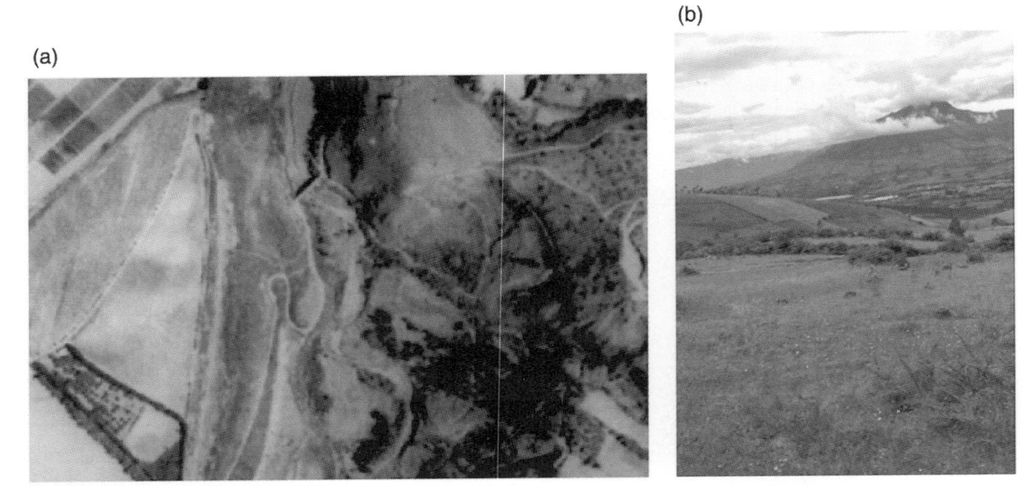

Fig. 4.14. (a) Difficult to classify areas adjacent to ravines (aerial photograph no. 13459 from 2000); (b) view of a zone used as a pasture (photo: Xavier Zapata Ríos).

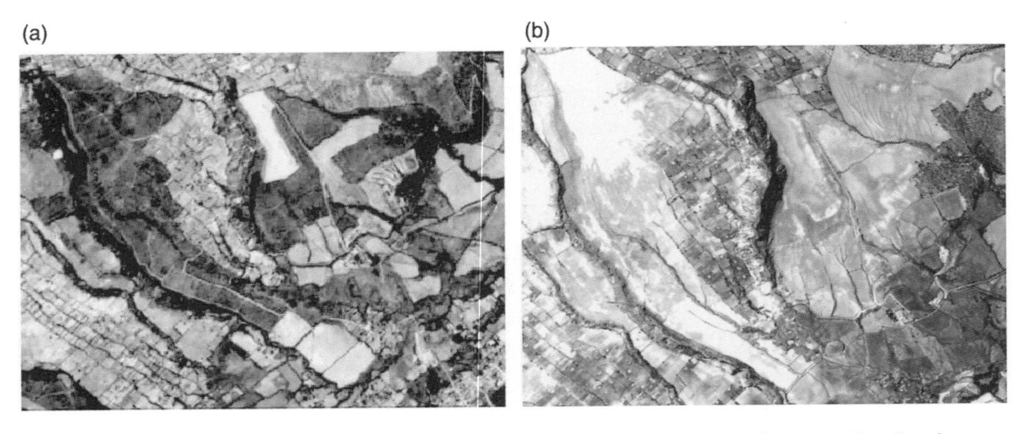

Fig. 4.15. (a) Hacienda Ocampo (aerial photograph no. 13460 from 2000) adjacent to the city of Cotacachi completely covered with eucalyptus forests; (b) Hacienda Ocampo (aerial photograph from 1963) used for crops.

Table 4.3. Land use, surface and percentage for each category by year (1963–2000).

Code	Category	1963 ha	%	1978 ha	%	1993 ha	%	2000 ha	%	Change ha
1	Páramo	4,096.0	18.7	4,093.0	18.7	4,098.0	18.7	4,112.0	18.8	16.0
2	Brush	4,677.0	21.4	3,946.0	18.0	3,915.0	17.9	3,474.0	15.9	–1,203.0
21	Páramo–brush land mix	959.0	4.4	1,092.0	5.0	1,234.0	5.6	1,516.0	6.9	557.0
3	Native forest	137.0	0.6	110.0	0.5	107.0	0.5	92.0	0.4	–45.0
4	Introduced forest	106.0	0.5	346.0	1.6	589.0	2.7	906.0	4.1	800.0
5	Pastures	190.0	0.9	174.0	0.8	154.0	0.7	152.0	0.7	–38.0
61	Cropland < 3 ha	3,961.0	18.1	4,341.0	19.8	4,908.0	22.4	5,190.0	23.7	1,229.0
62	Cropland 3–5 ha	146.0	0.7	317.0	1.4	616.0	2.8	1,144.0	5.2	998.0
63	Cropland > 5 ha	5,523.0	25.2	5,399.0	24.7	4,161.0	19.0	3,048.0	13.9	–2,475.0
7	Urban areas	83.0	0.4	97.0	0.4	152.0	0.7	191.0	0.9	108.0
8	Cotacachi mountain	530.0	2.4	530.0	2.4	530.0	2.4	530.0	2.4	0.0
9	Bodies of water	401.0	1.8	401.0	1.8	401.0	1.8	401.0	1.8	0.0
10	Greenhouses	0.0	0.0	0.0	0.0	14.0	0.1	68.0	0.3	68.0
11	Others	1,093.0	5.0	1,056.0	4.8	1,023.0	4.7	1,078.0	4.9	–15.0
	Total hectares	21,902.0		21,902.0		21,902.0		21,902.0		

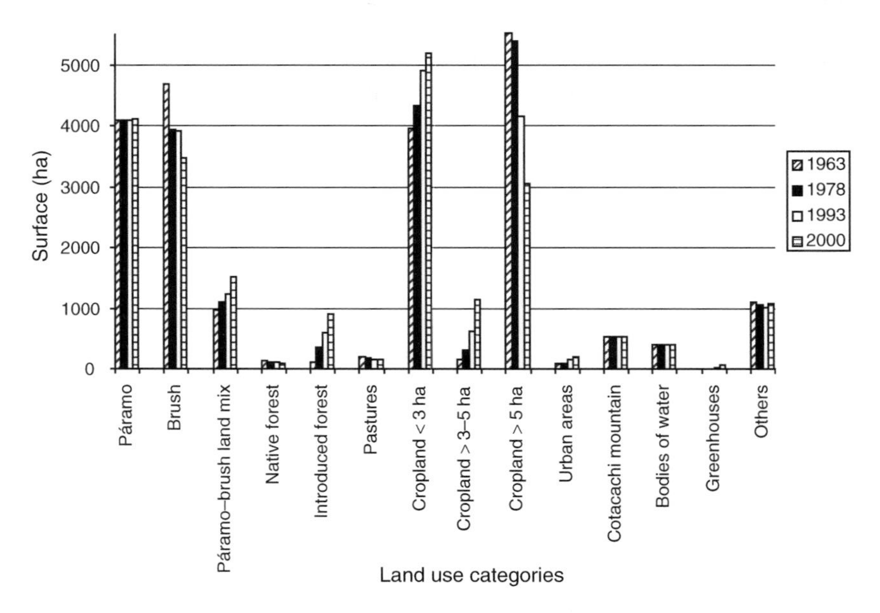

Fig. 4.16. Analysis of changes in land use, 1963–2000.

the Cotacachi-Cayapas Ecological Reserve in 1968 for the purpose of protecting and maintaining the diverse ecosystems within its borders. The reserve is located in the provinces of Imbabura and Esmeraldas. It covers 204,420 ha which range in altitude from 0 to 4939 masl. The reserve contains more ecosystems than any other reserve in the National System of Protected Areas (SNAP, Spanish acronym). According to the various studies of ecosystem diversity of avifauna, this is a priority conservation area (Sierra *et al.*, 1999b). Another reason the size of páramos has not varied is the fact that lower zones have good edaphological and climatic conditions for the production of potatoes and grains. Thus, peasants have not been interested in converting the high zones in the northern sierra to croplands (Recharte and Gearhear, 2001). In addition, hacienda owners in the northern part of Ecuador strictly controlled peasant access to páramo resources such as land, pasture and firewood. The páramo is currently used by commune members principally as grazing areas (Roffler, 1999), and this is why it is very difficult to find agriculture in areas above 3000 masl.

The loss of brush lands (code 2) is one of the most significant changes in the study area. This type of vegetation covered 4677 ha (21% of the total area) in 1963 and had diminished to 3474 ha in 2000 (16%). The change in this type of vegetation is most clearly seen in the Imantag zone in the northern part of the study area. The major cause of the loss of brush lands has been the growth of the agricultural frontier. Moreover, according to census data for 2001 (Instituto Nacional de Estadistica y Censos, 2001), 57% of the rural population uses wood as its sole source of fuel. Of the 1203 ha that have undergone some type of change, 10.6% corresponds to the 'others' category. We can also observe that the brush remnant in the zone is located in areas of difficult access, especially in ravines or on steep slopes.

It is interesting to observe the increase in the surface of the páramo–brush land mix category (code 21). From the original 4.4%, it had increased to 6.9% of the total by the year 2000. According to conversations with community members, these areas are used for grazing animals. The number of animals has decreased according to statements by community members, due to the migration of young people to the city and because of livestock theft.

The lands covered by páramo, brush and bushy páramo demonstrate a high level of change. The burning experienced by these lands during the dry season is the major reason. Community members believe that tender grass which is ideal for grazing sprouts anew after burning. Thus, it is common to see burning during dry periods, especially in areas above 3000 masl.

In 1963, introduced forest covered 106 ha (0.48% of the study area). By the year 2000, the forest cover had grown 8.5 times, to a total of 906 ha (4.1% of the area). The species most frequently introduced are eucalyptus and pine, with the former being by far the most common. Beginning in 1992, this species has been produced in the Ecuadorian sierra, principally for export.

A total of 43% of the study area was covered by crops (codes 61, 62 and 63) by the year 2000, holdings larger than 5 ha (code 63) covered 5523 ha of the area (25.2%) in 1963. By 2000, this figure had diminished to 3048 ha (13.9%). In other words, there was a net change of 2475 ha. Of this change, 41.3% (1023 ha) had become agricultural holdings of < 3 ha. A total of 19.7% of the net change had become holdings of between 3 and 5 ha (code 2). In addition, some 332 ha had been converted into eucalyptus forests. In the aerial photographs corresponding to the year 2000, we can still observe extensive agricultural lands that have not yet been divided into parcels. These haciendas are located, for the most part, in the lower zones of the study area; these are lands with gently rolling hills that provide better conditions for crops. The products of these haciendas, from both crop and livestock activities, satisfy demand in urban areas in cities of the province and the nation. Holdings < 3 ha (code 61) covered 18% of the study area in 1963. By the year 2000, this figure had increased to 1229 ha (23.70%). This change is due principally to agrarian reforms during the 1960s and 1970s. The reduction of smallholdings has been aggravated by inheritance practices, i.e. by the division of holdings into increasingly smaller parcels in order to leave something to all offspring. The increase in land covered by smallholdings and the constant parcelling of these represent a serious socioeconomic problem in that production on smallholdings does not, in the majority of cases, cover family needs nor lead to the production of crops for market.

In recent years in the Ecuadorian sierra, there has been an expansion of lands covered by greenhouses in which crops are grown, especially flowers for export. In the study area, there were 68 ha of greenhouses by the year 2000. There were no greenhouses in 1963 and 1978, in light of which it is clear that these structures have multiplied in recent years. The flowers for export industry began in 1980 and has spread rapidly and continuously since then (CIBEIDER, 1998).

The urban areas include the towns of Cotacachi, Quiroga and Imantag which together have increased 2.3 times since 1963. The growth in urban area has gone hand in hand with population growth. According to data from the Instituto Nacional de Estadística y Censos (Table 4.4), the population in the parishes of Cotacachi, Quiroga and Imantag increased from 17,338 to 25,223 between 1974 and

Table 4.4. Demographic growth in Cotacachi, Quiroga and Imantag.

Years	Cotacachi	Quiroga	Imantag	Total
1974	9,855	4,388	3,095	17,338
1982	10,659	4,728	3,723	19,110
1990	11,301	4,860	3,927	20,088
2001	15,002	5,561	4,660	25,223

Source: Instituto Nacional de Estadistica y Censos (2001).

2001 (Instituto Nacional de Estadística y Censos, 2001).

In terms of altitude, at > 4100 masl there are sand pits and rocks from snow-capped Mount Cotacachi. Between 3700 and 4100 masl, the landscape is dominated by grassland páramos. Between 3300 and 3700 masl, there is a transition area between brush land and páramo. The brush land is primarily above 3000 masl. Finally, all croplands, pastures, forests, greenhouses and urban areas are located below 3000 masl.

Conclusion

The study area has undergone a very dynamic change in soil use during the last four decades. The páramo area has not changed significantly in terms of size. However, brush land is losing ground to the expansion of agriculture. The landscape of the region is still dominated by large haciendas located in low zones. Agrarian reforms in Ecuador have led to a loss of hacienda lands which have been converted into holdings of < 5 ha. The export of lumber has brought about an increase in areas devoted to forest composed of introduced species such as eucalyptus. In general, the change in soil use is tied to the political situation, population growth and socio-economic conditions in the region. It is extremely important to identify clearly the factors that affect change in order to use them in a model that will predict future land use scenarios.

References

Acosta Solis, M. (1977) *Conferencias Fitogeográficas.* Instituto Panamericano de Geografia e Historia, Biblioteca Ecuador, Quito, Ecuador.

Cañadas, L. (1983) *El Mapa Bioclimático y Ecológico del Ecuador.* MAG/PRONAREG, Quito, Ecuador.

CIBEIDER (1998) Centro de Información y Documentación Empresarial sobre Ibero América.

Hofstede, R., Lips, J., Jonsma, W. and Sevink, Y. (1998) *Geografía, Ecología y Forestación de la Sierra Alta del Ecuador.* Abya Yala, Quito, Ecuador.

Hofstede, R. (2001) El impacto humano de las actividades humanas en el páramo. In: Medina, G. and Hofstede, R. (eds) *Los Páramos del Ecuador: Particularidades, Problemas y Perspectivas.* Abya Yala/Proyecto Páramo, Quito, Ecuador, pp. 161–185.

Instituto Nacional de Estadística y Censos (2001) *VI Censo de Población y V de Vivienda: Estadísticas del Cantón Cotacachi.* INEC, Quito, Ecuador.

Jørgensen, P.M., Ulloa, C., Madsen, J.E. and Valencia, R. (1995) A floristic analysis of the high Andes of Ecuador. In: Churchill, S.P., Balslev, H., Forero, E. and Luteym, J.L. (eds) *Biodiversity and Conservation of Neotropical Montane Forest.* New Botanical Garden, New York, pp. 221–237.

Korner, C. (1999) *Alpine Plant Life: Functional Plant Ecology of High Mountain Ecosystems.* Springer Verlag, Berlin.

Laegaard, S. (1992) Influence of fire in the grass páramo. In: Balslev, H. and Luteyn, J.L. (eds) *Páramo: an Andean Ecosystem under Human Influence.* Academic Press, London, pp. 151–170.

Marbberley, D.J. (1997) *The Plant-Book.* Cambridge University Press, Cambridge, UK.

Mena Vásconez, P. and Medina, G. (2001) La biodiversidad de los páramos en el Ecuador. In: Medina, G. and Hofstede, R. (eds) *Los Páramos del Ecuador: Particularidades, Problemas y Perspectivas.* Abya Yala/Proyecto Páramo, Quito, Ecuador, pp. 27–54.

Meyer, W.B. and Turner, B.L. II (1992) Human population growth and global land-use/cover change. *Annual Review of Ecology and Systematics* 23, 39–61.

Recharte, J. and Gearheard, J. (2001) Ecología política de los páramos en el Ecuador. In: Medina, G. and Hofstede, R. (eds) *Los Páramos del Ecuador: Particularidades, Problemas y Perspectivas.* Abya Yala/Proyecto Páramo, Quito, Ecuador, pp. 55–85.

Roffler, G. (1999) Dynamics of Páramo Vegetation Triggered by Burning and Grazing: the Effects of Pastoral Management Practices in Cayambe, Ecuador. PhD thesis, Washington State University, Pullman, Washington.

Sierra, R., Cerón, C., Palacios, W. and Valencia, R. (1999a) Criterios para la clasificación de la vegetación del Ecuador. In: Sierra, R. (ed.) *Propuesta Preliminar de un Sistema de Clasificación de Vegetación para el Ecuador Continental.* Proyecto INEFAN/GEF-BIRF/Ecociencia, Quito, Ecuador, pp. 29–54.

Sierra, R., Campos, F. and Chamberlin, J. (1999b) *Áreas Prioritarias para la Conservación de la Biodiversidad en el Ecuador Continental: un Estudio Basado en la Biodiversidad de Ecosistemas y su Ornitofauna.* Ministerio del Ambiente, Proyecto INEFAN/GEF-BIRF/EcoCiencia/Wildlife Conservation Society, Quito, Ecuador.

Ulloa, C. and Jørgensen, P.M. (1995) *Arboles y Arbustos de los Andes del Ecuador.* Abya Yala, Quito, Ecuador.

Valencia, R., Cerón, C., Palacios, W. and Sierra, R. (1999) Las formaciones naturales de la Sierra del Ecuador. In: Sierra, R. (ed.) *Propuesta Preliminar de un Sistema de Clasificación de Vegetación para el Ecuador Continental.* Proyecto INEFAN/GEF-BIRF/EcoCiencia, Quito, Ecuador, pp. 79–108.

Vanacker, V., Govers, G., Tacuri, E., Poesen, J., Dercon G. and Cisneros, F. (2000) Using sequential aerial photographs to detect land-use changes in the Austro Ecuatoriano. *Revue de Géographie Alpine* 3, 65–75.

5 Climate Change in Cotacachi

Robert E. Rhoades,[1] Xavier Zapata Ríos[2] and Jenny Aragundy[3]
[1]*University of Georgia, Department of Anthropology, 250 Baldwin Hall, Athens, GA 30605, USA*; [2]*SANREM–Andes Project, PO Box 17-12-85, Quito, Ecuador*; [3]*SANREM–Andes Project, Ciudadela Jardines del Pichincha, Pasaje B, N63-204, Quito, Ecuador*

Introduction

The purpose of this chapter is to analyse historical, climatological and local dimensions of climate change in the Cotacachi area. According to elderly indigenous people who live around the base of *Mama Cotacachi*, the Quichua name for the sacred mountain, the climate of the area has changed dramatically in their lifetimes. When recent visiting scientists first heard such folk declarations, they dismissed them as nostalgia on the part of older people for imagined rainy days of their youth or possibly an observation on a present-day seasonal drought. In our surveys, however, farmers consistently listed climate change more frequently than other factors as the single most important cause of agricultural change (see Campbell, Chapter 17, this volume). This observation was verified further by Pedro Loyo, former Agency Director of the Ibarra office of the National Council for Hydrological Resources, who noted that water sources (springs, rivers and water pools) throughout the region were starting to dry up and disappear (personal communication). Landowners who had water concessions from three decades ago were disappointed when they could no longer claim the same amount of water in renewing their concessions.

In an effort to reconcile local people's observations and beliefs with scientific and historical information, we analysed Cotacachi's climate change through a variety of sources: historical archives, historical photographs, landscape paintings, recent photographs taken by SANREM since 1996, aerial photographs, meteorological data, participatory workshops with local people using visual aids (three-dimensional participatory model of the landscape) and structured farmer interviews. We also conducted field visits and transects throughout the area, paying particular attention to streams, springs and rivers. In this chapter, we analyse this diverse information and point to impacts and consequences of climate change on the people of Cotacachi.

Evidence of Climate Change

Historical documents

Chroniclers, mountaineers, explorers and tourists provide written documentation over the past two centuries about Ecuador's Andean glaciers. The great German explorer, Alexander von Humboldt (1853, p. 21), in his travels through Ecuador in 1802 observed 'The Pichincha Mountain is located in the same direction and axis than the snow capped mountains Illiniza, Corazon, and Cotacachi.' In 1858–1859, another explorer mountaineer, M. Wagner (1870, p. 627–628) observed 'The low perpetual snow frontier for Cotacachi in May is 14,814 feet above sea level.' By 1867, another explorer J. Orton (1867), noted that 'Twenty-two summits are covered with perpetual snow . . . The snow limit at the equator is 15,800 feet (4815 metres above sea level (masl)). Cotacachi is always snow-clad.' We assume Orton observed Cotacachi throughout the year in making this observation and was not referring to seasonal snowfall. A decade later, in 1877, Dressel wrote that Cotacachi summit is 'covered with perpetual and compacted snow' (Hastenrath, 1981). Three years later, in 1880, British alpinist Edward Whymper – the first to ascend Cotacachi's peak – wrote that Cotacachi has 'permanent snow, in large beds, as low as 14,500 feet (4419 masl) . . . it is not likely that a crater lies buried beneath the glacier which at present occupies the depression between its peaks'. During the first half of the 20th century, Cotacachi continued to be an object of mountain climbing expeditions which fortunately also described and photographed the glacier zone (Bermeo, 1987, pp. 39–41; 126–129; Luis Bolaños, Club Nuevos Horizontes, 2004, personal communication). In 1989, a master's thesis student Christa Glee von Hillenbrandt noted 'The Cotacachi is the unique mountain in the northern part that has glaciers. There are moraines deposits at a limit as low as 3900 masl' (Hillenbrandt, 1989). When the SANREM project began work in Cotacachi in 1997, remnants of the glacier were documented through reconnaissance treks on the mountain. In 1998, Ekkehard and Hastenrath in a US Geological Survey documented the glacier with an area of 0.06 km^2 with the lowest glacier terminus at 4750 masl, although it is not clear if they were measuring seasonal snow or actual glacier (Ekkehard and Hastenrath, 1998, p. 3). According to Hillenbrandt and Ekkerhard's research, French scientists from INAHMI, aerial photos and personal interviews with Ecuadorian and French climbers, Cotacachi's glaciers might have disappeared at the end of the 1990s.

Paintings and photographs

Another source for measuring the decline of Cotacachi's glacier is landscape paintings and photographs. We were able to find these images in the published chronicles of explorers and mountaineers, tourist books and photograph archives. Edward Whymper (1892) in his book *Travels through the Majestic Andes of Ecuador* published the first artwork of Cotacachi mountain's glacier as it existed in 1880 (Fig. 5.1). In the Cotacachi market, we purchased from a street vendor an old photographic reproduction of Cotacachi mountain which showed that the glacier area was still quite extensive (Fig. 5.2). We estimate that this photograph was taken around 1890. Rafael Troya, an Ecuadorian landscape painter, painted Cotacachi volcano with its glacial zone in 1913 (Fig. 5.3). In 1978, B. Wuth published a photograph of Cotacachi in the book *Maravilloso Ecuador* (Vacas, 1978) (Fig. 5.4). The SANREM project's photographic archives show that in July 1997, there was still a small area covered with snow and glacier. By 2004, photographs by Xavier Zapata Ríos show that there are no glaciers, although occasionally the mountain will have a thin layer of seasonal snow (Fig. 5.5).

Fig. 5.1. East face of Cotacachi volcano (Whymper, 1892).

In addition, available aerial photos from 1963, 1978, 1993 and 2000 document almost four decades of changes in the snow and glacial zone (Military Geographical Institute, IGM) (Fig. 5.6). Although the glacial area in these aerial photos is not completely visible due to cloud cover, a steady loss of glaciated area appears to have taken place between March 1963 and October 2000. These same aerial photographs were useful for later discussions with local people who had personal knowledge of their landscape and changes in streams and springs in the area.

Fig. 5.2. East face of Cotacachi volcano (Anonymous, around 1890).

Fig. 5.4. South face of Cotacachi volcano (Vacas, 1978).

Fig. 5.3. East face of Cotacachi volcano (Troya, 1913).

Fig. 5.5. Southeast face of Cotacachi volcano (Photo: Xavier Zapata Ríos, 2004).

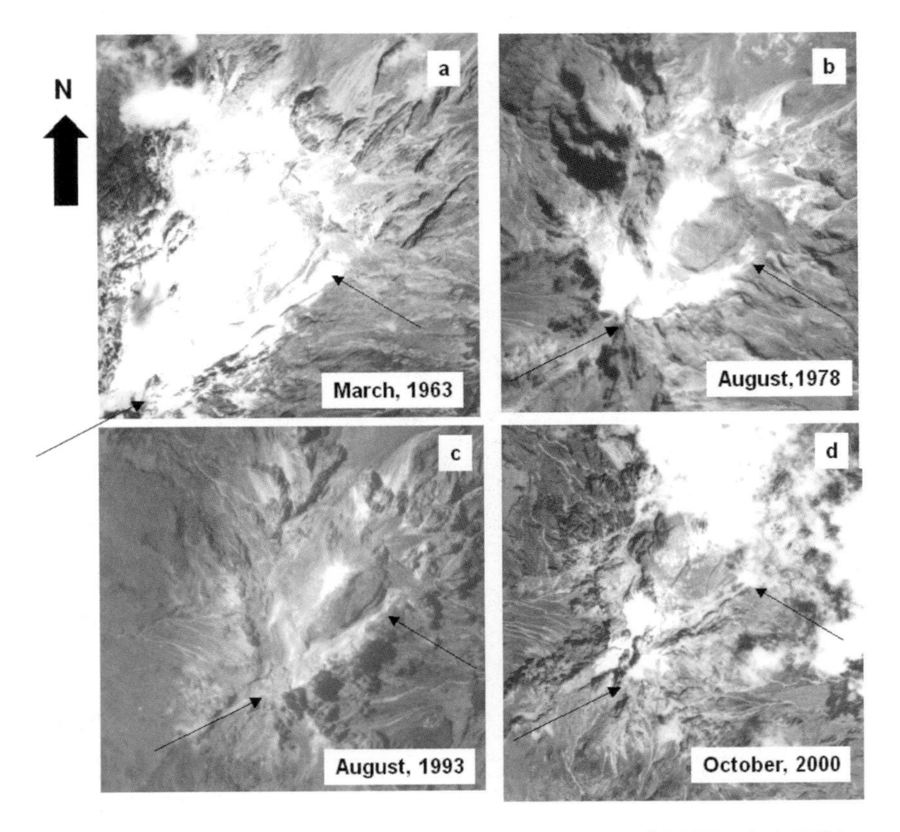

Fig. 5.6. Cotacachi summit. Aerial photos (IGM). (a) 1963, photo 2638. (b) 1978, photo 8134. (c) 1993, photo 31758. (d) 2000, photo 13473.

Meteorological data

Ecuador's National Institute of Meterology and Hydrology (INAHMI) provides the best available data for understanding changes in temperature and precipitation patterns. We selected data on annual temperatures from 15 meteorological stations in Ecuador. These data showed an increasing tendency in the mean, maximum and minimum annual temperatures and decreasing precipitation levels in the country, especially in the north (Cáceres, 2001). Figures 5.7 and 5.8 show mean annual precipitation and mean annual temperature variation over 40 years in the Cotacachi area. San Pablo and Hacienda Maria stations showed a decreasing tendency in precipitation, whereas Otavalo and Cotacachi stations showed an invariable precipitation tendency.

We recognize that INHAMI stations in the region are too dispersed and only permit regional level generalization, and any local climate trend analysis must await

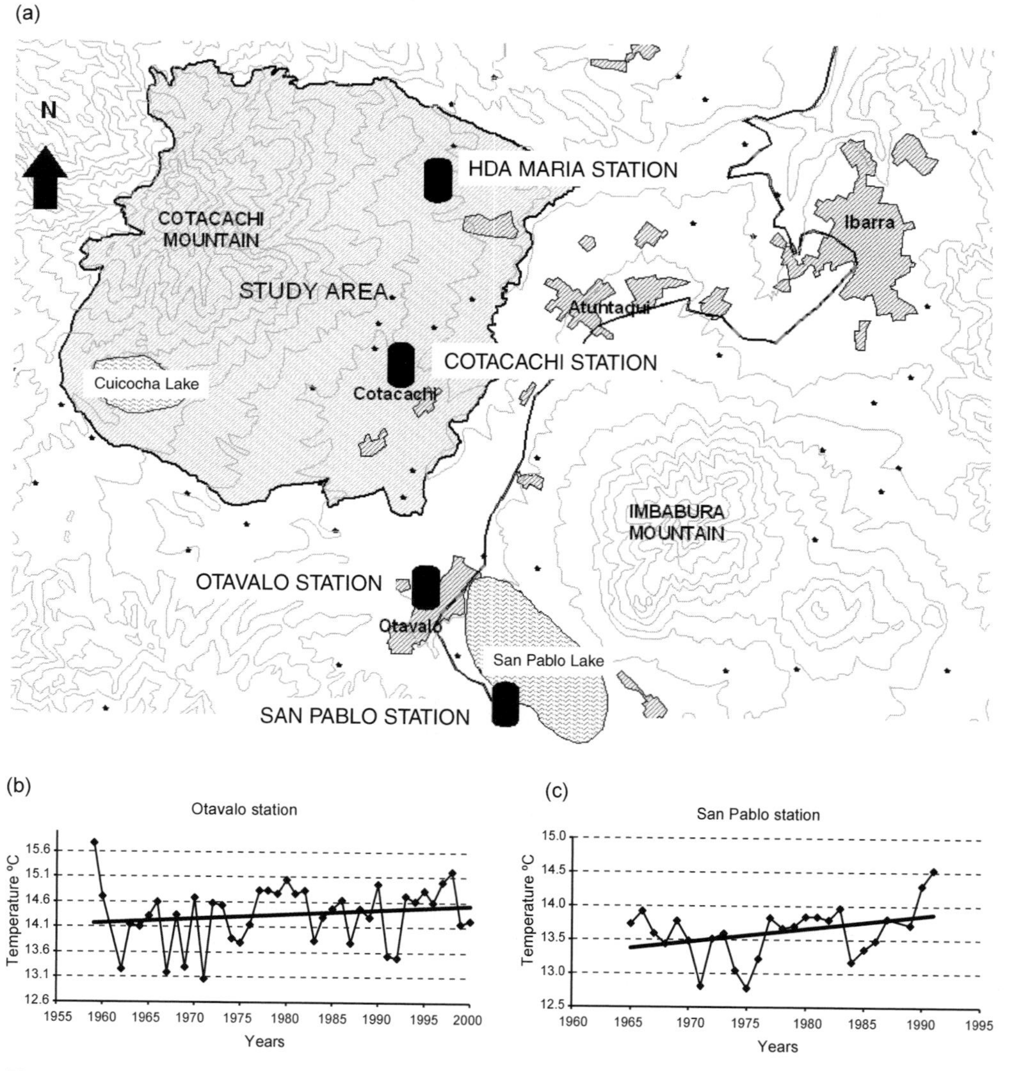

Fig. 5.7. (a) Location of INAHMI meteorological stations. Mean annual temperature for (b) Otavalo and (c) San Pablo stations.

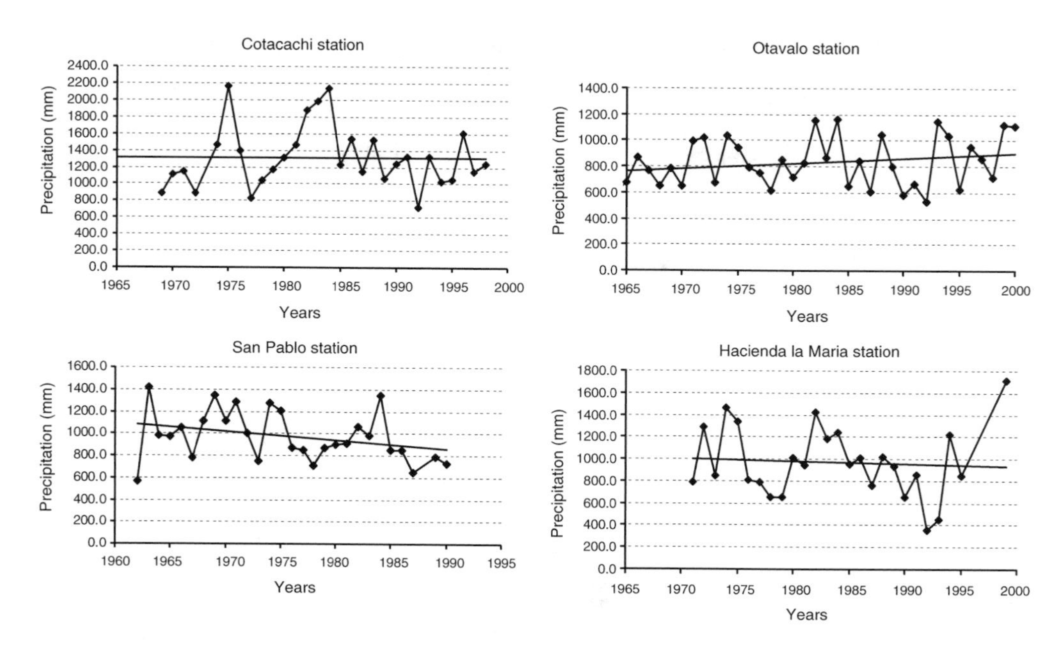

Fig. 5.8. Mean annual precipitation for Cotacachi, Otavalo, San Pablo and Hacienda Maria stations.

further verification. There is also a lack of rain and river stream flow data. SANREM presently has installed six pluviometer and five stream flow stations to measure water to help people manage this resource. The Cuicocha lake water level is also being monitored, but the time frame to date is too limited to draw solid conclusions.

Participatory workshop on climate change

During July 2004, we held a climate change workshop attended by 20 farmers from different communities in Cotacachi and conducted interviews with local farmers to gain an understanding of how they view climate change in Cotacachi. We constructed a three-dimensional physical model maqueta at a scale of 1:10,000 of the study area, including the Cotacachi peak (see Rhoades, Chapter 21, this volume). This replica of the Cotacachi mountain facilitated interaction and helped local people discuss specifics such as location of rivers and springs, or extent of the glacier (Fig. 21.1). In addition, we used aerial photographs from various years as a mechanism for informant recall of the past (Fig. 5.9). A number of field interviews with farmers and surveys with questions on agricultural change and climate provided useful information about climate (Campbell, Chapter 17, this volume).

Local people are very clear and specific in their claims that the climate is changing, although there is no consensus as to why. Elderly people and those from the more remote villages believe that it is the supernatural's punishment, while younger people with formal education are aware of global climate change effects. During 1997, we observed a ceremony at one of the highland water tanks conducted by a shaman who had been summoned by the villagers to appease *Mama Cotacachi* who, we were told, became angry that people were not behaving properly. One informant, the head of the water association, said it was because of uncontrolled burning on her 'skirt' (the local metaphor for the base around the conical shaped volcanic mountain where the villages are located). In her punishment, *Mama Cotacachi* had not provided rain and sufficient water for the communities below. In times of drought, children gather on the

Fig. 5.9. Discussion of the aerial photos (Photo: Robert E. Rhoades).

hillsides and 'call the rain' (*llamando a la lluvia*) in a ceremony (Fig. 5.10). In other ceremonies, sacrifices of animals are made or goods buried as an offering to appease *Pachamama* (Mother Earth).

The comments of local people in the climate change workshop can be divided into three main topics: seasons and rain patterns; glacier and snow on Cotacachi; and water availability.

Seasons and rain

While the disappearance of the glacial zone of Cotacachi mountain is the most visible and salient impact of climate change, the seasonal rain pattern seems to be more directly related to their day-to-day concerns. People emphatically state that there is less and less rain for farming and the rain that does fall is highly irregular and scarce. This presents confusion for timing of planting and other field preparation. Many referred to the rain 'playing' with the people. One 58-year-old female farmer from La Calera, a lowland community, notes: 'Currently it does not rain too much, and it is difficult to predict when rain comes. The weather is playing with us. Now there are only strong, dry winds.' Another female farmer's response was also typical: 'It doesn't rain as much anymore. It seems the weather is changing so that there is only a drizzle today. It used to be more abundant. The climate seems to be playing. The rains were harder and longer. Today the clouds are polluted and there are only strong winds and everything is dry.'

Glacier and snow on Cotacachi

The loss of glaciers and permanent snow from the Cotacachi mountain is one of the most significant landscape changes of Cotacachi perceived by local people. Thirty years ago, the mountain was permanently snow capped and the ice was considered a valuable local good. Ice was harvested and transported to the nearest cities such as Ibarra for food and medicine preservation. Typical comments were 'There was more snow in Cotacachi; it was whiter when it rained. They say that when there's snow you can see animal figures like at the Imbabura. There's figures of a rooster and lying cattle.'

Fig. 5.10. Rain ceremonies (*llamando a la lluvia*) (Photo: Natalia Parra).

Another woman responded 'What I remember of the Cotacachi peak was that it had more snow which came down to near the road which went to Intag'. One respondent said 'snow was used to make ice cream with crushed ice. The mestizos went to get it with donkeys.' Finally, one farmer from Morales Chupa offered a comparative viewpoint:

> I remember that the Cotacachi peak had snow before but today already it has hardly any. What we see is black. We used to climb to the paramo and observe that there was snow but now there is none. What I remember of the *cerro* is that it had snow and looked white. The 'pasillo' of Cotacachi now looks black.

He continued by noting that other peaks, Cotopaxi and Chimborazo, continue to have snow but Cotacachi is black.

Water availability

Some of the elderly participating in the workshop said that when they were young there were more earthquakes and a lot of water flowed down the rivers and streams. The rivers were remembered as being wide and springs plentiful. Today they say the rivers are more like *acequias* (canals) than rivers. A 49-year-old woman from the community of Turuco noted 'Rivers such as Pichavi and Yanayacu were wider 40 years ago, some people and animals died because they were swept away [by the river]. Now the rivers look like irrigation channels.' Another indigenous woman recalled that the Pichavi river was much wider in the past, about 8 m wide, and sometimes it would carry away people and houses. For about 20 years, it has been dry. Other participants pointed to former springs where they drew household water, but today these springs are no longer running.

A number of processes have affected water availability and use in the Cotacachi area over the past few decades. Following land reform in 1963 and 1974, indigenous people established their own farms and communities separated from former haciendas. In many communities, household potable water systems were installed, bringing water to a greater number of users. This, along with a significant increase in population, has led to a higher demand. Although the water systems for both irrigation and potable water have remained largely archaic and inefficient, a number of new systems (e.g. Cambugan, Chumavi and Imantag) have been constructed with outside aid.

Simultaneously, there has been an expansion of the urban zone in and around the town of Cotacachi and the growth of a viable floriculture and greenhouse industry with intensive use of water.

The increased demand for water and search for new sources have corresponded to the loss of Cotacachi's glaciers. Although it is likely that in the stage of rapid melting of the glacier there was an abundance of water in rivers and streams (thus accounting for local memories of abundant water), it is likely that supplies of water from the glacier are now drying up across the landscape. Based on our treks through the area and interviews with local people, we have identified visible consequences of glacier loss on water availability: (i) waterless streams from Chumavi and Caballito creeks (Fig. 5.11a and b); (ii) waterless streams from Alambi and Gualavi (Fig. 5.11c and d); (iii) streamflow decline in the Pichavi, Pichambiche and Yanayacu rivers (Fig. 5.11e); and (iv) decrease in the water level in Cuicocha lake (Fig. 5.11f). According to local farmers, Chumavi creek, which starts at 'Chumavi waterfall' located at 3780 masl above the Cuicocha lake in the southern part of the study area, once carried water which fell into Cuicocha lake. Although the streambed is 30 m wide, it is dry except for a small water spring in the riverbed at approximately 3600 masl. This is the source point for the Chumavi and San Nicolas water systems

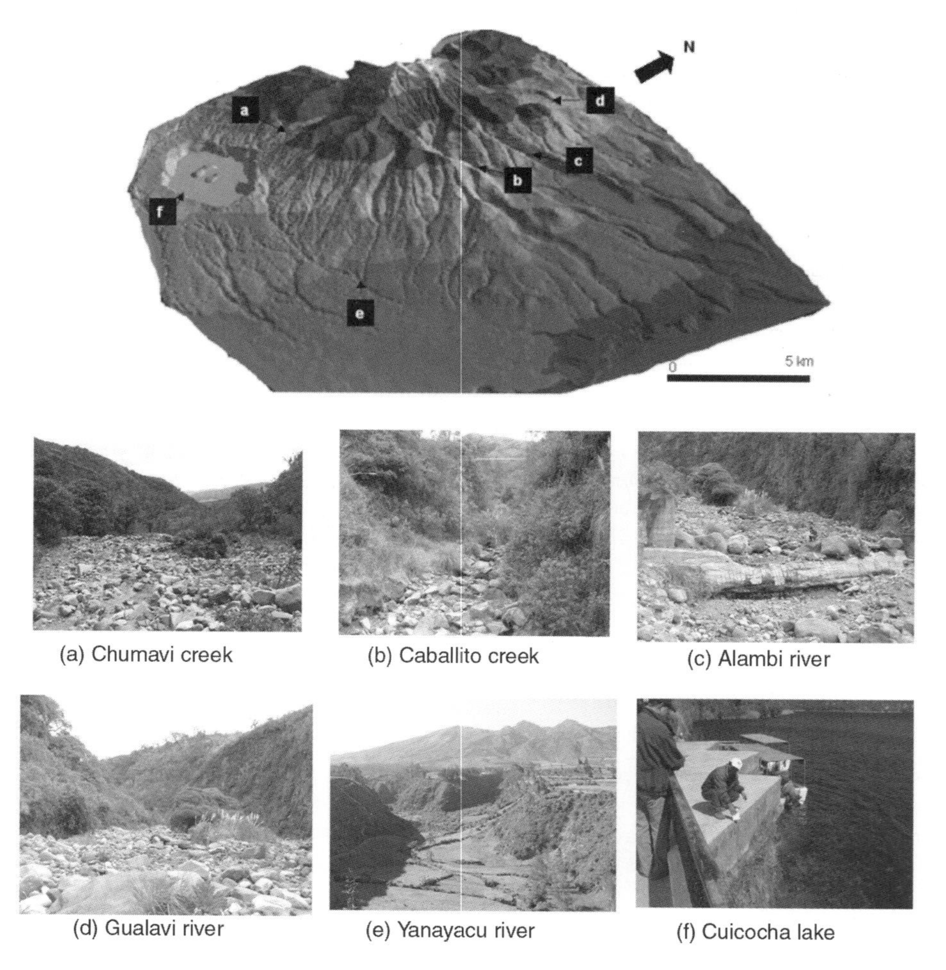

Fig. 5.11. Consequences of the glacier loss on water availability (Photos: Xavier Zapata Ríos).

which supply nine communities, including 487 families (Mayorga, 2004). Similar evidence is found for Caballito creek which is 10 m wide but has no running water. Another river, the Gualavi, located inside the Cotacachi Cayapas Ecological Reserve and within the hacienda El Hospital, is also dry.

Local people are clear that Cuicocha lake is also diminishing. A hotel was constructed on the edge of the lake 8 years ago. Since then, the water level has decreased about 5 m according to the tour boat concession operator on the lake. SANREM preliminary sensor data in the lake show that, during 2004, the water level has lowered approximately 40 cm.

Throughout the region, there is growing evidence of conflict over water sources. In the ‘La Marqueza’ system, water flow has dramatically decreased, leading to conflicts among users. Since no one has access to the same amount of water as before, legal issues over concessions are arising between hacienda owners, indigenous communities and the central government. Local authorities predict that conflict will increase dramatically in the coming years if nothing is done to reverse the situation (SANREM workshop on water, 20 January 2004).

The Alambi river, with a streambed of 30 m width, is located on the east side of the study area and directly under the now vanished large glacier described by Whymper and Wolf in 1880 and 1892. Thirty years ago, the Alambi ran full with water and was used to irrigate the northern part of the study area, including Imantag, Quitumba and Perafán communities. Evidence of its former use is an abandoned intake structure of the Imantag system located at 2800 masl. Due to vanishing water supply, however, during the 1980s the Imantag water authorities found another source and in 1993 received a 204 l/s water concession where they are building a new water system from the Sacha Potrero springs (Central Ecuatoriana de Servicios Agrícolas, 2000).

Conclusion

Glaciers in the Andes have been diminishing since the end of the little ice age between the 16th and 19th centuries. Over the past half century, however, the retreat of the glaciers has been dramatic (Garcia and Francou, 2002). It is predicted that in the next 15 years, 80% of the Andes glaciers will have vanished (The Geological Society, 2001). The shrinkage is not uniform as individual peaks in Ecuador demonstrate different stages and rates of glacial melting. Chimborazo, Cayambe, Antisana and Cotapaxi glaciers are in an ongoing melting process, while Cotacachi, Corazon and Sincholagua are examples of mountains that have already lost their glaciers in recent years.

As we have seen from the historical, visual (photographs, paintings and field visits) and climatological information, the local perception of climate change in Cotacachi cannot be written off as nostalgia or a misreading of their landscape. Local people are aware and are able to articulate climate change and its impacts. By combining different forms of evidence, we have been able to link biophysical evidence in the environment with human perception and response. By understanding local people’s awareness of weather and climatic change, we can also understand better their responses. In order to understand how humans will behave and adapt to climate change, it is essential to study people’s perceptions of climate and the environment in general. Since it forms the basis of decision making, local knowledge of climate should be incorporated into any strategy meant to mitigate the impact of climate change (Vedwan and Rhoades, 2001, p. 117).

References

Bermeo, J. (1987) *Ascensiones a las Atas Cumbres del Ecuador.* Colección Aventuras en las Montañas. Club de Andinismo ‘Nuevos Horizontes’. Impresores MYL, Quito, Ecuador.

Cáceres, L., Mejía, R. and Ontaneda, G. (2001) *Evidencias del Cambio Climático en Ecuador.* INHAMI, Quito, Ecuador. Available at: http://www.unesco.org.uy/phi/libros/enso/caceres.html

Central Ecuatoriana de Servicios Agrícolas (2000) *Gestión Social y Técnica del Agua en Imantag.* Serie Experiencias, CAMAREN, CESA, CICDA, Quito, Ecuador.

Ekkehard, J. and Hastenrath, S. (1998) Glaciers of Ecuador. In: Williams, R.S. Jr and Ferringno, J. (eds) *Sattelite Image Atlas of Glaciers of the World.* US Geological Survey Professional Paper 1386-I-3.

Garcia, M. and Francou, B. (2002) *El Corazón de los Andes.* Ediciones Libri Mundi, Quito, Ecuador.

The Geological Society (2001) Geology news: small Andean glaciers will have vanished in 15 years. 17 January 2001. Available at: http://www.geolsoc.org.uk

Hastenrath, S. (1981) *The Glaciation of the Ecuadorian Andes.* A.A. Balkema, Rotterdam, The Netherlands.

Hillenbrandt, C. (1989) Estudio Geovulcanológico del Complejo Volcánico Cuicocha-Cotacachi y sus Aplicaciones. Provincia de Imbabura. Tesis previa a la obtención del titulo de Magister en Geologia, Quito, Ecuador.

Humbolt, A. (1853) *Kleinere Schriften.* Vol. 1. Cotta, Stuttgart, Germany.

Mayorga, O. (2004) *Informe sobre el Sistema de Agua Entubada Chumavi.* Pontificia Universidad Católica del Ecuador y Proyecto SANREM–Andes, Quito, Ecuador.

Military Geographical Institute IGM (2004) Insituto Geográfico Militar. Archivo de fotografías aéreas.

Orton, J. (1870) The Andes and the Amazon; or Across the Continent of South America. Harper, New York.

Troya, R. (1913) Photo from the historical archive of the Central Bank of Ecuador. Central Bank of Ecuador, Quito, Ecuador.

Vacas, H. (1978) *Carchi e Imbabura in Maravilloso Ecuador.* Impreso en la industria gráfica Provenza, España, pp. 152–161.

Vedwan, N. and Rhoades, R. (2001) Climate change in the western Himalayas of India: a study of local perception and response. *Climate Research* 19, 109–117.

Wagner, M. (1870) *Naturwissenchafliche Reisen im tropischen Amerika.* Cotta, Stuttgart, Germany.

Whymper, E. (1892; reprinted 2001) *Viajes a Través de los Majestuosos Andes del Ecuador*, 3rd edn. Colección Tierra Incógnita No. 4. Ediciones Abya-yala, Quito, Ecuador.

Wolf, T. (1892) *Geología y Geografía del Ecuador.* Brockhaus, Leipzig, Germany.

6 Traversing a Landscape of Memory

Virginia D. Nazarea,[1] Rafael Guitarra[2] and Robert E. Rhoades[1]
[1]*University of Georgia, Department of Anthropology, Athens, GA 30602, USA*; [2]*UNORCAC, Cotacachi, Ecuador*

We shall not cease from exploration
And the end of all exploring
Will be to arrive where we started
And know the place for the first time.
T.S. Eliot

Introduction

One dominant theme running through our research in Cotacachi has been the need to compare, contrast and – when possible – combine local knowledge, also referred to as indigenous systems of knowing, on one hand, with the western scientific system, on the other. The former has been characterized as non-linear, qualitative, holistic, context-sensitive and, to a significant degree, spiritual, while the latter has been characterized as linear, quantitative, reductionistic, detached and self-consciously objective and secular. While it is important to understand these differences to counter simplistic and dangerous assumptions that the 'superior' western knowledge can be used to transform 'archaic' indigenous systems or, conversely, that scientific assessment and reasoning should be used to verify or legitimize local beliefs and practices, it is equally important not to forget the *sameness underlying the differences*. This is not to argue for human universals. Rather, it is to argue for a basic humanity that, to paraphrase Shylock in the Merchant of Venice, asks: if you prick us, do we not bleed, if you tickle us, do we not laugh, if you trample on our rights, do we not squeal? In the end, Cotacacheños desire for themselves and their children the same human rights and privileges as enjoyed by the 'developed' world and the scientists and development planners who represent it.

It is imperative, when we communicate cross-culturally and attempt to fathom different ways of knowing, to keep in mind the underlying sameness that responds to the same basic human exigencies and demands the same basic human decency. In the Sea Islands along the Georgia and Carolina coast, Gullah speakers say, 'some folks are gifted liars'. The folks they refer to are their favourite storytellers. According to Patricia Jones-Jackson (1987), 'lying' has more than one meaning in the Sea Islands. Aside from not telling the truth, it also means 'the telling of a good tale'.

This chapter is about the telling of tales among the *indigenas* or indigenous people of Cotacachi in northern Ecuador. The ethnoecology team collaborated with UNORCAC and local people to document the stories and folk tales told by the elders of Cotacachi, mainly Quichua speakers. These stories have been crafted and handed down through various stages of domination – from the Inca Empire, to the Spanish colonial period with its hacienda system and, finally, the post-colonial state. Many of the folk tales in Cotacachi talk about the *lobo* (wolf) and the *conejo* (rabbit), with the conejo perpetually running circles around the much larger but rather slow-witted lobo. The similarities between Cotacachi's conejo and lobo, with their many misadventures, and the American South's Brer Rabbit and his friends are striking. Jones-Jackson (1987, pp. 111–113) was discussing the parallels between Gullah tales in the American South and West African storytelling traditions, but she might as well have been talking about Cotacachi stories too, when she wrote:

> Both cultures delight in imbuing physically insignificant and seemingly helpless creatures with extraordinary mental acumen . . . Seldom does one find massive or ferocious animals . . . depicted as shrewd or insightful. They are continuously portrayed as stupid and as perpetually duped and humiliated by smaller animals who must from necessity live by their wits, not by their strength.

Trickster tales, as stories about the rabbit, the coyote, the raccoon and the raven, for example, abound in Native American folklore as well. In many cultures around the world, these stories elaborate origins, impart morals, vent frustrations, encode resistance and, obviously, entertain people of all ages. From an adaptationist standpoint, one has to suspect that any cultural feature that is so ubiquitous must perform a very basic, indispensable role. In the case of trickster tales present in the cultural repertoire of most marginalized groups, both venting and resisting functions must be of fundamental significance. The trickster is often at the centre of ploys that are not normally acceptable in society, but the audience is left with a sense of playfulness and innocence, not malevolence. The point is not the success of the trickster, since this is not always the case; the point is the lively and undefeated spirit he brings into the fray. In this regard, 'creative energy' and 'ambiguity' seem to be the key ingredients of trickster tales. As Blaeser (1994, p. 51) pointed out:

> Neither wholly this, nor wholly that; acting sometimes this way, sometimes another way – Trickster ambiguities continue to add up. Trickster is neither solely good, nor solely bad; neither completely wise, nor completely foolish . . . Not either/ or but either/and: Trickster mediates between supposed contradictory forces or elements retaining aspects of both, by revealing them to be coexisting parts of one whole, interconnected, often indistinguishable parts of the one. Ambiguity approaches truth in a way that clarity cannot.

Aside from lobo and conejo encounters, bird stories are also common in Cotacachi. There is, for example, a story about a white swan and a black condor, and how, to acquire the swan's coveted whiteness (and under the direction of the swan), the black condor submerged himself in the freezing lake until he perished. Then, there is the story about a bird who took on the female form and became a wife who, when overwhelmed with pregnancy and domestic responsibilities and a husband who could not help, reverted back into a bird and flew away. These animal personifications of human values, and sometimes assumption of human form, are embellished with every telling and encode resistance or transcendence in contexts that might otherwise overpower the weak. Other types of stories – the legends and myths of Cotacachi – describe the origins of their mountains and lakes, weaving in human frailties and strengths and acknowledging the spiritual forces of creation and destruction. Still others, such as the story about the *chificha* or the witch, seem to have evolved from school-learned Brothers Grimm fairy tales, but get infused along the way with a significant dose of local colour.

Indigenous tales endow the living and non-living components of the natural world with amusing, titillating and downright magical properties – properties that covertly encode resistance to different forms of subjugation. Although this realm is outside the more conventional focus of ethnoscience or local people's categorization and use of plants and animals in their environment, the current redirection towards understanding the social distribution and historical transformations of this knowledge calls for a closer examination of the boundaries of cultural significance (Nazarea, 1999). In other words, while the study and documentation of folk stories may be somewhat peripheral to the subject of ethnobiology and ethnoecology, we think it is time to venture into the more dynamic aspects of human engagement with the natural world; dynamic, i.e. in the sense of its vitality, and also in the sense of layers within layers of not just cultural, but also political, significance. If we claim that ethnoscience succeeds in achieving a deeper understanding of human perception, decision making and behaviour vis-à-vis the natural environment, is it not important to include human cognition, or fancy, that animates the features of the world around them to such an extent that the landscape *moves* with stories and memories that can be 'read' and 're-read'? As people from Cotacachi are wont to say, 'One story or myth may teach us many things'.

The themes that run through the stories we have collected so far encode struggles between dominant and subordinate groups and individuals, and teach not just resistance but persistence by being true to oneself, by thinking beyond the present and 'by keeping your wits about you'. These stories are customarily told at night, close to bedtime, or at wakes – ethereal contexts similar to what Ernst Bloch (1988, pp. 168–169) described:

> Towards dusk may be the best time to tell stories. Indifferent proximity disappears: a remote realm that appears to be better and close approaches. Once upon a time: this means in fairy tale language not only the past but also a more colorful and easier somewhere else . . . But the fairy tale does not allow itself to be fooled by the present owner of Paradise. Thus, it is a rebellious, burned child, and alert. One can climb a beanstalk up into heaven and then see how the angels make gold.

Historical reality and present reality coexist in the Cotacachi oral tradition. Through stories that they have maintained and transmitted, local knowledge and wisdom regarding human relations and the intimate–animate landscape persist. The community's elders tell stories not only to entertain youngsters but because they wish to transmit a particular cosmology that then becomes the foundation for an ideology of identity. *Memorias ancestrales* or ancestral memories, when told and retold, reinforce beliefs, teach lessons and shape destiny in spite of domination and conquest. The relationship between human beings and nature, transformation and transcendence, and the connection of everything to everything else are central to Cotacachi tales. The indigenous legends and myths of Cotacachi speak not of conservation but of coexistence, constituting an integrative cosmovision that prescribes 'sympathy' for, and empathy with, the sentiments of Pachamama or Mother Earth. Every feature of the environment is imbued with its own personality and every personality conveys its message of interrelationship and interdependence, of reward and punishment.

The Tales of Cotacachi

To provide a further glimpse of Cotacachi's oral tradition, we would like to share five stories which illustrate Cotacacheños understanding of society and their relationships with the environment. Two themes which run through all the folk tales, legends and myths shared by the *indigenas* of Cotacachi are those of creation and resistance. The first three tales presented below centre on indigenous accounts of how Cotacachi's landscape and its human inhabitants came to be. In 'God Father and Mother Water', a story similar to creation myths in many parts of the world, both male and female are accorded primacy in this creation role; not just the patriarchal male as is common in western religions. Such male–female dualism and synergy continues to play a dominant role in the culture of most Andean societies and is a central motif in beliefs surrounding agricultural practices. In the next folk tale about the origin of Lake Cuicocha, one of the most salient and sacred features of the Cotacachi landscape, a lesson in morality is conveyed. The evil *hacendado*, a common theme in Cotacachi folk tales, mercilessly sets his dogs upon a beggar (God) who in turn delivers justice by flooding the evil hacienda. One can imagine the feelings of an indigenous person when looking at the deep blue lake as a symbol of retribution for centuries of exploitation by the haciendas. Finally, the folk tale of the volcanic eruption is probably traceable to a devastating 19th-century volcanic eruption as it refers to the Spanish-founded town of Santa Ana de Cotacachi as well as the role of the Catholic saint, Virgen Santa Ana, who is said to be indigenous.

We have also included two stories/folk tales which illustrate indigenous resistance to the dominant society, symbolized in the past by the despised *hacendedo* of old Cotacachi. Could the conejo be the embodiment of the landless *huasipungero* who outsmarts and embarrasses the big, powerful, but always clumsy lobo (the *hacendado* or *mayordomo*)? In all probability, elders and parents who told these stories to the young were passing along a not so subtle message about place and identity. The tales seem to assert that indigenous people may be weak and powerless at the moment, but they possess a deep sense of self-determination and worth – a sense of belonging to the Cotacachi landscape that allows them to endure.

God Father and Mother Water

This is an old story that says that the God Father and Mother Water were both single and they were talking about creation. So they said, 'Let us create the earth, animals and plants.' 'But how can we create?' God Father asked. He told the Mother Water that, 'To do everything it would be better if we get married, we would be able to create more easily.' But the Mother Water did not accept the proposal of God Father, so the God Father pleaded with her for 3 days until she would agree with him.

After some days, Mother Water started to think and said to herself, 'In order to create all things, it is true that we need to get married and unite our powers.' She went to where the God Father was, apologized and told him 'Fine, I accept your marriage proposal.' And so they got married and started creating the land, hills, plants and animals of both sexes, male and female, who at that time knew how to speak.

When all these things were created, they said, 'Now something is missing that might have ideas and thoughts like us.' So they started talking about the creation of human beings. They mixed soil and water to form the man as well as the woman. So this was how our ancestors were formed and we are still growing. Our parents are God, Water and Earth; we are formed with these three elements that are in our bodies.

Miserly haciendas

They say that before, the haciendas were very stingy and miserly. Then, they say, God came converted into a beggar and asked for alms from the owner of the hacienda. Then the owner said, 'You stinking old dirty man! What do you want?' and he sent the young servant girl to release the dogs to bite him. Then, the beggar feeling the mistreatment said to the girl, 'Daughter, you have to leave the hacienda immediately tonight because something very grave is going to happen.' The girl obeyed and went to the house of a neighbour.

The next day she returned and saw that the hacienda was full of water forming a lake.

All the haciendas, from before till now, mistreat the poor people. That is why they say that there are several lakes in this province of Imbabura such as Lake Cuicocha, San Pablo, Yaguarcocha, Mojanda and Piñan.

The volcanic eruption

In this story told by my grandparents, this mountain called Huarmi Rasu (Cotacachi mountain) had exploded. They say that all the lava came down, breaking Mother Earth and making tremendous creeks. There, they say that the lava was coming down like running water where Topo Grande and Santa Barbara communities are located now. They told me also that the image of Virgen Santa Ana with her power diverted the lava to the other side. If the Virgen had not performed this miracle, the town of Cotacachi would have perished; thanks to her we are still living. Since this earthquake, we can see big stones in the boundary of Topo Grande and Santa Barbara communities. We call this sector Rumi Chupa (because it has the form of a tail with lots of stones).

The wolf who wanted to play the flute

Who wants to learn to play the flute? They say the rabbit is always playing the flute, playing beautiful, joyful music. One day, while he was playing, the rabbit met the wolf. The wolf admired the way the rabbit played his flute and asked, 'How did you learn to play such beautiful music?' To which the rabbit confidently replied, 'Well, do you want to learn?' The wolf then answered, 'Yes, yes, I want to.'

The rabbit seized the opportunity and said, 'This is very easy. All you have to do is wrap your entire body with straw and go alone with the flute and learn to play the music.' The rabbit just said this so the wolf would allow the rabbit to wrap him with straw.

The rabbit who is very cunning quickly lit the straw and watched as the wolf burned. At this time, the wolf shouted 'Is this how I'm going to learn to play the flute?' But he turned into ashes before he could get a reply. The rabbit then said, 'How smart I am, how stupid the wolf!'

The wolf who wanted to buy squash

They say the rabbit had a plot of squash. The squashes were very big, and seeing them the wolf approached the rabbit to ask him if he would sell his big squashes. So the rabbit says to the wolf, 'Very well, I can sell them to you, but you must go farther up because I have them up there.' The wolf quite happily said, 'OK, I can go farther ahead.' Then when they arrived there the rabbit said to the wolf, 'You can wait for me down here with your outstretched arms because I will send the squash down from up there.' Then the wolf says, 'Very well, I will wait for you.' The rabbit climbed and from above he threw an enormous rock instead of a squash. The rock fell on top of the wolf and he was squashed to death.

Persistence and Change

The study and documentation of folk stories allow us to venture into the more intimate and dynamic aspects of the landscape of human engagement with the natural world. In these distinctly personal and powerfully evocative tales, we see that the landscape *moves* with emotions and memories that can be 'read' and 're-read.' Moreover, the struggles between dominant and subordinate individuals and groups teach not just resistance but also creativity and persistence, underlining the importance of keeping one's wits in facing the trials and indignities of everyday life.

While each scientific discipline casts its own lens towards the Cotacachi landscape in an effort to understand its history, development and continuity, we should not forget that the indigenous people have their own interpretations of these same phenomena. Over the millennia, Cotacacheños have consciously and unconsciously modified and created the place upon which we train our scientific analysis. Through their long-term and intimate engagement with the landscape, they have imprinted their own patterns of meaning on the natural and man-made environments. The multifaceted characters in their tales may strike us as nothing more than imaginary diversions, but they persist and with great salience they populate the landscape and animate it with lessons of encouragement and hope (Figs 6.1 and 6.2). Thus, our scientific understanding – with its assumptions and selective perception – represents only one view of this reality. Scientific 'facts' need to be tempered by the understanding

Fig. 6.1. Landscape drawings with folk tale characters by Cotacachi children.

Fig. 6.2. Landscape drawings with folk tale characters by Cotacachi women.

of the landscape which the people themselves believe in and act upon.

As we can see from the foregoing examples, the storytelling tradition in Cotacachi has not yet disappeared. Its continued existence demonstrates that the culture is still vital, and viable. The folk tales show how local people look upon their landscape and seek their own cosmological answers to the origin and shaping of features of the landscape. The tales of Cotacachi, however, are not immune from roads, electricity and the insidious globalizing force, television; in other words, to what Connerton (1996) refers to as 'forced amnesia' and 'organized forgetting'. The storytelling tradition is waning as children attend schools and enjoy other diversions, much like our children, and as adults plunge into the market economy to be able to afford other diversions, much like us. The metanarrative of modernity that seduces us in our sameness compels us to relinquish what makes us whole. The seduction of 'organized forgetting' is so powerful that some authors have called for a *countermemory* that 'is essentially oppositional and stands in hostile and subversive relation' to this hegemony (Zerubavel, 1995). According to Zerubavel (1995, p. 10):

> The master commemorative narrative represents the political elite's construction of the past, which serves its special interests and promotes its political agenda. Countermemory challenges this hegemony by offering a divergent commemorative narrative representing the views of marginalized individuals or groups within the society. The commemoration of the past can then become a contested territory in which groups engaging in a political conflict promote competing views of the past in order to gain control over the political center or to legitimize a separatist orientation.

It is in this spirit that our project documented the stories in both Spanish and Quichua, and published these stories in a format that can be read and appreciated by schoolchildren (Nazarea and Guitarra, 2004) (Fig. 6.3). As formal education (mostly in Spanish), television (mostly in Spanish and English) and other forces of modernization disseminate homogenizing messages, both subliminal and direct, we believe it becomes all the more crucial to conserve these ancestral tales and memories and prevent their total obliteration. The indigenous people of Cotacachi are doing their part. It is, after all, their legacy and one that they cherish and protect. We are hoping that the conservation of these stories will not only enhance the use of Quichua in both oral and written form, but also strengthen the community's sense of self-determination, their sense of place.

Fig. 6.3. Virginia Nazarea (far left) presents a copy of the folklore booklet *Stories of Creation and Resistance* to two Cotacacheñas. This tri-lingual (English, Spanish and Quichua) book of folk tales is now being used in local schools (Photo: Robert E. Rhoades).

These stories are precious but fragile gifts from the elders to the children of Cotacachi. Hopefully, every generation will cherish and add to this legacy and their lessons will live on, impregnated in every mountain and lake as well as in every living creature with which humans share this planet. The problems of documentation and translation are too complicated to go into here; we fully realize that written and translated stories will never equal, or substitute for, a vibrant oral tradition in which stories are constructed in a stimulating verbal and non-verbal exchange between storyteller and audience. We are trying to freeze the tracks before they are obliterated in the landscape of memory, because the *conejo* in all of us will doubtless want to retrace these tracks again someday, *once upon a future time.*

References

Blaeser, K.M. (1994) Trickster: a compendium. In: Lindquist, M. and Zanger, M. (eds) *Buried Roots and Indestructible Seeds: the Survival of American Indian Life in Story, History, and Spirit.* University of Wisconsin Press, Madison, Wisconsin.

Bloch, E. (1988) *The Fairytale Moves on its Own Time. The Utopian Function of Art and Literature. Selected Essays.* MIT Press, Cambridge, Massachusetts.

Connerton, P. (1996) *How Societies Remember.* Cambridge University Press, Cambridge, UK.

Jones-Jackson, P. (1987) *When Roots Die: Engendered Traditions on the Sea Islands.* University of Georgia Press, Athens, Georgia.

Nazarea, V. (1999) A view from a point: ethnoecology as situated knowledge. In: Nazarea, V. (ed.) *Ethnoecology: Situated Knowledge/Located Lives.* University of Arizona Press, Tucson, Arizona.

Nazarea, V. and Guitarra, R. (2004) *Cuentos de la Creación y Resistencia.* Ediciones Abya-Yala, Quito, Ecuador.

Zerubavel, Y. (1995) *Recovered Roots: Collective Memory and the Making of Israel National Tradition.* University of Chicago Press, Chicago, Illinois.

Part II

Biodiversity Conservation and Use

Children of Cotacachi enjoy the harvest from the Ancestral Futures Farm (Photo: A. Shiloh Moates).

Although Cotacachi lies within one of the world's designated biodiversity 'hotspots' because of its high rates of endemic native species of flora and fauna and receives significant international funds for the protection of natural and agrobiodiversity, little is actually known about the number and status of different species, geographical distribution and rates of loss. Marcia Peñafiel, an Ecuadorian biologist, along with Marco Tipán, Lincoln Nolivos and Karla Vásquez present the results in Chapter 7 from their inventory and assessment of the conservation status of native biodiversity in Cotacachi's highland zone. Despite their documentation of degradation across the board of fauna and flora species, the area continues to have one of the richest distributions of mammals, birds, amphibians and floral species in Ecuador.

A striking land use pattern which is immediately visible as one enters Cotacachi from the Panamerican highway and glances towards *Mama Cotacachi* is the large blue-green forest blocks of eucalyptus plantations. Eucalyptus is at the centre of an international debate on ecological friendliness of plantations versus native forests. In Cotacachi, a ninefold increase in introduced tree plantation hectareage over the past 40 years has occurred, primarily on former hacienda land. Ashley Carse in Chapter 8 asks the question: how do Cotacacheños perceive tree plantations compared with their native forests in terms of ecological and economic costs and benefits? He discovers that native trees and introduced plantation trees serve different purposes and are valued for different reasons. Eucalyptus is valued for its accessibility, utility and monetary value, while native trees are perceived as ecologically sound and with different economic value. On another level, the larger tree plantations on hacienda lands seem clearly a strategy by mestizo landowners to keep land territorially marked and in commercial production with the lowest possible labour costs.

Kristine Skarbø in Chapter 9 turns our attention to agrobiodiversity which has undergone considerable 'genetic erosion' due to Green Revolution interventions and structural changes in the agrarian economy. Today, the region is characterized by a mix of native Andean landraces, historic Old World crops and new high value commercial crops. The reasons for change in agrobiodiversity turn out to be multiple and complex, ranging from market demand, climate change, infertile soils, declining farm size, pests and diseases, and globalization of the diet. She concludes that one of the positive forces working to bring back Andean crops is the strong local desire to rejuvenate Andean crops associated with indigenous cultural heritage.

A key decision-making group in the realm of agrobiodiversity is women. In Chapter 10, anthropologist Maricel Piniero looks at women's life histories and homegardens in La Calera, Cotacachi, how they classify their plants, and how the garden composition varies by age and socioeconomic class. Using interview and participatory mapping of past, present and future gardens, she shows us how homegardens help women cope with change and uncertainty. The maps highlight what women of different age and status see as important, not what development specialists try to impose in a uniform fashion. All women regardless of age and socioeconomic status want to increase cultivation for household subsistence, while higher income groups decrease tree planting since they now prefer to purchase cooking fuel and lower income women emphasize vegetables with market value as a source of household income. Piniero demonstrates as well that homegardens continue to function as *in situ* gene banks for local landraces and should be recognized for this role in biodiversity.

Juana Camacho in Chapter 11 focuses on the link between food habits, biodiversity and culture in Cotacachi. She argues that food is not only vital for sustenance but is a sensory and material interface between nature and culture. Local perceptions, linguistic classifications of food types and the effects of recent changes in the global economy have altered the indigenous food system. Increasingly, food aid and imported food grains (wheat and rice) have made substantial inroads in the local diets. As a marker of ethnic identity in Cotacachi, however, food remains a powerful symbol in a context where indigenous people are striving to rejuvenate their Andean culture. This also implies an interest in the preservation of native food species and landraces in the agricultural system even as external globalizing market forces make traditional farming uneconomical for most families. Maize, for example, is often grown for ceremonial purposes even when it is unprofitable to do so. Camacho concludes that while it is unrealistic to expect the people of Cotacachi to return to a diet based on historically important local foods, of both Andean and Old World origin, the widespread interest in maintaining both the biodiversity and cuisine of the Andes could be a strong incentive for the maintenance of landraces and native species.

7 Biological Diversity in Cotacachi's Andean Forests

Marcia Peñafiel,[1] Marco Tipán,[2] Lincoln Nolivos[3] and Karla Vásquez[3]
[1]*Alianza Jatun Sacha/CDC-Equador, Pasaje Eugenio de Santillán N 24-248 y Maurián, Quito, Ecuador*; [2]*Direccion Nacional de Recursos Naturales, DINAREN, Av. Amazonas y Eloy Alfaro, Quito, Ecuador*; [3]*Universidad Central del Ecuador, Quito, Ecuador*

Introduction

Few studies exist of the forest remnants found in Ecuador's high Andean ecosystems. Those that are available include brief descriptions of flora and fauna but contain no information on the interactions among species and their biological processes, which are being altered by human activities. Our review of the literature on biodiversity in the study area also revealed limited information on collections and lists of vascular plant and vertebrate species. In light of these gaps, the biodiversity component of the SANREM–Andes project undertook a study to learn about the conservation status of the flora and fauna in forest remnants of Cotacachi's Andean zone.

Fieldwork Sites

The study area within Cotacachi canton is located in the Andean buffer zone of the Cotacachi-Cayapas Ecological Reserve. Ecologically, the study area contains two distinct forest life zones: humid montane forest and very humid montane forest (Cañadas, 1983) or cloud forest (Acosta Solís, 1982; Holdridge, 1987). Our research is especially focused on plant species found in high montane evergreen forest ecosystems and highland grass ecosystems. According to Ecuador's zoogeographic division (Albuja *et al.*, 1980, 1991), the area pertains to the high Andean storey. The study area covers an altitudinal range extending from 2800 metres above sea level (masl; in the community of Iltaqui) to 3400 masl (border of the Cotacachi Cayapas Ecological Reserve), with an average temperature of 10 to 14°C, and between 1800 and 2800 mm of rainfall per year, with a relative humidity of 85%. As the altitude increases, the native Andean forest vegetation gives way to the grasses that are prominent in the highlands (Fig. 7.1).

Our study focused on three distinct sites within the Andean buffer zone of the Cotacachi Cayapas Ecological Reserve (see Fig. 7.1): Morochos páramo, Iltaqui and Topo Grande. Altitudinally, the sites ranged from the Rumihuasi ravine (2000 masl) to the high zone of Topo Grande at 3400 masl. During compilation and analysis of information, we reviewed topographical maps (1:50,000), forestry maps, area photographs and satellite images of the Cotacachi Andean zone, which gave us a basic understanding of the landscape characteristics of the area.

Morochos páramo

This site contains a native forest remnant at an altitude of 3500 masl near the Yanafaccha ravine towards the southeast, and other

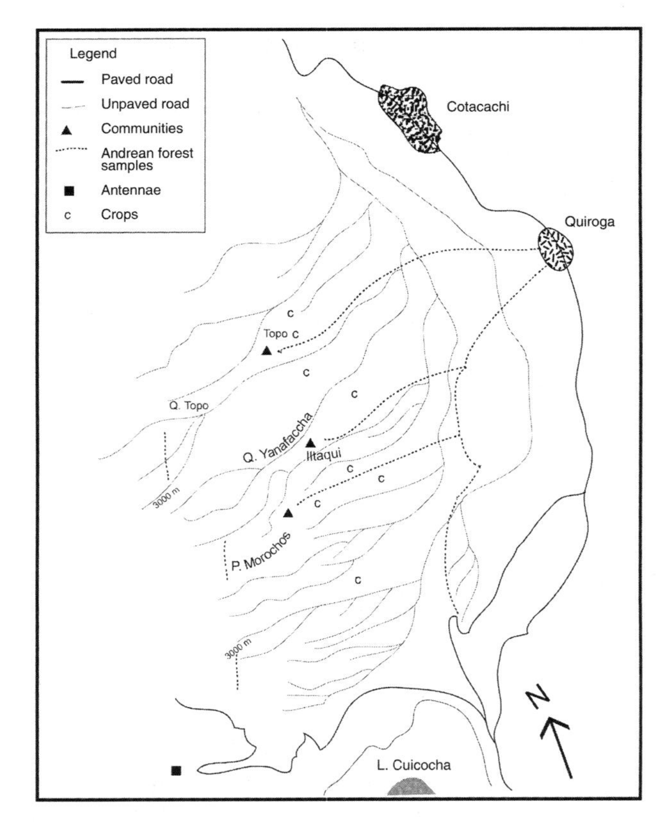

Fig. 7.1. Biodiversity study area.

remnants of native vegetation near two unnamed ravines towards the south. In this area, small fields of short cycle crops and bushy relicts are found. In addition, we studied and sampled the highlands vegetation zone between 3800 and 4000 masl.

Iltaqui

This site is influenced by the Iltaqui ravine with no surviving representative vegetation, but with extensive cultivated areas and bushy relicts, and the Yamafaccha or Seca ravine where at 3000 masl we observed mostly cultivated areas. Beginning at 3400 masl, vegetation representative of native forest is present.

Topo Grande

We walked the entire length of Topo Grande ravine, from 2500 to 3000 masl. We found no native forest remnants, but only cultivated areas and bushy relicts within the area. At about 3300 masl, we found a small remnant of representative native forest.

Methods

Flora

For the floristic inventory, we undertook observations and general collecting throughout the forest remnants in areas surrounding the communities, the ravines mentioned and the trails that join high parts and communities. In order to study the state of vegetation, we made ten transects measuring 50 m × 2 m each at the sampling sites, and recorded the following data: family, genus, species, altitude, habitat (tree, bush or grass), chest-level diameter (CLD), breast-height diameter (BHD) and phenological state (presence/absence of flowers and fruit).

After pressing, we treated plants with alcohol in order to preserve them until we got them to the National Herbarium in Quito (QCNE, Spanish abbreviation) where, after drying, we identified plants precisely by comparing them with specimens at the QCNE and a review of the pertinent literature. We used the similarity index to determine the degree of similarity of floristic composition and the wealth richness of species from the different sites.

Fauna

For the fauna survey, we used the methodology of rapid ecological evaluation (REE), modified and adapted according to the relief of the sampling areas and the criteria of Suárez and Mena (1994). We used specific methodologies for each vertebrate group.

Mammals

We listed mammals based on direct observation and capture (placement of Sherman traps and mist nets); indirect methods including sound recordings, identification of footprints, frequent signs of presence (faeces, feeding site, trails, nests and burrows or dens), and other signs indicating the presence of species in the study area. We undertook general observational walks through the area in order to include different habitats.

We inspected 20 traps in each place of sampling, once per day, always in the early morning. For bait, we used oatmeal, tuna, sardines and a number of essences including vanilla and cod oil. To complement information and confirm final lists, we interviewed guides and farmers from the area, with the aid of photos and illustrations from the scientific literature.

We collected only those specimens difficult to identify; for these, we conserved the dried skins and skeletons for study in a 70% alcohol solution. For all the individuals killed, as well as for some that were released, we took the following morphological measurements: total length (TL), head–body length (HB), length of hind leg (HL), length of tail (T) and weight (W). In addition, and according to the criteria of Larson and Taber (1987) and Tirira (1999), we recorded sex, sexual age and reproductive state.

For definitive laboratory identification, we reviewed mammal collections at the Ecuadorian Museum of Natural Sciences (EMNS) and the National Polytechnic University Museum (MEPN, Spanish abbreviation).

Birds

To evaluate birds, we identified sampling sites and at each one made tape recordings at dawn (5.30 a.m. to 6.30 a.m.), an hour when birds are very active. In addition to the standard recordings, and in order to obtain a more complete list, we observed and recorded at random sites around each sampling site.

We also made transects of 100 m long by 2.5 m high, covering an area of 250 m, in which we placed mist nets end to end. We identified and photographed individuals captured and recorded data; the latter included indications of the type of observation, the forest stratum and the trophic niche.

Amphibians and reptiles

Data were obtained at study points through quadrants, a technique which makes it possible to record species that live in holes, as well as through sightings during walks along trails to get a complete list of amphibians and reptiles at each point of study. In addition, we undertook day and night-time reconnaissance during routine movement along trails and in open areas at each point.

We took all specimens captured in quadrants and on trails to camp (the amphibians in plastic bags and the reptiles in cloth bags) to review their taxonomy with the aid of scientific literature (Coloma, 1991, 1995). Specimens difficult to identify we placed in a 10% clorethone solution and then put them in a fixing solution of 10% formaldehyde and, finally, preserved them in a 70% alcohol solution. For each specimen, we recorded the time of capture, the altitudinal storey in which it was found, the height above ground level, the kind of habitat and climatic conditions. These data made it possible to determine vertical distribution in the different forest stories.

Results and Discussion

Flora

We recorded 1009 individuals representing 94 species of vascular plants in the study area; these plants are members of 75 genera and 44 families. *Oreopanax ecuadorensis* Seeman (*Araliaceae*) is the species we recorded most frequently; it was present in all transects, and we registered 193 individuals. Next in order of frequency is the 'pucunero', *Siphocampylus gigateus* (Cav. G. Don) (*Campanulaceae*), present in most transects, for which we recorded 67 individuals. These species were followed by the 'colca', or *Miconia papillsa*, for which we found 66 individuals, and *Gynoxys fuliginosa*, with 52 individuals. The species mentioned have a CLD > 5 cm and were most representative of this Andean forest.

Native forest

The forest is characterized by the presence of trees higher than 8 m and with a BHD > 10 cm. The trees are covered with mosses and epiphytes such as bromeliads, orchids and ferns. The constant deforestation to which the sampling sites have been subject is endangering the native forest; thus, we could see only small patches of surviving forest in the study area, especially in the ravines (Peñafiel, 2003).

The dominant species in this type of vegetation are 'pumamaqui', *O. ecuadorensis* Seeman (*Araliacae*), 'peralillo', *Vallea stipularis* L. f. (*Elaeocarpaceae*), and 'matachi', *Weinmannia pinnata* L. (*Cunnoniaceae*); residents use the wood of these species for crafts, firewood and charcoal.

Brush forest (Bosque de matorral)

This type of vegetation is found at sites that have lost their native vegetation and are now in a process of recovery. Predominant species are no more than 5 m tall and their CLD is < 4 cm. Among the most representative species are 'chilca' *Baccharis arbutifolia* (Lam.) M. Vahl, *Ageratina pseudochilca* (Benth.) R.M. King & H. Robinson, *G. fulignosa* (Kunth) Cass and *Badillos salicina* (Lam.) R.M. King & H. Robinson of the *Asteraceae*. The 'colca' *M. papillosa* (Desr.) Naudin and 'pucachaglla' or 'Inca's earrings' *Brachyotum ledifolium* Triana, a member of the *Melastomataceae* family were also found. We also identified the 'romerillo' *Hypericum laricifolium* Juss. (*Clusiaceae*) and the *Ericaceae* family represented by the following species: the 'mortiño,' or blackberry, *Vaccinium floribundum* Kunth, with an edible berry used in making the traditional 'colada morada', served on All Souls' Day (2 November), and the 'taglli' *Pernettya prostrata* (Cav.) DC. Members of the *Rosaceae* family include the 'cerote' *Hesperomeles obtusifolia* (Pers.) Lindley and the 'wild blackberry' *Rubus adenotrichos* Schltdl.

Grasslands (páramo)

The páramos, or highlands, studied are characterized by grasses; *Poaceae* is the most conspicuous family as it sprouts with ease after the burnings to which páramos are subject. The most representative species are: *Calamagrostis macrophylla* (Pilger) Pilger, *Anthoxanthum odaratum* L., 'sigse' *Cortaderia nitida* (Kunth) Pilger, and the 'straw' *Stipa ichu* (Ruiz & Pav). Kunth also lists *Orthrosanthus chimboracensis* (Kunth). Baker lists (*Iridaceae*), *Rhynchospora macrochaeta* Steudel ex Boeckeler (*Cyperaceae*) and other genera pertaining to the *Scrophulareaceae* and *Gentianaceae* families, 'Maigua' *Epidendrum evectum* Hook.f., *Pleurotallis grandiflora, Masdevallia bonplandii* H.G. Reichb., *Elleanthus arpophyllostachys* (Rchb.) (f.) Rchb. f, *Pleurothallis* sp. R. Brown, *Pleurothallis macra* Lindley and *Pleurothallis sclerophylla*.

Crops are evident in all communities. Those most frequently produced include: the broad bean (*Vicia faba* L.), peas (*Pisum sativum* L.), ocas (*Oxalis tuberosa* Sav.), mellocos, potatoes and others.

We found the greatest diversity in the community of Morochos, with a value of 0.6720 and a total of 528 individuals, followed by the community of Topo Grande, with a rating of 0.5623; the lowest rating was that found in the Yanafaccha-Iltaqui ravine. Table 7.1 contains a list, in alphabetical order, of the plant species recorded in the course of the study.

Fauna

Mammals

We recorded a total of 23 mammal species in the study area, belonging to eight orders, 12 families and 16 genera. This number includes species that are known to inhabit the area. The total number of species recorded represents 47% of those found in the high Andean zoogeographic storey and 6% of the mammal species existing in Ecuador. The most representative orders are the Rodentia, with the *Akodon*, *Thomasomys* and *Microryzomys* genera; and the Lagomorpha represented by the species *Sylvilagus brasiliensis* (Fig. 7.2). Table 7.2 contains the site, type of sighting, relative abundance, habitat and trophic niche of the mammals listed.

REPRESENTATIVE WILD SPECIES. Although the sites we studied (the páramo of Morochos, Iltaqui and Topo Grande) and surrounding areas have been altered by humans, we found evidence of the presence of small mammals, albeit restricted to small remnants of brush and forest. In these remnants, they have sufficient resources to feed and reproduce. These habitats are thus the area's wildlife refuges for small mammals, but are not permanent refuge sites for large mammals.

The genus *Akodon* (country mouse) is common in temperate and high storeys in the Andes (Albuja, 1991). We captured the mouse in the study area in native forest remnants, near cultivated fields, in bushy relicts and near brush lands on the edges of farms near the sampling sites. The wild rabbit *S. brasiliensis* was also frequently found throughout the study area, especially in grasslands.

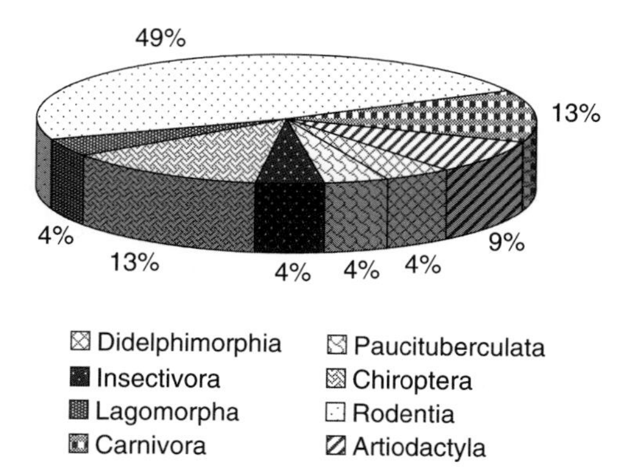

Fig. 7.2. Total mammals recorded in the study area.

STATE OF CONSERVATION OF LARGE MAMMALS. The destruction of native forests and the alteration or loss of habitats and microhabitats leads to the isolation and, consequently, the extinction of the large mammal population which, unlike species of other groups, needs larger areas in order to survive. The Red Book of the IUCN (International Union for the Conservation of Nature, 2001) and Agreement on International Trade in Endangered Species (CITES, 2000) provide an assessment of the state of conservation of large mammal species in the study area. Of all the species registered within the mammal group, the mountain cat *Oncifelis colocolo* has a neotropical distribution. In Ecuador, its habitat ranges from the forests of the high montane evergreen formation to the páramos of the Andes. It has rarely been seen in the study area, possibly because it has been greatly affected due to the reduction and alteration of its natural habitat. This, in turn, is due to the fact that currently part of the remaining natural remnants are burned to expand croplands. The páramo wolf *Pseudolopex culpaeus* was listed on the basis of faeces in the forest remnant and on the trail leading to Cuicocha lagoon. The deer *Odocoileus virginianus* was recorded based on hoofprints in the Morochos páramo; farmers from the area also report infrequent sightings within forest remnants.

Table 7.1. List of plant species recorded in the communities of the study area.

Family	Scientific name	Common name	Morochos	Iltaqui	Topo Grande	Total
Asteraceae	*Achyrocline alata* (Kunth) DC			7		7
	Ageratina pseudochilca (Benth.) R.M. King & H. Robinson	Chilca	10	1		11
	Baccharis arbutifolia (Lam.) M. Vahl	Chilca	6		25	31
	Baccharis genistelloides (Lam.) Pers.		2			2
	Baccharis latifolia Ruiz Lopez & Pavon	Chilca negra	2	11	32	45
	Barnadesia arbórea Kunth	Espino chivo	2		5	7
	Badilloa salicina (Lam.) R. King & H. Robinson		16	12		28
	Bidens andicola Kunth	Ñachag	1	5	5	11
	Chuquiraga jussieui J.F. Gmelin	Flor andinista			1	1
	Gynoxys hallii Hieron		4	7		11
	Gynoxys fuliginosa (Kunth) Cass.		32	1	19	52
	Hypochoeris sessiliflora Kunth	Achicoria	3			3
	Liabum igniarum (Kunth) Less	Santa María	1	3		4
	Perezia pugnes (Bonpl.) Less		1			1
	Pentacalia sotarensis (Hieron) Cuatrec.				1	1
Actinidaceae	*Sauravia bullosa W.*		1		3	4
Amarylidaceae	*Stenomesson* sp. Herbert			1		1
Apiaceae	*Arracacia moschata* (Kunth) DC.				2	2
	Azorella pedunculata (Spreng.) Mathias & Constance		1	1		2
Aquifoliaceae	*Ilex* sp.		1			1
Araliaceae	*Oreopanax ecuadorensis* Seemann	Pumamaqui	114	53	26	193
Araceae	*Anthurium pulchrum* Engler				1	1
Berberidaceae	*Berberis* sp.	sp1		1		1
Boraginaceae	*Tournefortia scabrida* Kunth		1		5	6
Bromeliaceae	*Tillandsia* sp. L	Guaycundo	3	2	1	6
	Puya clava-herculis Mez & Sodiro	Achupalla	1		8	9
Loganiaceae	*Buddleja bullata*			1		1
Campanulaceae	*Siphocampylus giganteus* (Cav.) G. Don.	Pucunero	44	23		67
Caesalpinaceae	*Cassia tomentosa*		1			1
	Senna multiglandulosa				1	1
Cyperaceae	*Rhynchospora macrochaeta* Steudel ex Boeckeler		1	1	12	14
Corrariaceae	*Coriaria ruscifolia* subsp. *microphylla* (Poir.) L. Skog	Shanshi	2		13	15
Columeliaceae	*Collumelia oblongata*			1		1
Clusiaceae	*Hypericum laricifolium* A. L. Juss.	Romerillo	2		23	26
Cunnoniaceae	*Weimania* cf. *pinnata* L.	Matacachi		1		1

Table 7.1. *Continued.* List of plant species recorded in the communities of the study area.

Elaeocarpaceae	*Vallea stipularis* L.f.	Peralillo	18	4	19	41
Ericaceae	*Cavendishia bracteata* (Ruiz Lopez & Pavon ex J. St.) Hil.) Hoerold	Zagalita			3	3
	Dysterigma enpetrifolium		1		3	4
	Pernettya prostrata (Cav.) DC.	Taglli	2	9	21	32
	Vaccinium floribundum Kunth	Mortiño		12	6	18
Scrophulariaceae	*Calceolaria crenata* Lam.	Zapatitos	3	3	1	7
Euphorbiaceae	*Euphorbia laurifolia*	Lechero		1	1	2
Gentianaceae	*Halenia weddelliana* Gilg		1			1
Iridaceae	*Orthrosanthus chimboracensis* (Kunth.) Baker				1	1
Lorantaceae	*Dendrophtora chrysostacha*		3		2	5
	Tristerix longebracteatus (Desr.) Barlow & Wiens	Popa macho	1		5	6
Melastomataceae	*Brachyotum ledifolium* Triana	Pucachaglla	32	10	2	34
	Miconia crocea (Desr.) Naudin	Colca	44	6	4	54
	Miconia papillosa (Desr.) Naudin	Colca	39	1	26	66
	Miconia sp. Ruiz Lopez, Hipólito & Pavon, Jose Antonio				5	5
Myricaceae	*Myrica pubescens* Humb. & Bonpland, Aime Jacques Alexandre		3		5	8
Myrtaceae	Sp 1			22		22
Orchidaceae	*Cyclopogon* sp. Presl			1		1
	Oncidium pentadactylon Lindley			1		1
	Elleanthus arpophyllostachys (Rchb. f.) Rchb. f.				1	1
	Epidendrum evectum Hook.f.	Maygua			1	1
	Pleurothallis sp. R. Brown				1	1
	Pleurothallis macra Lindley	sp3			1	1
	Pleurothallis sclerophylla Lindley	sp4			1	1
	Masdevallia bonplandii H.G. Reichb.				1	1
Oxalidaceae	*Oxalis lotoides* Kunth	Chulco		3		3
Papaveraceae	*Bocconia frutescens* L.	Pucunero		1		1
Passifloraceae	*Passiflora mixta* var. *eriantha* (Benth.) Killip	Taxo silvestre	5	3		8
	Passiflora alnifolia Kunth	Taxo	1			1
Papilonaceae						
Faboideae	*Dalea coerulea* (L.f.) Schinz & Thell.	Izo	3			3
	Lupinus pubescens Benth	Ashpa chocho			3	3
	Otholobium mexicanum (L.f.) J.W. Grimes	Trinitaria	8	2	7	17
Piperaceae	*Peperomia fruticetorum* C. DC.		1		6	7
	Piper barbatum Kunth		9	2	26	37
	Piper andiculum Kunth				6	6
	Piper bogotense H.B.K		1	2	6	9
	Peperomia hartwegiana Miq.				1	1
Plantagonaceae	*Plantago australis* L.			1		1
	Plantago linearis Kunth			1		1

Continued

Table 7.1. *Continued.* List of plant species recorded in the communities of the study area.

Family	Scientific name	Common name	Morochos	Iltaqui	Topo Grande	Total
Poaceae	*Poa annua* L.		1			1
	Anthoxanthum odaratum L.	Pasto	16			16
	Calamagrostis macrophylla (Pilger) Pilger	Pasto			3	3
	Cortaderia nitida (Kunth) Pilger	Sigse		5		5
	Stipa ichu (Ruiz Lopez & Pavon) Kunth	Paja		21		21
Polypodiaceae	*Campyloneurum amphostenon* (Kunze ex Klotzsch) Fee	Calahuala	12		1	12
Polygalaceae	*Monnina obtusifolia* Kunth		12	1	2	15
	Monnina crassifolia (Bonpl.) Kunth	Higuillan		1		1
Phytolacaceae	*Phytolacca bogotensis* Kunth	Atugzara			1	1
Polygonaceae	*Rumex obtusifolius* L.				1	1
Rosaceae	*Hesperomeles obtusifolia* (Pers.) Lindley	Cerote	47		43	90
	Margyricarpus pinnatus Lamark Kuntze	Niguitas		2		2
	Prunus cerofita	Capulí		1		1
	Rubus nubigenus Kunth	Mora silvestre	2	2	4	8
	Rubus adenotrichos Schldl.	Mora silvestre	1	6	1	8
Rubiaceae	*Arcytophyllum thymifolium* (R. & P.) Standley	Tillin		3	2	5
Solanaceae	*Datura stramonius*	Chamico	1	5		6
	Solanum caripense Dunal	Tzimbalo	2	3		5
	Solanum brevifolium Dunal				3	3
	Solanum crinitypes		2		1	3
	Solanum oblongifolium Dunal			2		2
Urticaceae	*Phenax rugosus* (Poiret) Weddell		1			1

Birds

Bird diversity in the study area, with two types of life zones – high montane evergreen forest and páramo – diminishes considerably compared with that of forests farther down (Sierra *et al.*, 1999). Thus, we sighted 48 species distributed in nine orders, 23 families and 43 genera. The most representative orders are the Passeriformes, which account for 57% of sightings, and Apodiformes with 21% (Fig. 7.3). Within these orders, the most representative families are Thraupidae and Trochilidae, respectively. Of the 48 species sighted, 94% were seen on the páramo of Morochos, possibly due to the existence of more of the zone's representative vegetation which constitutes

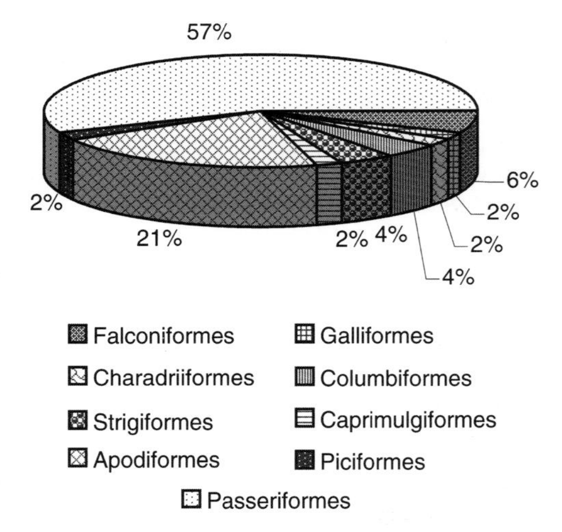

Fig. 7.3. Total birds recorded in the study area.

Table 7.2. Mammals recorded in the localities of the study area.

Scientific name	Common name	Locality	Type of record	Relative abundance	Habitat	Trophic niche
DIDELPHIMORPHIA						
Didelphidae						
Didelphis alviventris	Raposa	IL, TG	In	F	BN, ra, cu	Om
PAUCITUBERCULATA						
Caenolestidae						
Caenolestes fuliginosus	Ratón marsupial	PM	Co	R	BN, ra, cu	Om
INSECTIVORA						
Soricidae						
Cryptotis thomasi	Ratón topo	PM	Co	F	BN	I
CHIROPTERA						
Phyllostomidae						
Anoura geoffroyi	Murciélago longirostro	PM, IL, TG	Co	F	BN, ra	Nc
Sturnira erythromos	Murciélago andino	PM, IL, TG	Co	F	BN, ra	Fr
Vespertilionidae						
Histiotus montanus	Murciélago orejudo andino	PM, TG	Co	R	BN	I
LAGOMORPHA						
Leporidae						
Sylvilagus brasiliensis	Conejo silvestre	PM, IL, TG	Ma, V, Hu, In	C	BN, ra, P	H
RODENTIA						
Muridae						
Akodon mollis	Ratón de campo	PM, IL, TG, Pm	Co, Li	C	BN, ra, P	Fr
Akodon sp.	Ratón de campo	PM, IL, TG, Pm	Co, Li	F	BN, ra	Fr
Microryzomys altissimus	Ratón	PM, IL, TG, Pm	Co, Li	F	BN, ra, P	Fr
Microrysomys minutus	Ratón	PM, IL, TG	Co, Li	R	BN, ra	Fr
Thomasomys aureus	Ratón	TG	Co	R	ra	Fr, Om
Thomasomys baeops	Ratón	PM, IL, TG, Pm	Co	F	BN, ra, P	Fr
Thomasomys cinereus	Ratón	PM	Co	F	BN	Fr
Thomasomys rhoadsi	Ratón	PM	Co	R	BN	Fr
Thomasomys sp. 1	Ratón	PM, IL	Co	R	BN, ra	Fr
Thomasomys sp. 2	Ratón	PM, IL	Co	R	BN, ra	Fr
Cavidae						
Cavia aperea	Sacha cuy	PM, IL, HG	Hu	F	BN, ra	H
CARNIVORA						
Canidae						
Pseudolopex culpaeus	Lobo de pàramo	PM, IL, TG	V	F	BN, ra, P	Cr
Mustelidae						
Conepatus semistriatus	Zorro hediondo	PM, IL, TG	Hu, In	F	BN, ra	Cr
Felidae						
Oncifelis colocolo	Gato pajero	PM, Pm, IL, TG	Au, Hu	R	BN, ra, P	Cr

Continued

Table 7.2. *Continued.* Mammals recorded in the localities of the study area.

Scientific name	Common name	Locality	Type of record	Relative abundance	Habitat	Trophic niche
ARTIODACTYLA						
Cervidae						
Odocoileus virginianus	Venado de cola blanca	PM	In	R	BN	H
Mazama rufina	Soche, cervicabra	PM, Pm	In	R	BN, P	H

Relative abundance: C = common (6–10 recordings); A = abundant (> 10 recordings); F = frequent (2–5 recordings); R = rare (one recording).
Type of record: V = visual; Hu = footprints or other signs; Co = collected; Li = liberated; In = information; Au = auditory.
Trophic niche: H = herbivore; Fr = frugivore; Om = omnivore; I = insectivore; Cr = carnivore; Nc = nectarivore.
Habitat: BN = natural forest remnants; ra = brushy relict; cu = crops; P = grasslands (páramo).
Locality: PM = páramo de Morochos 3300–3500 masl; Pm = Paramo de Morochos 3800 masl; IL = Iltaqui; TG = Topo Grande.

habitats or refuge sites for the local bird fauna. Table 7.3 contains the sites, type of sighting, habitat and trophic niche of each of the birds listed.

REPRESENTATIVE WILDLIFE. Among the most representative of birds is the 'huarro' (*Geranoaetus melanoleucus*). The state of conservation of accipiters is poor because there are almost no nesting areas left for these birds of prey. The destruction of wildlife also reduces their chances for survival. In the study area, this bird was sighted only in the páramo of Morochos.

The Andean turkey (*Penelope montagni*) is a large, gregarious bird that prefers humid forest habitats. At one of the study sites (páramo of Morochos), the birds were observed in pairs in April and May, months which coincide with their nesting period from March to June. The 'torcaza' (*Columba fasciata*) was observed in very active groups or flocks in late afternoon throughout the study area. The Troquilidos, members of the order Apodiformes (hummingbirds), were one of the most representative groups in our study area.

The species *Aglaectis cupripennis* was often sighted in bushes and brush in the three study sites. Among birds that are partial specialists (Sierra *et al.*, 1999) and found in páramo life zones or in the high montane forest, the species *Aglaeactis cupripennis* and *Grallaria quitensis* were sighted.

The study area contains a wide variety of Passeriform species; of these, the Thraupidos is the family with the largest number of species in the area. Among the most representative, the following are outstanding:

- Scarlet-bellied mountain tanager (*Anisognathus igniventris*) was the most frequently sited on the páramo of Morochos and in Topo Grande. These birds hide in bushes near the forest edge or in those surrounded by grasslands. The bird was observed in pairs and small groups.
- The blue-and-black tanager (*Tanagara vassorri*) was observed in pairs and small groups. This very active species was seen especially in the páramo of Morochos.
- The great thrush (*Turdus fuscater*) was sighted frequently in open areas, especially near crops.

STATE OF CONSERVATION OF BIRDS. None of the species sighted in this study is listed in IUCN's Red Book of fauna, nor in the CITES appendices. Nevertheless, it can be assumed that, in light of the constant alteration to which the area is subject due to the expansion of the agricultural frontier, the

Table 7.3. Birds recorded in the localities of the study area.

Scientific name	Common name	Locality	Type of record	Habitat	Trophic niche
Accipitridae					
Geranoaetus malanoleucus	Aguila pechinegra, Huarro	PM, IL, Pm	V	BN, P	Cr
Buteo polyosoma	Gavilán	PM, IL, TG	V	Ra, P	Cr
Falconidae					
Falco sparverius	Quilico	PM, IL, TG, Pm	V	cu, BN	Cr
Cracidae					
Penelope montagnii	Pava de monte	PM, Pm	Au	BN	Fr
Laridae					
Larus serranus	Gaviota andina	PM, IL	V	BN, ra	I
Columbidae					
Columba fasciata	Tortola orejuda	PM, IL, TG	V, Au	BN, ra, cu	Fr
Zenaida auriculata	Torcaza	PM, IL, TG, Pm	V, Au	BN, ra, cu	Fr
Tytonidae					
Tyto alba	Lechuza	PM, IL, TG	Au	Ra	Cr
Strigidae					
Otus sp.	Buho	IL	Ca	Ra	Cr
Caprimulgidae					
Caprimulgus longirostris	Chotacabras alifajeado	TG	V, Au	BN	I
Apodidae					
Streptoprocne zonaris	Vencejo cuelliblanco	PM, TG	V	BN, ra, P	I
Trochilidae					
Patagona gigas	Colibrì gigante	PM, IL, Pm	V	BN	Nc
Aglaectis cupripennis	Colibrì rayito brillante	PM, IL, TG	V, Co	ra, BN	Nc
Pterophanes cyanopterus	Colibrì alizafiro grande	PM, Pm	V	BN	Nc
Eriocnemmis luciani	Zamarrito colilargo	PM	V	BN	Nc
Eriocnemmis vestitus	Colibrì zamarrito luciente	PM, TG	V	BN	Nc
Lesbia victoriae	Colacintillo colinegro	PM, IL, TG, Pm	V	BN, ra	Nc
Lesbia nuna	Colacintillo coliverde	PM, TG, Pm	V	BN, ra	Nc
Colibrì coruscans	Colibrì orejivioleta-ventriazul	PM	V	BN	Nc
Metallura tyrianthina	Colibrì metalura tiria	PM, IL, TG, Pm	V, Co	BN, ra, P	Nc
Picidae					
Piculus rivolii	Carpintero dorsicarmesì	PM	V	BN	Fr, I
Furnariidae					
Cinclodes fuscus	Cinclodes alifrajeado	PM	Au	BN	Fr, I
Schizoeaca fuliginosa	Colicardo barbiblanco	PM	Au	BN	Fr, I
Synallaxis azarae	Pues-pues	PM	V	BN	I
Formicariidae					
Grallaria quitensis	Gralaria leonada	PM	V	BN	Om
Rhinocryptidae					
Scytalopus latrans	Tapacola	PM, IL, Pm	V, Au	ra, BN	Fr
Acropternis orthonyx	Tapaculo ocelado	PM, IL	V, Au	BN	Fr
Tyrannidae					
Agriornis montana	Arriero piquinegro	PM, TG	V, Au	BN, ra	I
Corvidae					
Cyanolyca turcosa	Urraquita turquezsa azulejo grande	PM	V	ra	Om
Turdidae					
Turdus fuscater	Mirlo	PM, IL, TG, Pm	V, Au	ra, BN, P, cu	Om

Continued

Table 7.3. *Continued.* Birds recorded in the localities of the study area.

Scientific name	Common name	Locality	Type of record	Habitat	Trophic niche
Hirundinidae					
Notiochelidon cyanoleuca	Golondrin azul y blanca	PM, TG	V, Au	BN, ra	
Troglodytidae					
Troglodytes solstitialis	Sotorrey montañes	PM, IL, TG, Pm	V, Co	ra, P	I
Parulidae					
Myioborus melanocephalus	Candelita de anteojos	PM	V	BN, ra	I
Thraupidae					
Conirostrum cinereum	Picocono conereo	PM, Pm	V	BN	Fr
Diglossa humeralis	Pinchaflor negro	PM, IL, TG, Pm	V, Co, Li	ra, BN	Nipa
Diglossa cyanea	Pinchaflor enmascarado	PM	V	ra	Nc, Fr
Diglossa lafresnayii	Pinchaflor satinado	PM	Au	BN, ra	Nc
Diglossa sittoides	Pinchaflor	PM	Au	ra	Nc
Anisognathus igniventris	Tangara-ventriflama montana	PM, IL, TG	V	BN, ra	Fr
Catamblyrhynchus diadema	Gorradiadema	PM	V	BN	Fr
Chlorornis riefferii	Tangara carirroja	PM, TG	V, Au	BN	Fr
Tangara vassorii	Tangara azulynegra	PM	V	BN, ra	Fr
Cardinalidae					
Pheuticus chrysogaster	Picogrueso amarillo sureño	PM	V	ra	Fr
Emberizidae					
Catamenia analiss	Semillerito colifajeado	PM, TG		BN, ra	Fr
Phrygilus unicolor	Frigilo plomizo	PM	V	ra	Fr
Atlapetes rufinucha	Matorralero nuquirrufo	PM, TG	V	BN, ra	Fr, I
Zonotrichia capensis	Gorrión	PM, IL, TG	V	BN, ra, cu, P	Om
Buarremon torquatus	Matorralero cabecilistado	TG	V, Au	BN	

Type of record: V = visual; Au = auditory; Ca = hunting; Co = collected; Li = liberated.
Trophic niche: Fr = frugivore; Om = omnivore; I = insectivore; Cr = carnivore; Nc = nectarivore; Nipa = mixed feeding nectar–small insects.
Habitat: BN = natural forest remnants; ra = brushy relicts; cu = crops; P = grasslands (páramo).
Locality: PM = Páramo de Morochos 3300–3500 masl; Pm = Paramo de Morochos 3800 masl; IL = Iltaqui; TG = Topo Grande.

birds may move to other areas or disappear as the few remaining habitats are destroyed.

Amphibians and reptiles

We recorded 11 species of reptiles and amphibians, a figure which represents 1.37% of the total in Ecuador and 22% of the number found in the high Andean life zone (Coloma, 1991, 1995).

We detected the presence of nine species of amphibians at the sampling sites; these are members of the Anura order (two families and two genera), *Eleutherodactylus* and *Gastroteca*. We listed only three reptile species, all belonging to the order Sauria (two families and three genera), *Pholidobolus, Proctoporus* and *Stenocercus* (Fig. 7.4). In Table 7.4, we provide the site, type of sighting, habitat and trophic niche of the amphibians and reptiles sighted.

The largest number of mammals and birds were sighted on the Páramo of Morochos due, possibly, to the presence of well-conserved native forest remnants.

REPRESENTATIVE WILDLIFE SPECIES. The most representative group of amphibians (altitudinal range: 3300–3700 masl) is the

Leptodactylidos, with the *Eleutherodactylus* genera displaying specific adaptations to the harsh páramo climate. This is especially evident in terms of reproductive mode which consists of direct reproduction, i.e. the tadpole phase takes place within the egg, taking advantage of the vitellus and then hatching as small adults. The species occupy specific ecological niches, including bromeliads, achupallas and cavities in trees, where they feed and reproduce.

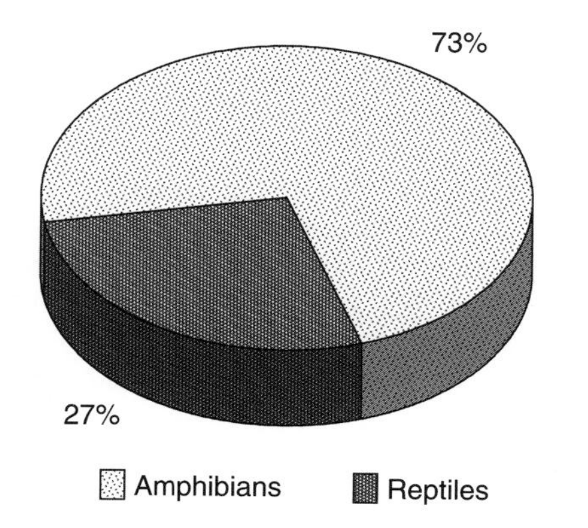

Fig. 7.4. Total amphibians and reptiles recorded in the study area.

Another important group sighted is the *Hylidos*, represented by *Gastroteca riobambae*. This species has a unique reproductive strategy consisting of laying fertilized eggs in a skin fold (marsupial pocket) on the female's back where they develop to the tadpole stage, after which the female places them in small pools where they complete their metamorphosis. They live primarily in bromeliads and clumps formed by lichens.

Among sightings during this study, the most representative species are:

- *Eleutherodactylus unistrigatus*, which displays a compex of variations, known as *unistrigatus*, in terms of colouring, morphology and size according to the

Table 7.4. Amphibians and reptiles recorded in the localities of the study area.

Scientific name	Daily activity	Vertical distribution	Locality	Type of record
AMPHIBIA				
Leptodactylidae				
Eleutherodactylus unistrigatus	N	a/s	PM, IL, TG	Co
Eleutherodactylus buckley	N	a/s	PM, IL, TG, Pm	Co
Eleutherodactylus curtipes	N	a/s	IL, TG	Co
Eleutherodactylus themelensis	N	a/s	PM	Co
Eleutherodactylus orcesi	N	a/s	IL	Co
Eleutherodactylus sp. 1	N	a/s	PM, IL	Co
Eleutherodactylus sp. 2	N	a/s	PM, IL	Co
Hylidae				
Gastroteca riobambae	N	s	IL	Au
REPTILIA				
Teiidae				
Proctoporus unicolor	D	s	PM, IL, TG	Co
Pholidobolus montius	D	s	PM, IL, TG	V, Co
Tropiduridae				
Stenocercus guentheri	D	s	PM, IL, TG	Co

Daily activity: D = diurnal; N = nocturnal.
Vertical distribution: a = altitudinal store; s = ground.
Type of record: V = visual; Co = collected; Au = auditory.
Locality: PM = Páramo de Morochos 3300–3500 masl; Pm = Paramo de Morochos 3800 masl; IL = Iltaqui; TG = Topo Grande.

sites they are found in, as a result of which they are the most abundant species in the majority of Andean páramos.

- *Eleutherodactylus curtipes*, also frequent in the Andean páramo, is of medium size with a conspicuous white line along its upper maxillary.
- *Eleutherodactylus orese*, a small amphibian with a dark back, is seldom seen in the páramos.

The reptiles are represented by the Tropiduridae family, specifically the *Stenocrecus* genus with a wide geographical distribution (500–4000 masl). At sampling sites, the 'guagsa', *Stenocercus guentheri*, which is found as high as 4000 masl, was sighted. The species reaches 15–17 cm in length, deposits its eggs on clumps of lichens, and eats only beetles and grasshoppers.

The other family listed is the Teiidae. Its representative species, the *Pholidobolus montium*, or 'mining lizard', typical of the high páramos, is also found in the temperate zone. This lizard prefers cover, i.e. a life beneath the litter layer where it deposits its eggs. *P. montium* is a generalist in terms of diet, eating different types of invertebrates and organic matter.

STATE OF CONSERVATION OF AMPHIBIANS AND REPTILES. In general, the state of conservation of amphibians is critical, especially that of species inhabiting areas at 2500 masl and above where a number of factors have caused habitat fragmentation, including the expansion of the agricultural frontier, and in particular the loss of the ozone layer. Unfortunately, no precise studies exist that quantify and indicate the degree to which amphibian populations have disappeared, and thus the information presented here offers a preliminary view of their conservation status, based on Coloma's (1991) criteria.

Sensitive or indicator species

Among the mammals listed, the following can be taken as indicators of the good state of conservation of the various habitats in the study area: carnivores such as the páramo wolf (*P. culpaeus*) and mountain cat (*Ocifelis colocolo*), which require habitats that are unaltered or only slightly altered. They are thus species sensitive to a change in habitat due to human intervention.

Sensitive bird species which may be indicators of the state of conservation of different habitats in the study area include: Andean guan (*P. montagnii*), sapphire-vented puffleg (*Eriocnemis luciani*), ocellated tapaculo (*Acropternis orthonyx*), plush-capped finch (*Catamblyrhynchus diadema*), grass-green tanager (*Chlorornis riefferii*) and the stripe-headed brush finch (*Baurremon torquatus*). These species are sensitive to changes in their habitats, especially those due to human intervention. As for the amphibians, today it is assumed that all species in habitats above 1000 masl are sensitive species.

Endemism

Of all the vertebrates recorded in this study, some of the small mammals, particularly the 'country mice' *Thomasomys baeops* and *T. rhoadsi*, are considered endemic to Ecuador.

Uses of fauna

Human communities in the study area do not use all the fauna resources. Instead, there is sporadic hunting of deer (*O. virginianus*), rabbits (*S. brasiliensis*) and Andean guan (*P. montagnii*).

Impacts and threats to animal populations

As is true for many areas throughout Ecuador, the study area contains natural forest remnants which are disappearing, due to habitat destruction caused by uncontrolled fires for extending the agricultural frontier. Burning affects fauna and lengthens dry periods by diminishing vegetation.

In addition, it negatively affects ecological soil processes and increases habitat fragmentation.

Conclusion and Recommendations

In the study area, grazing lands have expanded with burning. This, combined with the deforestation of the few patches of remaining forest in order to use these areas for the production of broad beans, potatoes, mellocos and maize, has produced in three decades a repopulation of bushes and grasses such as the Asteraceas and the Poaceas. All of the study areas' ravines have been deforested in large part in lower areas near communities where *B. latifolia* Ruiz López & Pavón, *Poa annua* L., *Margyricarpus pinnatus* Lamark Kuntze, and other species predominate.

Of the eight watersheds, only four have water in the high areas, a situation illustrating the way the use and abuse of natural resources is bringing about the loss of water and the erosion of soil, two resources necessary for the sustenance of the farm income-dependent families who live in these communities. Thus, reforestation with native species is recommended; these species also add an aesthetic note as they provide habitats for orchids, lilies and bromeliads, which depend on this type of forest. With such forests, residents would be able to attract foreign tourists and bring in additional income.

A comparison of the results of this study with others such as the vegetation inventory of Cuicocha lake and its surroundings (CITES, 2000), and with maps of the last three decades indicates that vegetation communities have recovered from the 1960s when they suffered as a result of invasion of mountainsides by indigenous communities rebelling against feudal overlords. Today a similar invasion is occurring. To avoid further degradation, inappropriate cultivation practices must be stopped and local people's knowledge must be improved.

In surrounding areas, there are steep slopes and ravines where shade, soil types, and wind and other climatic characteristics that affect the formation of microhabitats have provided for the greatest variety and abundance of plant species. The conservation of the native flora of watersheds should be a priority as biodiversity is richest in these microhabitats, as is evident in the upper portion of the Yanafaccha watershed where a natural orchidarium has formed, with five different species, which in the future could become a tourist attraction and, thus, a source of income for residents.

The results of this study do not suffice to evaluate the state of conservation of species for the entire area included in the SANREM–Andes project. For such an evaluation, more sites have to be studied in order to provide broader, more complete information. Based on this study, however, we offer the following recommendations as regards fauna.

- Despite the general degradation of plant life in the study area, we listed 23 mammal species in our study, i.e. 47% of the total registered in the country for the high Andean zoogeographic storey. In addition, we listed 48 bird species, or 3% of the total existing in the country, and eight species of amphibians and three of reptiles, or 1.37% of the total in the country.
- Among the bird species listed in this study, none is listed in the IUCN Red Book, nor in CITES' appendices. One can speculate that with constant alteration suffered due to the expansion of the agricultural frontier, birds may move to other areas or disappear due to the destruction of their remaining habitats.
- The majority of forest amphibians of the Andean summits and páramos are seriously affected by the decline and extinction of populations due to habitat fragmention, expansion of the agricultural frontier and climate change.
- The destruction and alteration of native forests and the loss of habitats and microhabitats result in the isolation and subsequent extinction of local fauna.

- Residents near the study area do not use most of the fauna resources. They hunt small mammals such as rabbits and sacha cuy on a sporadic basis.
- Appropriate practical education in the management of these ecosystems through improved practices and new alternatives is essential.
- Environmental education campaigns should be undertaken in all communities near the Cotacachi-Cayapas Ecological Reserve in order to raise the consciousness of people regarding the role of forests in maintaining ecological resources.

References

Acosta-Solís, M. (1982) Los pastizales naturales del Ecuador. Conservación y aprovechamiento de los páramos y sabanas. *Revista Geográfica* 17, 87–9.

Albuja, L. (1991) Lista de vertebrados del Ecuador, Mamíferos. *Revista de Información Técnico-Científica* 16, 3.

Albuja, L., Ibarra, M., Urgilés, J. and Barriga, R. (eds) (1980) *Estudio Preliminar de los Vertebrados Ecuatorianos.* Escuela Politécnica Nacional, Quito, Ecuador.

Cañadas, L. (1983) *El Mapa Bioclimático y Ecológico del Ecuador–MAG.* PRONAREG, Quito, Ecuador.

CITES (2000) *Lista Roja de Especies Amenazadas.* UICN-Unión Mundial para la Naturaleza, Gland, Switzerland.

Coloma, L.A. (1991) *Anfibios del Ecuador: Lista de Especies, Ubicación Altidudinal y Referencias Bbliográficas.* EcoCiencia, reportes técnicos, Quito, Ecuador.

Coloma, L.A. (1995) *Ecuadorian Frogs of the Genus* Colostethus *(Anura: Dendrobatidae).* Miscellaneous Publication No. 87. University of Kansas, Natural History Museum, Lawrence, Kansas.

Holdridge, L.R. (1987) *Ecología Basada en Zonas de Vida.* Instituto Interamericano de Cooperación para la Agricultura, San José, Costa Rica.

International Union for the Conservation of Nature (2001) *IUCN Red List Categories and Criteria: Version 3.1.* IUCN Species Survival Commission, IUCN, Gland, Switzerland.

Peñafiel, M. (2003) *Flora y Vegetación de Cuicocha.* Ediciones Abya-Yala, Quito, Ecuador.

Sierra, R., Campos, F. and Chamberlin, J. (1999) *Areas Prioritarias para la Conservación de la Biodiversidad en el Ecuador Continental. Un Estudio Basado en la Biodiversidad de Ecosistemas y su Ornitofauna.* Ministerio de Medio Ambiente, Proyecto INEFAN/GEF-BIRF, EcoCiencia and Wildlife Conservation Society, Quito, Ecuador.

Suárez, L. and Mena, P.A. (1994) *Manual de Métodos para Inventarios de Vertebrados Terrestres.* Fundación EcoCiencia, Quito.

Tirira, D. (1999) *Mamíferos del Ecuador.* Museo de Zoología, Centro de Biodiversidad y Ambiente, PUCE-Sociedad para la Investigación y Monitoreo de la Biodiversidad Ecuatoriana, Quito.

1.

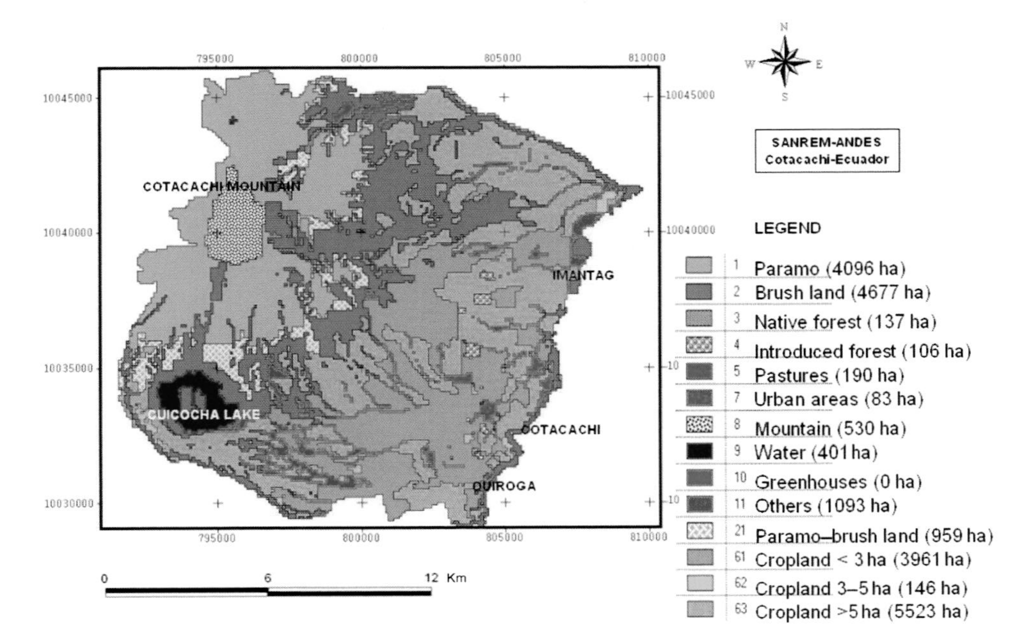

2.

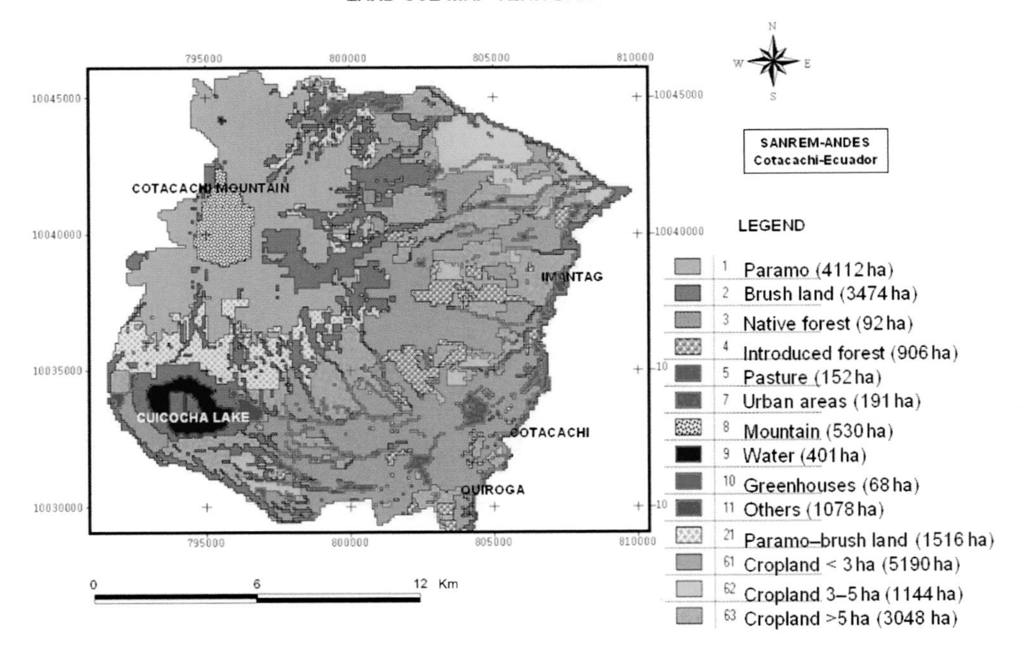

Plate 1. Land use map 1963.

Plate 2. Land use map 2000.

3.

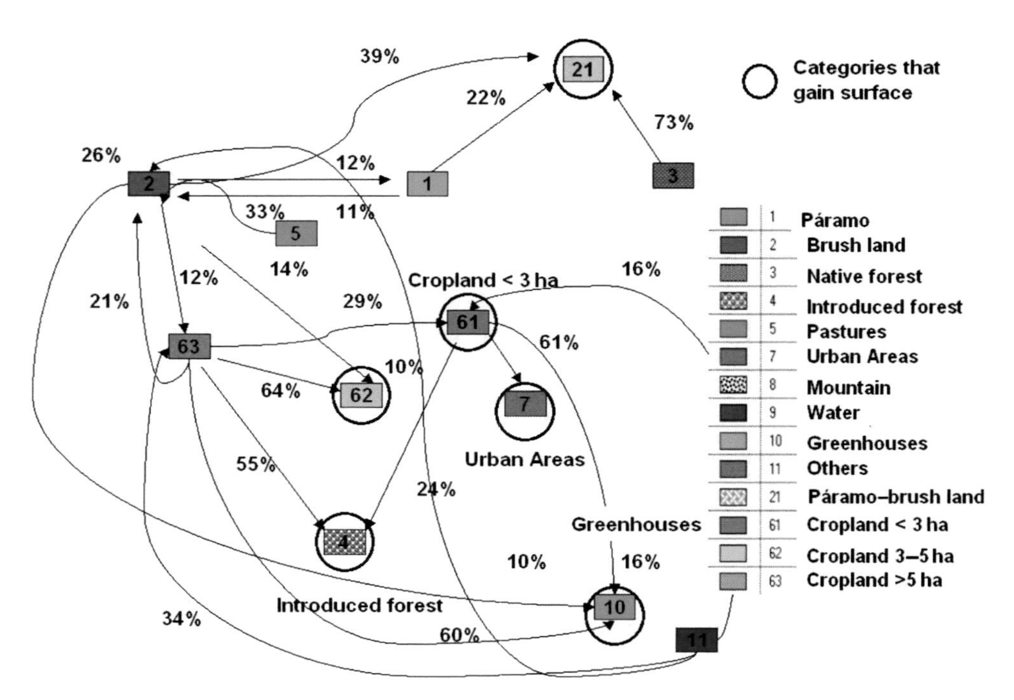

4.

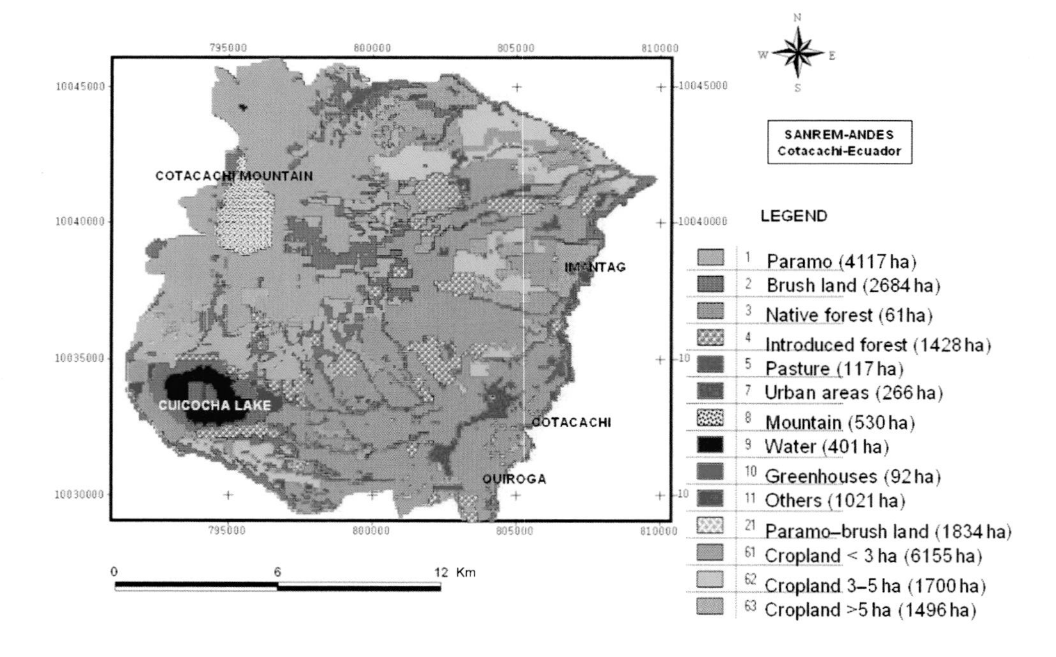

Plate 3. Dominant tendencies of land use categories.

Plate 4. Land use map 2030.

8 Trees and Trade-offs: Perceptions of Eucalyptus and Native Trees in Ecuadorian Highland Communities

Ashley D. Carse
University of North Carolina, Chapel Hill, Department of Anthropology, CB# 3115, 301 Alumni Building, Chapel Hill, NC 27599-3115, USA

Introduction

From a distance, the volcano-studded landscape of the northern Ecuadorian highlands brings to mind an irregular patchwork quilt. The mosaic of gold, brown and green patches becomes, upon closer inspection, individual plots of local agricultural staples *papas y maíz* – potatoes and maize – occasionally interrupted by the darker greens of forest. While, from a distance, these patches might be imagined as remnants of native forest – last refuges from an agricultural expansion moving ever upward and outward – this is generally not the case. For the most part, these are forest plantations of eucalyptus species planted in recent decades.

The land use change analysis of the Cotacachi Andean zone by Zapata Ríos *et al.* (Chapter 4, this volume) demonstrates two important forest cover trends in the study area between 1963 and 2000: (i) a decrease in the land area covered in native forest; and (ii) an increase in the land area covered in forest plantations (Fig. 8.1). Because native forests and forest plantations differ both ecologically and economically, these trends represent tensions between conservation and development, often competing goals in Ecuador, a biodiversity hotspot in the economically struggling Andean region of South America.

The increasingly dominant presence of forest plantations, most commonly planted with eucalyptus species, in the landscape of the northern sierra, coupled with the heated debate surrounding their promotion in international development efforts, compelled me to investigate how the people in highland Quichua communities perceive introduced trees in comparison with native Andean species. In this study, I address the following question: how do people in two northern highland Quichua communities near Cotacachi, Ecuador perceive eucalyptus and native tree species in terms of their ecological and economic costs and benefits?

This chapter begins with a historical description of the processes of deforestation and establishment of forest plantations in Cotacachi. Then, to provide epistemological, geographical and cultural context for the study, the approach, methodology and field sites are detailed. I argue that native and introduced plantation tree species often serve distinct purposes and, consequently, are valued for different reasons in the study communities. While many community members recognize the potential ecological risks of monocultures of introduced tree

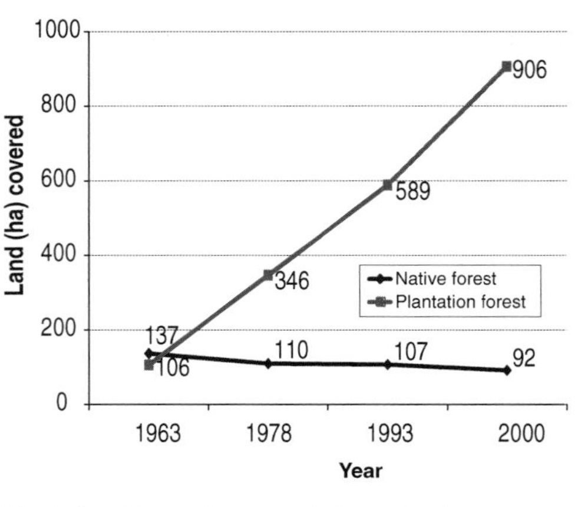

Fig. 8.1. Native forest and plantation forest cover (ha) in the SANREM–Andes study area in 1963, 1978, 1993 and 2000 (Source: Zapata Ríos *et al.*, Chapter 4, this volume).

species, they appreciate the trees' accessibility, utility and monetary value. Native trees, in contrast, are perceived to be very important ecologically and to provide economic benefits distinct from those of plantation species, but are also considered less accessible due to diminished quantities and reduced geographic distribution. While the notion of reforestation with native tree species is overwhelmingly popular in the communities, the form that future projects might take is less clear. This case study of perceptions of native and introduced trees demonstrates the difficulties and trade-offs inherent in the pursuit of conservation and development objectives in rural Andean communities.

Cotacachi's Native Forests and Forest Plantations

Agricultural expansion and deforestation

Agricultural expansion and deforestation in the northern Ecuadorian Andes have a long history, beginning during the precolonial period and continuing to the present day (Sarmiento, 2002). Geographer William Denevan (1992) argues that long before European contact, dense populations of indigenous agriculturalists significantly modified the physical environment, producing anthropogenic landscapes across the Americas. In Ecuador, recent decades have seen national policy play an increasingly important, if often indirect, role in driving environmental change. The Ecuadorian land reforms of the 1960s and 1970s returned marginal highland areas to indigenous communities as communal property, but left most of the fertile lowlands in the hands of *mestizos*, people of mixed indigenous and European descent, especially the wealthy owners of large tracts of land known as *haciendas* (Kenny-Jordan, 1999). In the 1980s and 1990s, the lands received by indigenous communities during previous decades were divided into individual family plots, which continue to be subdivided into smaller plots today as each child inherits a share of his or her family's land. Sven Wunder connects this demographic pressure with deforestation in rural Ecuador, finding 'a direct link between the division of plots after an inheritance and the clearing of new lots' (1996, p. xvi).

Native forest has diminished on the western slopes of the Cotacachi volcano. In Chapter 4 of this volume, Zapata Ríos *et al.* conclude that the total land area covered in native forests declined from an already meagre 137 ha in 1963 to 92 ha in 2000, a decrease of 33% over the period. Figure 8.1 juxtaposes this decline with a sharp increase in forest plantations, primarily eucalyptus species, in the study area. If the high-altitude mix of brush and small trees called *matorral* is also included in the analysis, the decrease in native flora becomes even more striking. The land area covered in *matorral* was determined to have declined from 4677 ha in 1963 to 3467 ha in 2000, a 26% decrease. Ethnographic research conducted in indigenous communities as part of this investigation bears out macrolevel claims of shifting land use. People link changing use of trees and forest products within their own households to the fact that native forests are now *lejos*, or 'far away', while eucalyptus are close, accessible and plentiful.

Counter to the common discourse of population-driven deforestation in the developing world, aggregate forest cover is expanding in many parts of the northern sierra. With few exceptions, however, the forests that increasingly cover the slopes of the inter-Andean valleys are not native, but plantation monocultures of fast-growing eucalyptus and pine often situated on *hacienda* and community lands formerly cultivated with grains and other agricultural crops.

Eucalyptus introduction and plantations

Eucalyptus globulus, commonly known as the Southern or Tasmanian Blue Gum, was the first species of the genus cultivated outside Australia and remains the most popular variety in Ecuador and the world today (Doughty, 2000). The genus was reportedly introduced to Ecuador via France by Garcia Moreno in 1865 (Doughty, 2000). Nearly 30 years later, in 1894, the nation's first eucalyptus plantations were established and have been expanding erratically since (Instituto Ecuatoriano Forestal y de Áreas Naturales y Vida Silvestre, 1996). Plantation growth was sluggish and inconsistent in the first half of the 20th century. Not until the 1950s and 1960s did significant quantities of the trees begin to appear in the Ecuadorian sierra. By the 1970s, approximately 17,700 ha of eucalyptus alone had been planted across the country, with the sierra region accounting for approximately 90% of the total plantation area (Food and Agriculture Organization, 2003). The real boom, however, has taken place in the past 30 years. By 2000, the land area covered in eucalyptus plantations had jumped to 81,000 ha, a remarkable 458% increase in land area in < 30 years. The establishment of government initiatives such as 1993's *Plan Nacional de Fomento a las Plantaciones Forestales* (PLANFOR), which aims to promote forest plantations by providing incentives to cover the costs of establishment and maintenance, coincided with this era of plantation growth.

In the northern sierra province of Imbabura and, more specifically, the region surrounding the city of Cotacachi, the expansion of plantations has far exceeded rapid growth at the national level. The SANREM–Andes land use analysis of the Cotacachi study area indicates that the amount of land dedicated to eucalyptus-dominated plantation forestry increased by a whopping 855%, from 106 to 906 ha, in the 1963–2000 period (Zapata Ríos *et al.*, Chapter 4, this volume). Figures 8.2 and 8.3 depict one local example of this trend: a large section of the *Hacienda Ocampo* situated just outside of the city of Cotacachi. In Fig. 8.2, a photo taken in 1963, the light parcel of land outlined in black is cultivated with grain crops, most likely wheat and barley. Figure 8.3, a photo taken in 2000, depicts the same parcel of land decades later. Now dark and outlined in white, it is cultivated with *E. globulus*.

Fig. 8.2. Hacienda Ocampo in 1963. The land outlined in black is cultivated with grain crops, most likely wheat and barley (Zapata Ríos *et al.*, Chapter 4, this volume).

Fig. 8.3. Hacienda Ocampo in 2000. The same parcel of land – outlined in white – is now cultivated with eucalyptus (Zapata Ríos *et al.*, Chapter 4, this volume).

When asked about the transition from agriculture to plantation forestry, one local *hacienda* owner explained that eucalyptus trees were often first planted in response to changing precipitation patterns that left insufficient water for grain cultivation.

Local opinions of changing forest cover near Cotacachi are complex and often contradictory. Few people take explicitly 'good' or 'bad' positions on the impacts of ever-increasing forest plantations and diminishing native forests. More commonly, people articulate the trade-offs inherent to 'losing' environmentally friendly native trees and gaining 'useful' eucalyptus trees. A community member might, for example, praise the eucalyptus for its remarkable wood production capacity and utility, but criticize it for harming the environment and decreasing the availability of previously plentiful agricultural day labour. Forest plantations require minimal labour relative to the agricultural crops they replaced on local *haciendas*. As a result, the need for day labourers from indigenous communities declined dramatically following the introduction of eucalyptus plantations. Environmentally, descriptions of eucalyptus trees are mixed but generally negative. An individual might describe them as consuming large quantities of water and depleting the soil of nutrients – all while benefiting the local climate. Meanwhile, native forests are viewed as protecting the soil and water or, by some, as occupying land that would be better used for agricultural production. Among these seeming contradictions, one thing is certain: the eucalyptus – fast-growing, adaptable and useful for both domestic and industrial purposes – continues to be planted heavily at a variety of scales, from expansive plantation-style tracts to small windbreak rows on individual *campesino* plots. The trees satisfy local and regional demand for construction materials and fuelwood as well as international demand for pulp and paper.

The potential ecological and social costs of eucalyptus trees have been increasingly debated within scientific and development circles in recent decades (Shiva and Bandyopadhyay, 1983; Saxena, 1994). At times, however, the most vocal critics have included not only academics, researchers and non-govermental organization (NGO) professionals, but also people living near eucalyptus plantations. While the controversy surrounding eucalyptus promotion has an obvious ecological component, the trees have also come to symbolize an increasingly contentious market-oriented development trajectory. On the debate surrounding the Indian Social Forestry Programme's pro-eucalyptus policies, John B. Raintree writes, 'While most of the argument was couched in ecological terms, many of the underlying issues in the debate were socio-economic in nature' (1996, p. 1). Eucalyptus plantations, then, are not simply stands of introduced trees with fiercely debated impacts on ecosystems, but also a form of land use with profound social, economic and political ramifications. The common pro- and anti-eucalyptus claims that characterize the debate are summarized in Table 8.1.

The economic benefits of eucalyptus trees clearly outweigh their potential ecological and social costs for many plantation owners and foresters. However, how is the impact of extensive planting of introduced tree species perceived within neighbouring communities? This and other questions surrounding the eucalyptus require further local ecological and social research.

The Study

The approach: economic ethnoecology

This study takes an ethnoecological approach that situates perceptions of the local environment within a broader political–economic framework. Two principles underlie this approach: (i) ecological problems are socially defined and inextricably linked to economics and culture; and (ii) politics and power dynamics have to be taken seriously in order to understand ecological problems.

In the social sciences, human understanding of the environment is increasingly recognized to be subjective and dynamic rather than absolute and static. According

Table 8.1. Pros and cons of eucalyptus plantations.

Pro-eucalyptus claims[a]	Anti-eucalyptus claims[b]
1. Fast-growing wood producers meet domestic and international demand.	1. Intensive water and nutrient consumers.
2. Produce regular quantities of high-quality wood due to scientific plantation production techniques.	2. Decrease floral and faunal diversity. Do not provide food resources for animals.
3. Economic benefits for the rural poor: cash crops for smallholders and plantation owners meet domestic and international demand.	3. Economic benefits do not accrue evenly: wealthy land owners benefit disproportionately from government assistance and subsidy programmes; wage labour employment declines when forest plantations replace agriculture.
4. May reduce runoff, flooding and waterlogging in regions with erratic and severe rainfall because they absorb large quantities of water.	4. Out-compete other species for water: extensive root systems tap water inaccessible to other plants.
5. May reduce erosion and gully formation caused by rainfall on formerly barren slopes.	5. Deplete groundwater and lower water table.
6. Decrease deforestation pressure by providing alternative local sources of fuelwood, construction material, etc.	6. Poor organic fertilizer: leaf litter is less beneficial to the soil than that of other trees.
	7. Toxic: allelopathic properties suppress growth of neighbouring plants.

[a]Most often made by: foresters, government officials and policy makers, plantation owners.
[b]Most often made by: environmentalists, social activists, local people, some NGOs and social scientists.

to Martin Fishbein (1967), there is a causal link between beliefs and behaviour: beliefs develop attitudes, attitudes lead to intentions, and intentions determine behaviour. If environmental perceptions, beliefs and knowledge are indeed linked to action, then studies like this one, which seeks to examine local, cultural perceptions and attitudes toward forests and environmental change, are critical to effective natural resource management. An ethnoecological approach is ideally suited to exploring perceptions of the environment. Harold Conklin, who coined the term ethnoecology, writes, 'Ethnoecological factors refer to the ways in which environmental components and their interrelations are categorized and interpreted locally' (1961, p. 6).

A better understanding of these factors is important, practically speaking, because natural resource management regimes initiated and/or controlled by external stakeholders – typically national governments, businesses, development organizations and wealthy landowners – directly impact local-level economies, ecosystems and cultures. This is especially true in poor, developing areas such as the Andes, where a disconnection between those responsible for officially 'managing' resources and the local actors that depend upon these resources for sustenance has historically had profoundly negative consequences. To take one example, development and conservation projects have long failed to consider local opinions and cultural knowledge, decreasing the likelihood of local appropriateness, equitability and success. In a study of anthropogenic landscape change in highland Ecuador, Fausto Sarmiento addresses the distance between those planning and those impacted by projects, labelling ethnoecology, 'an essential contributor to the current discourse about conservation-within-development' (2002, p. 213).

Highland Quichua people do not live in a cultural vacuum. Thus, anthropologists and other social scientists conducting local-level research must not ignore the significance of external political, economic and social factors. In investigating linkages between cognition and action, it is important to remember that external factors may constrain or motivate local action. Political ecology research (Blaikie, 1985; Stonich, 1993; Peet and Watts, 1996) has demonstrated that, while human-driven ecological problems have cultural components and occur at the local level, they are often precipitated by the ideology disseminated, policy created, and actions taken at the regional, national and international levels.

Field sites

The Quichua communities of Morochos and San Pedro, field sites for this investigation, lie high on the flanks of the Cotacachi volcano, outside and well above the city of Cotacachi. Both communities are located at an altitude of 2700 m or higher in the *zona alta* or 'high zone', but are situated within different sections of the volcano's western slope, which is broken up into a northern, central and southern zone. The locations of communities on the Cotacachi volcano are significant because indigenous communities in the southern zone, including Morochos, and the central zone, including San Pedro, are more likely to maintain aspects of Quichua culture than those in the north. In the southern and central zones, for example, men generally wear their hair long, while in the northern zone men are more likely to wear their hair in a short *mestizo* style. Although seemingly superficial, this detail is representative of cultural difference between the zones. Language is also significant. Most people in the southern and central zones speak Quichua in their communities. Children learn both Quichua and Spanish in primary school and usually speak Quichua at home. In the northern zone, in contrast, Quichua tends to be limited to the elderly; few young people learn to speak the language. The zones also differ in terms of agricultural production. Many farmers in northern zone communities focus on the large-scale production of a few cash crops for sale at market, while those in southern and central zone communities primarily grow diverse crops for autoconsumption (see Skarbø, Chapter 9, this volume). In many ways, then, the communities of Morochos and San Pedro are more alike than different. Some key community characteristics are summarized in Table 8.2.

Participants in both Morochos and San Pedro report that, on average, the nearest native forest is a distance of 3 km from their homes. It is important to note that the 'native forest' described in this case is generally the high-altitude mix of brush and small trees also labelled *matorral*. When large trees are needed, they are often harvested in nearby eucalyptus plantations. This is an important distinction between the communities. Morochos, considered to be a relatively wealthy and well-organized community, owns a large eucalyptus plantation. Although the plantation is primarily managed for outside sale, community members may collect fallen branches for fuelwood and harvest trees for construction with the permission of the community government. San Pedro, on the other hand, does not own a community forest. Fuelwood and wood for construction come either from the small number of trees grown on personal plots, from the relatively distant and sparse native forest, or, most commonly, from a neighbouring *hacienda* eucalyptus plantation. As Wunder points out, 'Access to trees and wood resources may not fully coincide

Table 8.2. Morochos and San Pedro communities compared.

Variable	Morochos	San Pedro
Elevation	2700 m	2780 m
Population	~800 people	~520 people
Zone	High, southern	High, central
Nearest eucalyptus forest	Community-owned plantation	Hacienda plantation
Nearest native forest	3+ km	3+ km
Males/females interviewed	12M/8F	11M/9F
Average age of participants	41.90 years	43.95 years

with land tenure, particularly when wood is abundant: large landowners often give access to peasants to graze animals and collect fuelwood and other non-timber products' (2000, p. 41). Such a relationship seems to exist between some plantation owners and nearby indigenous communities in the study area.

Methodology

How can perceptions of the ecological and economic costs and benefits of trees be determined? To approach this question, exploratory focus groups were conducted in the Quichua communities of Cercado, Morochos and San Pedro. Focus groups provided an understanding of: (i) the costs and benefits most commonly associated with trees of both types; and (ii) the appropriate local language and terminology for describing ecological and economic impacts. After creating an interview template based on exploratory data, 40 semi-structured interviews were completed, 20 in Morochos and 20 in San Pedro. Interviews were carried out in either Spanish or Quichua, depending on the preference of the interviewee. The author attempted roughly to balance the number of males and females and speak with community members of all ages. A bias may exist in the sample because the translators also served as community 'gatekeepers,' often soliciting those closest within their own social networks to be interviewed first. An attempt was made to minimize the bias by interviewing people at random on the street and at community events.

In addition to collecting basic demographic data through surveys and asking a series of open-ended questions, perceptions of soil, water and tree utility were measured using semantic differential scaling, a common instrument in cognitive social science research (Bernard, 2002). Participants were first given a target item, which, in this case, was either 'eucalyptus' or 'native trees'. They were then asked to describe the target item in terms of paired adjectives situated at opposite ends of a line segment. The paired adjectives had previously been assigned numeric values of 1 and 5, with values of 2, 3 and 4 assigned to the segments between them. The adjectives used were chosen based on information gathered during the focus groups. In describing soils, for example, participants were given a line segment delimited by 'damages the soil' and 'sustains the soil' (Fig. 8.7). Each participant was asked to indicate which point along the line – and, therefore, which corresponding value – best described eucalyptus or native trees in terms of soil costs and benefits. The process was repeated for water consumption and tree utility for both native and eucalyptus trees. In addition to semantic differential scaling and open-ended interview questions, additional data were collected using a 'free listing' technique (Bernard, 2002, pp. 282–285). Participants were asked to list all of the varieties of native tree species known to them in order to indicate breadth of knowledge on the subject, the varieties most well known in the communities and differences in knowledge between generations.

Results

Ecological costs and benefits

Perceptions of the ecological costs and benefits associated with the presence of eucalyptus (Fig. 8.4) and native trees were measured in three categories: water consumption, soil impact and a general environment category. These categories were deemed important for investigation based upon data gathered during several exploratory focus groups conducted with the aid of translators prior to individual interviews in Morochos and San Pedro. As mentioned previously, the perceived costs and benefits associated with water and soil were measured using the semantic differential scaling method with the values 1 and 5 attached to paired adjectives. The general environment category, addressed through a series of open-ended questions, provided a wealth of data about perceived climatic impacts, possibly due to

Fig. 8.4. Eucalyptus poles and boards for construction await purchase in Cotacachi (photo: Ashley D. Carse).

participants' association of the 'environment' and climate.

Eucalyptus and native trees: water consumption

Once the darling of development projects worldwide and considered beneficial to dry, environmentally degraded areas, the eucalyptus has been heavily criticized in recent years for its reportedly extravagant water consumption. Many environmentalists, social activists and people living near plantations claim that the trees consume more water than native trees and out-compete other trees, plants and crops for water and nutrients near the surface. They also maintain that the roots penetrate to depths inaccessible to other plants and lower the water table (Doughty, 2000). Foresters and policy makers tend to dispute these claims, arguing that they are based upon emotion and politics, not science.

Studies of water consumption by eucalyptus trees often reach contradictory conclusions. As a result, both opponents and supporters are able selectively to utilize scientific evidence to bolster their claims. There is general agreement, however, that eucalyptus species consume water intensively in areas where soil moisture is freely available and atmospheric humidity demand is high (Zobel *et al.*, 1987). In such areas, the transpiration rates of eucalyptus trees are extremely high because they do not exhibit stomatal regulation, thus 'pumping' moisture from the ground to the atmosphere more quickly than many other trees (Calder, 1992). For this reason, the trees have a history of being used as 'water pumps' to drain marshes and swamps. While it is not surprising that the image of introduced trees guzzling valuable groundwater has provoked strong reactions from environmentalists, some research in arid areas – where measuring water consumption levels is most salient – indicates that water depletion by eucalyptus trees is less significant (Zobel *et al.*, 1987).

In the Ecuadorian highlands, thorough scientific research on the comparative water use of eucalyptus and native trees has yet to be carried out. As Ian R. Calder points out, such local research is very important:

> To quantify fully the effects of eucalypt plantations in a particular environment requires research; in particular, studies are required to determine the comparative water use and the mechanisms which control the use of water in regions of different climate and soil type from the major vegetation types; these must include commonly planted *Eucalyptus* and other tree species, natural and degraded forests and agricultural crops.
>
> (1992, p. 174)

Not surprisingly, the majority of locally specific, in-depth research has been undertaken in the Australian home habitat of the eucalyptus. In order to provide the information necessary for better environmental policy making in settings such as the Ecuadorian sierra, it is critical that both basic ecological and social data are available to decision makers.

If rapid growth and wood production are key benefits of eucalyptus trees, then high levels of water consumption are a concomitant cost. In terms of water consumed per unit of wood produced, the eucalyptus is an efficient – if thirsty – plant. A study in India, for example, reported that eucalyptus trees consumed 0.48 l of water/g of wood produced versus 0.55–0.88 l/g for a selected group of native trees (Prabhaker, 1998). However, such a straightforward comparison of wood production misses two key points. First, for many poor, rural people, forests serve purposes beyond wood production, often including food, medicine and assorted traditional uses. Second, although eucalyptus trees may indeed use water more efficiently in wood production than many native trees, they do not necessarily consume water less intensively. Since eucalyptus trees produce wood mass so quickly, even the most efficient require large quantities of water.

In the study communities of San Pedro and Morochos, participants believe that the eucalyptus consumes more water than native trees. On a scale of 1–5, with 1 signifying consume 'a little' water and 5 'a lot' of water, eucalyptus trees received an average score of 4.90 in Morochos, 4.26 in San Pedro and 4.58 in both communities combined (Fig. 8.5). In fact, 85% of participants believe that eucalyptus trees consume 'a lot' of water. Meanwhile, native trees received average scores of 1.40 in Morochos, 1.88 in San Pedro and 1.61 in the communities combined. In stark contrast to eucalyptus, 72% of participants reported that native tree species consume only 'a little' water. Significantly lower average scores in the two communities reflect a widely held belief that native trees are 'protectors' of water, consuming only minimal quantities, while eucalyptus trees are heavy water consumers.

Eucalyptus, water consumption and agriculture

Tall rows of slender eucalyptus trees often line the edges of agricultural plots in the Quichua communities of the northern Ecuadorian Andes. Primarily planted as windbreaks to protect topsoil and crops, these trees serve a variety of domestic purposes and can also provide a small source of income. In general, however, individual Quichua farmers have neither the land nor capital to engage in plantation-scale tree cultivation. The perceived usefulness of eucalyptus trees as windbreaks is decreasing among community members, who increasingly see them as damaging the very soil and crops they have been planted to protect.

In San Pedro and Morochos, 50% of participants, when asked about water consumption by eucalyptus trees, framed their replies in terms of the very same example. The following explanation, by a 33-year-old Morochos man, was typical:

> You cannot plant them [eucalyptus trees] around the edges of the field because

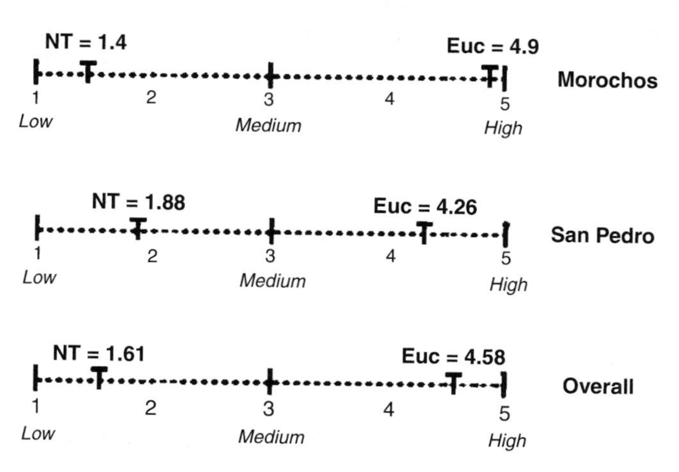

Fig. 8.5. The average perceived water consumption of eucalyptus ('Euc') versus native tree ('NT') species in the communities of Morochos, San Pedro and overall.

> they damage the crops. And they consume more water, as well, so there is no ground moisture to be consumed by the corn. So, they [the corn plants] cannot produce. It is bad.

Community members stress that crops – particularly maize – growing near eucalyptus trees do not grow as quickly or yield as much as those more distant from the trees' root systems (Fig. 8.6). They attribute stunted growth and reduced crop production to high water and nutrient consumption by the eucalyptus, which has extensive root systems that compete vigorously with neighbouring plants for available water. Empirical research supports these claims, indicating that competition from eucalyptus trees can decrease the agricultural output of crops up to 10 m away (Jagger and Pender, 2000). Moreover, 19 of the 20 participants providing this example identified the trees as consuming 'a lot of' water. The perceived link between eucalyptus trees, water consumption and diminishing agricultural output appears to coincide with an increasingly popular sentiment in Morochos and San Pedro that the trees, while extremely useful, should be relegated to areas where no crops are cultivated.

Research substantiates *campesino* claims that planting eucalyptus trees can decrease soil moisture and damage agricultural production. Hodnett *et al.* (1980) found that land previously occupied by agricultural crops will, once forested, see a marked increase in evaporative losses under both dry and wet conditions. Trees will intercept more rainfall and utilize their root systems, deeper and more extensive than those of previous crops, to exploit more water resources. Such a scenario is especially relevant in San Pedro because the eucalyptus-covered tracts of *hacienda* land surrounding the community were cultivated with grain crops several decades ago. According to Hodnett *et al.*'s analysis, the evaporative loss caused by the introduction of trees in this formerly agricultural land would have depleted soil moisture and lowered the area's water table, impacting water-needy farmers in San Pedro and in other communities situated near the eucalyptus plantation.

Eucalyptus and native trees: soil impact

Participants were asked to rate the soil impact of eucalyptus and native trees on a semantic differential scale like that used for water consumption. The paired adjectives 'damages the soil' and 'sustains the soil' were situated at opposing ends of a line

Fig. 8.6. The dashed white semi-circle indicates an area of stunted maize growth attributed to high water consumption by neighbouring eucalyptus trees (Photo: Ashley D. Carse).

segment and assigned numeric values of 1 and 5, respectively (Fig. 8.7). Like water, soil is of critical importance in the indigenous communities near Cotacachi due to its importance in agricultural production. People in the study communities conceptualize the soil impacts of trees in three broad categories: (i) soil moisture; (ii) soil nutrients; and (iii) soil erosion/ stabilization. Not surprisingly, these categories, which will be further discussed later, also play important roles in the international eucalyptus debate.

In the communities of San Pedro and Morochos, as illustrated in Fig. 8.7, there is a general perception that eucalyptus trees damage the soil and that native trees sustain the soil. On a scale of 1–5 – again, with 1 signifying 'damage' and 5 'sustain' – eucalyptus trees received average scores of 1.55 in Morochos, 2.00 in San Pedro and 1.78 in both communities. The negative soil impacts commonly mentioned include: a lack of organic material benefiting soil fertility, high levels of nutrient absorption and soil desiccation. Native trees, in contrast, are regarded very positively, receiving average scores of 4.90 in Morochos, 4.47 in San Pedro and 4.69 in both communities. While 62.5% of participants identify eucalyptus as damaging the soil, a mere 2.5% believe that native species damage the soil.

What is the significance of these quantitative values? In some cases, they reflect a grappling with conflicting priorities. As participants considered the soil-related costs and benefits of eucalyptus, what often emerged in their responses were complex sets of trade-offs. Few responses were purely positive or negative. Rather, they were generally coloured in shades of grey according to individual experience and education. The individual analyses of participants not only reveal the complexity of local human–environment relationships, but also show how community members weigh costs and benefits in determinations of preferred land use scenarios. Do the organic fertilizer, ecological benefits and non-wood products provided by native trees outweigh the fact that they are less financially productive than agriculture or eucalyptus cultivation? Do the erosion protection, wood and potential small income generation provided by eucalyptus windbreaks outweigh the cost of intensive nutrient and water uptake near crops? Not surprisingly, mixed responses like that of a 20-year-old man from San Pedro were common:

> There are two things. On one hand it hurts and on the other hand it maintains the soil. When I say damages the soil: the roots of the eucalyptus are long and go close to the crops, which damages the crops. When I say maintains the soil: trees that are grown always help to protect against the wind, so there is no erosion.

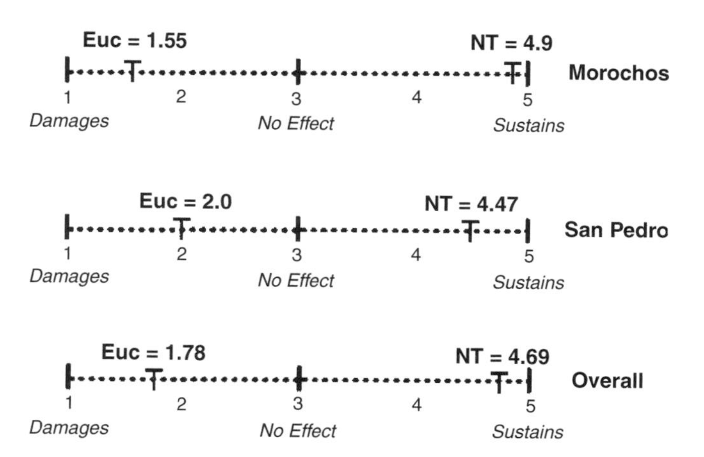

Fig. 8.7. The average perceived soil impact of eucalyptus ('Euc') versus native tree ('NT') species in the communities of Morochos, San Pedro and overall.

When asked where he would place eucalyptus trees along the soil impact line segment, this man chose the middle value of 3, indicating that, for him, the positive and negative impacts cancel each other out. In fact, 40% of participants in San Pedro, but fewer in Morochos, responded to the question with similarly mixed responses, usually indicating that eucalyptus trees damage or sustain the soil depending upon location.

Trees and soil nutrients

Critics of eucalyptus cultivation argue that the genus is nutrient intensive and depletes the surrounding soil. Proponents counter that the trees do not impoverish the soil any faster than other fast-growing species (Doughty, 2000). This may be true, but in comparison with relatively slow-growing Andean species, the soil is more likely to be impoverished of nutrients under intensive eucalyptus cultivation.

The views of participants varied greatly with regard to the relationship between the two tree types and soil nutrients. Native trees were repeatedly identified as providing *abono natural*, or 'natural fertilizer,' by depositing their leaves. Eucalyptus, in contrast, were generally seen as: (i) depleting soil quality through nutrient absorption; (ii) not building up soil fertility with their leaves; and (iii) stunting the growth of nearby plants – especially crops – through nutrient absorption. In the words of one San Pedro man, 'They damage the soil a lot, absorbing all of the nutrients and leaving it infertile, causing poor development of other plants and crops.'

Trees and soil erosion

Clearly, eucalyptus trees are perceived negatively in San Pedro and Morochos in terms of their impacts on soil moisture and nutrients. When it comes to erosion prevention, however, opinions about these introduced trees are much more positive. Although the impacts of trees on erosion were discussed less frequently (mentioned specifically in only 15% of interviews) than nutrients and water use, people who did mention them spoke of the protection afforded soil by both native and eucalyptus trees. Speaking of the eucalyptus, a 36-year-old man from San Pedro said, 'Around the crops they begin to absorb the nutrients, but when we plant them in the uncultivable parts, for example, on the hillsides, they help to protect against erosion.' When asked about the soil impacts of eucalyptus and native trees, those who did discuss soil erosion/stabilization mentioned that the root systems of all tree varieties provide protection against soil erosion. There is little difference, then, in the benefits that eucalyptus and native trees are perceived to provide in terms of soil erosion prevention. Figures 8.8 and 8.9 depict land covered in native Andean vegetation and eucalyptus forest.

Fig. 8.8. Native trees and shrubs above Morochos (Photo: Ashley D.Carse).

Fig. 8.9. Eucalyptus trees and erosion near San Pedro (Photo: Ashley D. Carse).

Trees and the 'environment': climatic impacts

> Our grandparents told us, 'where there are trees, there is water. Where there are no trees, we have droughts because they have been cut down.' They say that we have no clouds today because they have left to seek out trees elsewhere.
>
> (Man from San Pedro, 36)

Water is of great importance in the agriculture-dependent communities near the city of Cotacachi; rainfall, irrigation and drinking water are all critical issues. In both formal interviews and informal conversations in Cotacachi, Morochos and San Pedro, people – both *mestizo* and *indigena* – consistently lamented haphazard weather patterns, insufficient rainfall and dwindling groundwater. Among the most important ecological benefits of trees is, in participants' words, their ability to 'attract' clouds, rain, wind and cooler weather and to 'purify' the air. In terms of these climatic effects, there appears to be little difference in descriptions of eucalyptus and native tree species – all trees are considered beneficial.

When asked about the relationship between trees and the environment in general terms, participants most often speak of the relationship between forests and the climate. Community members tend to associate the term 'environment' with climatic factors such as weather and air quality, excluding soil, water and other key components of the western environment concept. This seems peculiar because the Quichua concept of *Pachamama* (roughly analogous to Mother Nature) represents the interplay between all components of the natural world. Consequently, an intriguing avenue for future anthropological research may be exploring the apparent distinction made by Quichua community members between the western concept of *medio ambiente*, as 'environment' is often translated, and the *Pachamama* of Quichua cosmovision.

Economic costs and benefits

The perceived economic utility of native and eucalyptus trees was measured, as with water consumption and soil impact, using a semantic differential scale. Participants were asked to place eucalyptus and native trees on a line segment with the paired adjectives 'not useful' and 'very useful' assigned to opposing ends with numeric values of 1 and 5 (Fig. 8.10). The middle of the line segment, assigned a value of 3, was labelled 'medium useful.'

Despite the widespread popularity of the eucalyptus in recent decades and its local reputation as a useful and valuable tree, native tree species received only slightly lower utility scores in Morochos

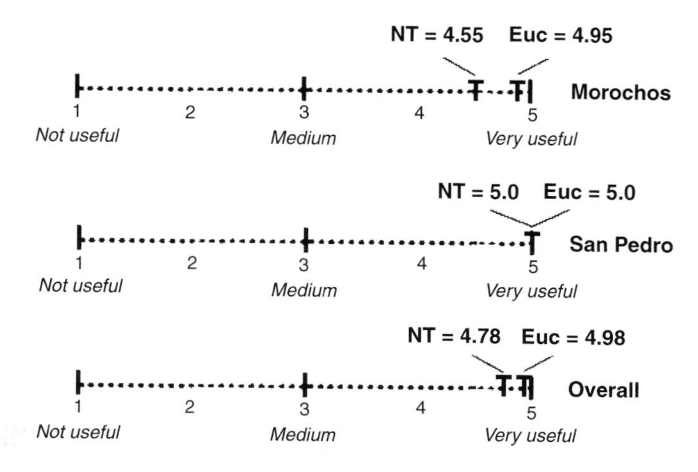

Fig. 8.10. The average perceived utility of eucalyptus versus native tree species in the communities of Morochos, San Pedro and overall.

and San Pedro. On a scale of 1–5, eucalyptus trees received very high average scores: 4.95 in Morochos, 5.00 in San Pedro and 4.98 in both communities. Native trees received only slightly lower average scores of 4.55 in Morochos, 5.00 in San Pedro and 4.78 in both communities. The close proximity of these utility scores is underlined by the fact that 87.5% of those interviewed consider both eucalyptus and native trees to be 'very useful,' which is somewhat surprising given low levels of reported native tree use. While community members report using native trees in fewer ways and less often than in the past, people clearly still consider these species to be useful, especially for medicine, food and assorted traditional uses. It is important to note here that actual use is not necessarily determined by preference alone. In Morochos and San Pedro, local availability and ease of access are also critical factors. While eucalyptus trees – planted on individual, community and *hacienda* lands – are nearby and widely available, native tree varieties are sparse and usually found several kilometres or more uphill from communities. The comparison of the past and present uses of native trees that follows demonstrates that the decreased accessibility of native forests and the diminished use of native trees are correlated.

Native trees: past and present uses

Across gender and age groups, Morochos and San Pedro community members report using native trees less regularly and in fewer ways than in the past. It has often been argued that the expansion of forest plantations, which provide alternative sources of fuelwood and construction material, lowers deforestation pressure on native forests. While the results of this study indicate that the inhabitants of Morochos and San Pedro are indeed using fewer native trees today than in the past, eucalyptus substitution is not the only explanation. As native forests recede from communities, use patterns change due to decreased availability and access. By asking participants to compare the past and present uses of native forest products within their own households (Fig. 8.11), the shifting relationship between livelihoods and native forests can be better understood.

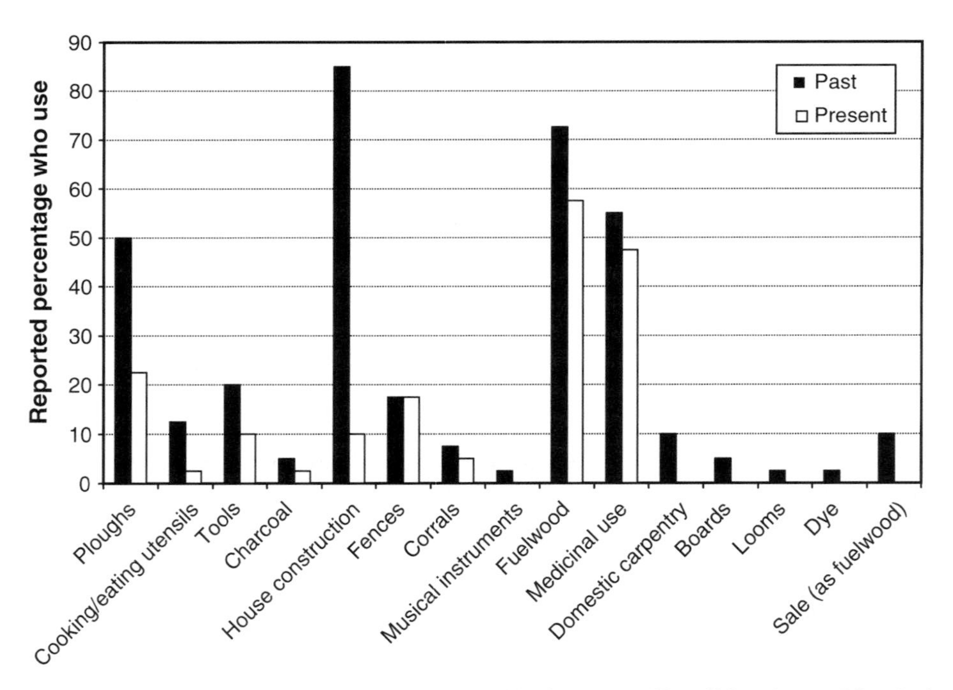

Fig. 8.11. Past and present uses of native trees reported in the communities of Morochos and San Pedro.

Figure 8.11 disaggregates shifting native tree use in the study communities. Over participants' lifetimes – their own self-defined 'pasts' and 'presents' – the percentages using native trees within their households at present is consistently less than or equal to corresponding past percentages. For example, in house construction – the category with by far the greatest disparity between past and present use – 85% of respondents reported using native trees in the past, compared with a mere 10% today. It is easy to understand why: not only are eucalyptus trees fast growing and widely available, but they are also straight with minimal branching off the trunk. The combination of these traits makes the trees ideal for the long wooden poles used for framing houses in Quichua communities near Cotacachi. Native trees, on the other hand, grow more slowly and tend to be shorter and more twisted than eucalyptus. Nevertheless, a handful of dissenting participants speak of the advantages of native trees over eucalyptus for house construction. A man from San Pedro said, 'Wood from the native trees is very resistant and durable for many more years than the eucalyptus. Wood from some native tree species lasts 200 years, while the eucalyptus hardly lasts 60 years.' The main point here is that both native and eucalyptus trees are considered useful, but in different ways. While eucalyptus trees are most often used for house construction, sale and fuelwood, native species are regularly used for traditional medicine, ploughs and agricultural implements, and fuelwood.

Eucalyptus plantations and native forests: who benefits?

A central question surrounding the push for wood- and pulp-producing eucalyptus plantations in the global south has been: who benefits from plantation forestry efforts, especially those supported by government subsidies and technical assistance or funding from various international organizations? Raintree (1996) and other advocates argue that much of the debate surrounding eucalyptus species is misplaced. Opponents' charges are focused on the ecological impacts of the genus itself, they argue, while the principal problems are rooted in social and management-related issues. A common social critique is that development efforts have emphasized a plantation forestry approach that benefits the wealthy landowners rather than the poor. National governments often support large-scale eucalyptus plantations because they provide a cash crop with an international market and a means of meeting domestic wood demand. The high yields and resulting financial benefits of intensively cultivated, fast-growing plantation species are potentially substantial; however, as with 'Green Revolution' agriculture, so are the potential costs. Even if the most optimistic economic claims proved true and economic benefits were better distributed across social strata, the fact remains that forest plantations are not equivalent substitutes for native forests in terms of the ecological services and diversity of products provided to poor, rural people.

According to Instituto Ecuatoriano Forestal y de Áreas Naturales y Vida Silvestre (INEFAN), in 1993, 67% of native forest harvested in Ecuador was used for fuelwood, with the remaining 33% going to industrial uses (Wunder, 1996). Meanwhile, of wood harvested from plantations, 21% was used for fuelwood and the remaining 79% for industrial uses (Wunder, 1996). While these aggregate figures fail to account for a multitude of non-wood forest uses, the wide disparity in native and plantation wood use that emerges does highlight significant differences between distinct forest types. What types of products does each provide and who are the primary consumers of those products? Robin Doughty writes, 'Opponents argue that forest plantations close down options for other resources at the local level, including food, medicines, and forage, which rural folk would extract from native woodlands' (2000, p. 144). Funding allocated for the financial or technical support of plantation forestry, like that provided by the Ecuadorian government to support 'reforestation' with introduced species, is most efficient when the objective is the improvement or expansion of large-scale cash cropping. These efforts can be very successful in terms of boosting income among landed individuals or groups. However, as a means of replacing the diverse, widely distributed benefits, in terms of both household economics and the local ecosystem, provided by communally owned native forests or generating income for the landless poor, these policies are markedly less effective. Those on both sides of the eucalyptus debate are increasingly recognizing that native and eucalyptus forests serve different, but equally important, environmental, economic and cultural purposes. The question, then, becomes: what type of forest and, more specifically, varieties of trees do people want in or near their communities?

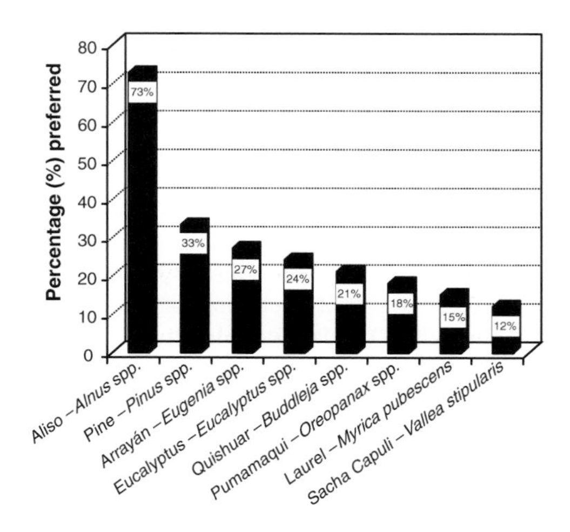

Fig. 8.12. The preferred tree varieties for reforestation, as reported by interview participants in the communities of Morochos and San Pedro.

Reforestation preferences

Large tracts of formerly agricultural lands in the northern sierra have been converted to forests of introduced tree species during the plantation boom of the past decades. Under the aforementioned PLANFOR national forestry plan, which aims to reforest hundreds of thousands of hectares for both commercial and ecological purposes, the Ecuadorian government provides financial and technical assistance to owners of forest plantations (Food and Agriculture Organization, 2003). The plan emphasizes reforestation with profitable plantation species such as eucalyptus and pine over native species. Despite such efforts to increase Ecuador's overall forest cover, if native forests alone are considered, the process of deforestation continues (Food and Agriculture Organization, 2003).

When asked which tree varieties they prefer for reforestation, inhabitants of both San Pedro and Morochos emphatically favoured native over exotic plantation species (Fig. 8.12). The native *aliso* tree (alder in English) was by far the most popular, listed by 73% of participants. The eucalyptus was fourth most popular, mentioned by 24% of participants. While 39% favour including pine, eucalyptus or both in reforestation efforts, less than half of these support planting plantation species alone without native species. Many participants mentioned that both eucalyptus and native species should be planted in order to receive their distinct benefits. In most cases, native trees were selected for their perceived ecological benefits while eucalyptus and pine were selected based upon their perceived utility and potential economic returns. As this case study has demonstrated, however, native trees are also valued for their diverse uses.

Finally, Food and Agriculture Organization (FAO) research indicates that gender is a critical variable in determining reforestation preferences in indigenous Ecuadorian communities (Kenny-Jordan, 1999). Men ostensibly prefer exotic trees that can be sold for cash, such as pine, eucalyptus and cedar. Women, on the other hand, are said to prefer native trees such as the *aliso* and *quishuar*. In this study, no significant difference emerged between the reforestation preferences of men and women. Preferences were similar across gender, age and community lines. A nearly equivalent 73% of females and 72% of males listed the *aliso* tree as a preferred reforestation variety. Similarly, 27% of females and 22% of males listed eucalyptus as a preferred reforestation variety. Gender appears to be of little significance for determining preferred tree varieties for reforestation in the study communities.

Summary of Findings

Connected shifts in tree availability and use

Zapata Ríos *et al.* (Chapter 4, this volume) conclude that, in the Cotacachi study area, the land area covered in plantation forest continues to increase, while that covered in native forest diminishes. In the study communities of Morochos and San Pedro, people recognize these trends and connect them to their own changing patterns of forest use. Eucalyptus trees, considered especially useful for house construction, firewood and sale, are close, accessible and easily available to community members. Despite the fact that native forests provide benefits not available from plantation monocultures – particularly traditional medicine, food and wood ideal for ploughs and kitchen implements – they are exploited less regularly as they 'move away' or are 'lost' in the eyes of community members. The oldest community members are most likely to walk the kilometres uphill necessary to gather and return with products of the native forest, but many say that they are no longer physically able to make the arduous journey. While participants appreciate the benefits of the fast-growing, readily available and easily accessible plantation trees, many miss the numerous products provided by the larger, more diverse native forests of the past.

Ecological costs and benefits

Eucalyptus trees are perceived by inhabitants of Morochos and San Pedro to have negative ecological impacts, including high water consumption and groundwater depletion, reduced agricultural production and soil nutrient depletion. Perceived benefits are limited to those considered common to all trees, namely preventing soil erosion and improving the local climate. Native trees, on the other hand, are seen first and foremost – before economic benefits provided or domestic utility – as ecologically beneficial plants that 'protect' groundwater, prevent soil erosion, improve the local climate, sustain the soil and provide 'natural fertilizer' by dropping leaves. In general, eucalyptus trees are thought to damage the local environment while native trees benefit it.

Local environmental discourse: shaped by generation and outside influence

When asked about the importance of reforestation projects, a 55-year-old woman from San Pedro replied, 'We don't participate in any of the projects that come. I think that the younger generation has a lot of interest in reforestation because I have seen them plant native trees around their crops.' As we spoke over lunch the next day, two young people knowledgeable about native plants expressed the same view from their own perspective. They explained that the lack of interest shown by elders towards reintroducing native trees and protecting forests is the result of past struggles for survival. Due to extreme poverty and nutritional deficiencies in the past, many members of the older generation – who, incidentally, often know the most about native plants – place an emphasis on productive agriculture over native forests. For these community members, food and human survival come before reforestation or conservation efforts. Perceptions among the younger generation, especially people in their 20s and 30s, are quite different. Even young people with very little knowledge about native plants are quick to affirm the importance of native trees and voice support for reforestation projects. Karl Zimmerer writes:

> Indeed the local knowledge apparent in the everyday discourses of . . . peasants about erosion has rarely been manifest as a strictly self-contained and self-referential dialogue. Their commonplace observations suggest strongly that such expressions have influenced, and been influenced by, ideas on erosion of other social groups in the region.
>
> (1996, p. 113)

The local discourse surrounding the ecological costs and benefits of forests in

San Pedro and Morochos is clearly subject to exterior influence. Environmental topics are common fare at the ubiquitous *talleres*, or 'workshops,' conducted by various local, national and international development and conservation organizations in Cotacachi and the surrounding indigenous communities. It would not be surprising, then, if the statements of young participants were influenced by the anti-introduced/pro-native plant rhetoric typical at these meetings. Moreover, increasingly market-savvy indigenous communities recognize that native forests and plants can be a successful means of attracting tourism and development funds, which might also be reflected in discourse.

Economic costs and benefits

While eucalyptus trees are exploited much more frequently than native trees in Morochos and San Pedro, they are considered only slightly more useful. This apparent disparity may be the result of two factors. First, native trees exist in smaller quantities and are less accessible than eucalyptus, so perceived utility may not correspond to actual use. Second, the discourse common among NGOs and various international organizations in Cotacachi today emphasizes the ecological and cultural benefits of native species over eucalyptus trees, which may lead even those who do not use them to speak of them as useful. While eucalyptus trees may provide a small source of income for community members who grow them on their plots, the introduction of eucalyptus plantations on *hacienda* lands has decreased the availability of agricultural labour and income for many people in the community. As these jobs disappeared, household members, especially men, became more likely to travel farther and stay longer in cities such as Quito and Otavalo where wage labour is more readily available (Flora, Chapter 18, this volume). While, in terms of monetary income, the impact of eucalyptus plantations on nearby communities has been negative, the overall economic impact is more ambiguous.

Reforestation

Ninety-five per cent of participants in Morochos and San Pedro indicated that they believe reforestation projects 'are important.' While nearly everyone asked agreed that planting more trees is positive, the preferred locations for and varieties of trees to be planted are disputed. Some advocates of reforestation believe that trees should be planted in and around the community for accessibility, while others believe that they should be planted above the communities in order to regulate water flows and not monopolize productive agricultural land. The native *aliso* tree – which is ecologically beneficial and has uses ranging from medicine to handicrafts – is by far the most popular variety for reforestation, listed as preferred by 73% of participants. Support also exists for planting more pine and eucalyptus. Pine was listed by 33% and eucalyptus by 24% of participants. Based on interview data collected in San Pedro and Morochos, ideal future reforestation efforts would simultaneously provide useful forest products, income generation, ecological benefits and biodiversity protection.

Future Research Questions

The results of this study, which takes an ethnoecological and economic perspective on issues of conservation and development in two indigenous communities, provide a foundation for understanding indigenous perceptions of trees and forest-related changes in the landscape as they relate to local livelihoods. From this analysis, two especially interesting paths for further study emerge.

1. Political ecology of eucalyptus plantations in the Ecuadorian sierra. A belief exists among some members of the study communities that the appearance of eucalyptus trees on neighbouring *hacienda* lands once cultivated with grain crops – an event that coincided with the agrarian reform era in Ecuador – was a calculated move by *hacienda* owners to prevent expropriation efforts by indigenous groups under new land reform laws. Future studies might examine the role that legal and political manoeuvring by various stakeholders played, and continues to play, in the rise and spread of forest plantations in the Ecuadorian sierra.

2. Role of discourse in shaping perceptions of the environment in the Ecuadorian sierra. This study explores perceptions of the ecological and economic costs and benefits of eucalyptus and native trees. An important question that emerges from this analysis is: what forces shape people's perceptions of the environment in San Pedro and Morochos? With a multitude of NGOs and international organizations sponsoring projects and workshops on conservation and development issues in and around Cotacachi, local people have been exposed to a considerable amount of environmental rhetoric. Future studies might shed light upon the role that local, regional, national and international discourses – in addition to factors such as culture and personal experience – play in shaping local perceptions of the environment.

References

Bernard, H.R. (2002) *Research Methods in Anthropology: Qualitative and Quantitative Approaches*, 3rd edn. AltaMira Press, Walnut Creek, California.

Blaikie, P. (1985) *The Political Economy of Soil Erosion in Developing Countries*. Heinemann, London.

Calder, I.R. (1992) Water use of eucalypts – a review. In: Calder, I.R., Hall, R.L. and Adlard, P.G. (eds) *Growth and Water Use of Forestry Plantations*. John Wiley and Sons, Chichester, UK, pp. 167–179.

Conklin, H. (1961) The study of shifting cultivation. *Current Anthropology* 2, 27–61.

Denevan, W.M. (1992) The pristine myth: the landscape of the Americas in 1492. *Annals of the Association of American Geographers* 82, 369–385.

Doughty, R.W. (2000) *The Eucalyptus: a Natural and Commercial History of the Gum Tree*. Johns Hopkins University Press, Baltimore, Maryland.

Fishbein, M. (1967) *Readings in Attitude Theory and Measurement*. John Wiley and Sons, New York.

Food and Agriculture Organization (2003) FAO Forestry Department Ecuador Country Profile. Available at: http://www.fao.org/forestry/foris/webview/forestry2/index.jsp?siteId=5081&sitetreeId=18308&langId=1&geoId=208

Hodnett, M.G., Bell, J.P. and Gupta, D.K. (1980) Phase I of the soil moisture studies in the Betway catchment. *Institute of Hydrology Report* (unnumbered). Wallingford, UK.

Instituto Ecuatoriano Forestal y de Áreas Naturales y Vida Silvestre (1996) *Principales Estadísticas Forestales del Ecuador*. Instituto Ecuatoriano Forestal y de Áreas Naturales y Vida Silvestre, Quito, Ecuador.

Jagger, P. and Pender, J. (2000) *The Role of Trees for Sustainable Management of Less-favored Lands: the Case of Eucalyptus in Ethiopia*. EPTD Discussion Paper No. 65. International Food Policy Research Institute, Washington, DC.

Kenny-Jordan, C.B., Carlos, H., Mario, A. and Andrade, M. (1999) *Pioneering Change: Community Forestry in the Andean Highlands*. Food and Agriculture Organization of the United Nations, Quito, Ecuador.

Peet, R. and Watts, M. (eds) (1996) *Liberation Ecologies: Environment, Development, Social Movements*. Routledge, New York.

Prabhaker, V.K. (1998) *Social and Community Forestry*. Satish Garg, New Delhi.

Raintree, J.B. (1996) *The Great Eucalyptus Debate: What is it Really All About*? Reports submitted to the Regional Expert Consultation on Eucalyptus, Vol. II. RAP Publication, Bangkok.

Sarmiento, F.O. (2002) Anthropogenic change in the landscapes of highland Ecuador. *Geographic Review* 92, 213–234.

Saxena, N.C. (1994) *India's Eucalyptus Craze: the God That Failed*. Sage, New Delhi.

Shiva, V. and Bandyopadhyay, J. (1983) Eucalyptus – a disastrous tree for India. *Ecologist* 13, 184–187.

Stonich, S.C. (1993) *'I am Destroying the Land!': the Political Ecology of Poverty and Environmental Destruction in Honduras*. Westview Press, Boulder, Colorado.

Wunder, S. (1996) *Los Caminos de la Madera: una Investigación de los Usos Domésticos y Comerciales de los Productos de la Madera, y su Relación con el Proceso de Deforestación.* IUCN, Quito, Ecuador.
Wunder, S. (2000) *The Economics of Deforestation: the Example of Ecuador.* St Martin's Press, New York.
Zimmerer, K.S. (1996) Discourses on soil loss in Bolivia: Sustainability and the search for socio-environmental 'middle ground'. In: Peet, R. and Watts, M. (eds) *Liberation Ecologies: Environment, Development, and Social Movements.* Routledge, London, pp. 110–124.
Zobel, B.J., Van Wyk, G. and Stahl, P. (1987) *Growing Exotic Forests.* John Wiley and Sons, New York.

9 Living, Dwindling, Losing, Finding: Status and Changes in Agrobiodiversity of Cotacachi

Kristine Skarbø
Norwegian University of Life Sciences, N-6200 Stranda, Norway

Introduction

For thousands of years, the Andean region has been a stage for rich crop domestication and cultivation (National Research Council, 1989). Varied climatic and environmental microniches along the slopes of the Andes allowed a diversity of crops and cultivars to develop (Zimmerer, 1996). Following the Spanish conquest in the 16th century, Old World crops were introduced to Andean fields, setting the conditions for an even greater complex of agricultural genetic resources available to farmers.

Over the centuries, many Andean crops have been looked down on by South Americans of European and mixed ancestry who disdained the Native culture. The conquistadores and their successors developed a cuisine distinct from the indigenous people (Weismantel, 1988; National Research Council, 1989). However, while many 'Inca' crops were ignored by the ruling class, the Native masses continued to grow their own food on small plots. Season after season they kept the native plants and old heritage alive in their gardens and fields.

Genetic erosion of these Andean crops is a contemporary concern (Frankel, 1973; Rhoades and Nazarea, 1998). Studies have shown that intraspecific genetic diversity is declining in Andean root and tuber crops (Tapia, 2001) and some species have fallen out of cultivation since the Inca Empire ceased (National Research Council, 1989; Hernández Bermejo and León, 1994). This situation has led to increasing global attention given to the neglected Andean crops, referred to by some as 'The Lost Crops of the Inca' (National Research Council, 1989). Researchers, impressed by nutritious properties and environmental robustness of these species, are evaluating their potential for crop improvement and export to other parts of the world (Sperling and King, 1990; Izquierdo and Roca, 1998; Scott *et al.*, 2000; Scheldeman *et al.*, 2003).

This chapter centres on the current state of agrobiodiversity within Cotacachi's Andean zone. First, I will introduce the study area and the five communities selected for in-depth study. Second, I will access the status of agrobiodiversity in Cotacachi. Third, an analysis of causes and effects of genetic erosion will be undertaken. Next, trends and patterns of change will be examined. Finally, I will attempt to project what the future might bring for Cotacachi's agrobiodiversity. This study was carried out in Cotacachi during 2003–2004 and included semi-structured interviews with 45 farmers in five communities, a survey of 50 children from the same communities, several workshops, participatory observation and in-depth interviews with farmers as well as local organizations.

 Development with Identity: Community, Culture and Sustainability in the Andes (ed. R.E. Rhoades)

The Study Area

Five of Cotacachi's 43 communities, were selected for the study: Quitugo (2500 metres above sea level (masl)) and El Batan (2480 masl) from the lower zone; San Pedro (2780 masl) and Peribuela (2600 masl) from the middle; and Ugshapungo (3200 masl) from the upper zone. These communities also represent the three latitudinal geographical zones (northern, central and southern) of the Andean part of the canton. Different agroecological zones and developmental patterns in crop composition and agricultural calendars are found across these communities.

An important historical change in rural Cotacachi has been in average farm size (Zapata *et al.*, Chapter 4, this volume). Farmers in Quitugo, El Batan and San Pedro benefited little from land redistribution in the agrarian reforms of 1963 and 1974, only obtaining property rights to their own plots, and consequently land holdings remain small. Alternatively, Ugshapungo and Peribuela are situated on former hacienda land. Since farms in these two communities are generally larger than in the three previously mentioned communities, the possibilities for producing crop surplus for commercialization is greater. Peribuela farmers devote most of their land to cash crops. Ugshapungo farmers produce tubers for sale, but still meet a large part of subsistence need with home produce. Even if Quitugo, El Batan and San Pedro farmers were to maximize production, most do not have enough land to feed their families throughout the year. To supply food and other needs, men migrate and seek jobs as construction workers and women make handicrafts, keep little stores, clean houses or bake bread for sale (see Flora, Chapter 18, this volume).

Status of Agrobiodiversity in Cotacachi

Crops grown

The overall number of crop species is high in the Cotacachi area; a total of 61 different food crop species were identified. Half of these (34) are native species while the rest are introduced from the Old World. However, crop species are not farmed equally by farmers and there is a tendency towards farming maize, beans and potatoes (Table 9.1). Nineteen species traditionally make up the majority of the crops; including cereals, tubers, roots, legumes and cucurbits. Maize (*Zea mays*) originated in Mexico, but early on found its way to the Andes, where an extensive genetic diversity has developed over the millennia. The common bean (*Phaseolus vulgaris*) was probably domesticated independently in the Andes and in Central America. Cotacachi farmers grow both climbing and bush beans. Most potatoes grown in Cotacachi belong to the species *Solanum tuberosum* subsp. *andigena*, but fast-growing *Solanum phureja* and weedy *arapapas* (*Solanum* sect. *petota*) are also found. In addition to different species of potatoes, a number of other Andean tubers and root crops are still grown in Cotacachi. Roots include mashwa (*Tropaeolum tuberosum*), arracacha (*Arracacia xanthorrhiza*) and sweet potato (*Ipomoea batatas*), and tubers ulluco (*Ullucus tuberosum*) and oca (*Oxalis tuberosa*). Two species of cucurbits are grown; the *zambo* (*Cucurbita ficifolia*), domesticated in Mexico, and the zapallo (*Cucurbita maxima*) which originated in the southern Andes (National Research Council, 1989). Together with lupins and the pseudocereal quínua, these crops constituted the basis of a diverse Andean diet 500 years ago. Old World crops integrated later include wheat, barley, rye, fava beans, lentils, peas and chickpeas, and a number of vegetables and fruits.

Vegetables and herbs

Except for the roots and cucurbits mentioned above, Andean farmers did not traditionally cultivate many vegetables. Of the 15 vegetable crops (divided between ten species) registered in Cotacachi, only three are of South American origin, and of these only the *ají* or Andean chilli pepper (*Capsicum baccatum*) is deeply rooted in the Cotacachi campesinos' cuisine. To

Table 9.1. Crops grown in Cotacachi[c].

Type of crop	Origin	English common name[a]	Local Spanish name	Scientific name	Portion of farmers (%) ($n = 45$)
Cereals	Native	Maize	Maíz	*Zea mays*	100
	Introduced	Wheat	Trigo	*Triticum aestivum*	38
	Introduced	Barley	Cebada	*Hordeum vulgare*	38
Pseudo-cereals	Native	Quinua	Quinua	*Chenopodium quinoa*	54
Legumes	Native	Common bean	Fréjol	*Phaseolus vulgaris*	83
	Native	Pearl lupine	Chocho	*Lupinus mutabilis*	54
	Introduced	Fava bean	Haba	*Vicia faba*	51
	Introduced	Pea	Arveja	*Pisum sativum*	64
	Introduced	Lentil	Lenteja	*Lens culinaris*	9
Tubers	Native	Potato	Papa	*Solanum tuberosum* subsp. *andigena*	80
	Native	Chaucha potato	Papa chaucha	*Solanum phureja*	9
	Native	Wild potato	Arapapa	*Solanum* sect. *Petota*	2
	Native	Ullucu	Melloco	*Ullucus tuberosus*	28
	Native	Oca	Oca	*Oxalis tuberosa*	36
Roots	Native	Sweet potato	Camote	*Ipomoea batatas*	4
	Native	Mashwa	Mashua	*Tropaeolum tuberosum*	7
	Native	Arracacha	Zanahoria blanca	*Arracacia xanthorrhiza*	2
Cucurbits	Native	Squash	Zambo	*Cucurbita ficifolia*	51
	Native	Winter squash	Zapallo	*Cucurbita maxima*	18
Vegetables	Native	Andean pepper	Ají	*Capsicum baccatum*	7
	Native	Sweet pepper	Pimiento	*Capsicum annuum*	7
	Native	Tomato	Tomate riñon	*Lycopersicon esculentum* var. *esculentum*	9
	Introduced	Cabbage	Coles	*Brassica oleracea* var. *capitata*	40
	Introduced	Repollo cabbage	Col verde/silvestre/ Repollo	*Brassica oleracea* var. *capitata*	11
	Introduced	Broccoli	Brocolia	*Brassica oleracea* var. *italica*	4
	Introduced	Cauliflower	Coliflor	*Brassica oleracea* var. *botrytis*	2
	Introduced	Onion	Cebolla blanca	*Allium cepa* var. *cepa*	36
	Introduced	Red onion	Paiteña	*Allium cepa*	11
	Introduced	Lettuce	Lechuga	*Lactuca* spp.	27
	Introduced	Carrot	Zanahoria	*Daucus carota*	22
	Introduced	Red beet	Remolacha	*Beta vulgaris*	18
	Introduced	Leaf beet	Acelga	*Beta vulgaris* var. *cicla*	9
	Introduced	Radish	Rabano	*Raphanus sativus*	7
	Introduced	Zucchini	Zuquini	*Cucurbita pepo* subsp. *melopepo*	2
Fruits	Native	Tree tomato	Tomate de árbol	*Cyphomandra betacea*	49
	Native	Andean blackberry	Mora de Castilla	*Rubus glaucus*	33

Continued

Table 9.1. *Continued.* Crops grown in Cotacachi.

Type of crop	Origin	English common name[a]	Local Spanish name	Scientific name	Portion of farmers (%) ($n = 45$)
	Native	Avocado	Aguacate	*Persea americana*	13
	Native	Pacay	Guava	*Inga* sp.	11
	Native	Passion fruit	Granadilla	*Passiflora ligularis*	11
	Native	Banana passion fruit	Taxo	*Passiflora mollissima*	9
	Native	Goldenberry	Uvilla, capuli	*Physalis peruviana*	13
	Native	Walnut	Nogal/tocte	*Juglans neotropica*	7
	Native	Cherimoya	Cherimoya	*Annona cherimola*	4
	Native	Mountain papaya	Babaco	*Carica* × *heilbornii*	4
	Native	Mountain papaya, Chamburo	Chilehuaca	*Carica pubescens*	2
	Native	Guava	Guayaba	*Psidium guajava*	2
	Introduced	Lemon	Limón	*Citrus limon*	22
	Introduced	Orange	Naranja	*Citrus sinensis*	13
	Introduced	Mandarin	Mandarina	*Citrus reticulata*	11
	Introduced	Peach	Durazno	*Prunus persica*	9
	Introduced	Strawberry	Frutilla	*Fragaria* × *ananassa*	9
	Introduced	Plum	Claudia	*Prunus domestica*	7
	Introduced	Apple	Manzana	*Malus sylvestris*	4
	Introduced	Fig	Higo	*Ficus carica*	4
	Introduced	Banana	Plátano	*Musa* sp.	2
	Introduced	Lime	Lima	*Adelia ricinella*	2
	Introduced	Grape	Uva	*Vitis vinifera*	2
	Introduced	Pomegranate	Granada	*Punica granatum*	2
Medicinal plants/herbs[b]	Native	Roman chamomile	Manzanilla	*Anthemis nobilis*	
	Native		Limoncillo	*Siparuna* sp.	
	Native	Lemon balm	Toronjil	*Melissa officinalis*	
	Native	Lemongrass	Hierbaluisa	*Cymbopogon citratus*	
	Native	Lemon verbena	Cedrón	*Lippia citriodora*	
	Introduced	Mint	Hierbabuena	*Menta piperita*	
	Introduced	Parsley	Perejil	*Petroselinum crispum*	
	Introduced	Coriander	Culantro	*Coriandrum sativum*	

[a]For several Andean plants, there is no consensus on the English name. Here names from the National Research Council (1989) are employed.
[b]Medicinal plants are not registered on all farms, thus percentages are not included. The listed herbs are the eight most commonly grown.
[c]In total, 64 different food crops divided into 61 species were documented. Nineteen species traditionally make up the majority of the fields; an additional ten vegetable species, 24 fruit species and eight herbs are grown. About half of the species (34) are of native origin. Horticultural crops make up the majority of introduced species; many of these are only cultivated by a few farmers at present.
Sources: Fieldwork; National Research Council (1989); Rehm and Espig (1991); Smartt and Simmonds (1995).

complement and enrich their diet, people gathered a variety of wild plants which were also highly appreciated for their medicinal value (Rodriguez Cuenca, 1999). Quite a few people in Cotacachi still go out to find *bledo* (*Amaranthus blitum*), *ataco* (*Amaranthus quitensis*), *berro* (*Nasturtium officinale*), *achocha* (*Cyclanthera explodens*) and *paico* (*Chenopodium ambrosiodes*), either growing as weeds among other crops or in *quebradas*, steep ravines that rift the landscape. Old World vegetables such as onions, cabbage and carrots have been incorporated into the local cuisine, and lately a new wave of exotic vegetable crops is being cultivated. The highest number of vegetable crops is registered in Peribuela (13), and the lowest number is found in Ugshapungo (6). Also species number per cultivator is highest for Peribuela (5.5). Overall, each vegetable cultivator on average has 2.3 species. People also grow medicinal plants around their houses, some of which were introduced by Spaniards (Table 9.1).

Fruits

Among a total of 24 fruit species, 12 are native and 12 introduced. Despite the richness of native fruit species, these have not, as one woman describes, occupied a large part of the campesinos' land:

> Earlier they almost only had them in the haciendas, the fruit trees. For example, we could have some trees of avocados or guavas; those were all we had in the communities, whereas in the haciendas they had mandarins, oranges, lemons, plums, all of those. But recently we are starting to plant them in the communities as well.
>
> (Female farmer, 41, El Batan)

Small-scale Cotacachi farmers are now becoming more interested in planting their own trees. Nevertheless, the most abundant fruit production is found in Peribuela, where tree tomatoes now cover large areas earlier devoted to maize and beans. Together with *mora*, or Andean blackberry, a native *Rubus* species, and lemon, the tree tomato (*tomate de árbol*) is the most popular fruit among the cultivators. Each fruit cultivator on average grows 3.7 species, but this number varies from one to 22.

Crop composition: differences in time and space

> Now we only sow maize, the most fundamental.
>
> (Female farmer, 40, Quitugo)

Even if this woman in Quitugo plants both beans and other seeds in her maize field, her statement points to the lack of diversification in Andean crops today. Maize, beans, potatoes and peas dominate, while plants of many other species, in particular native roots and tubers, play a marginal role (Fig. 9.1). The once more common root crop mauka (*Mirabilis expansa*, locally called *biso*) is no longer cultivated by a single respondent. This trend of crop simplification is most profound among commercial farmers in Peribuela where monoculture of a few crops is the rule. The number of crops grown varies widely between farmers, and some of this variation can be explained by differing agroecological zones; grain and tuber crops are often considered *cultivos del cerro* ('crops of the hill') and most of the current cultivators reside in the higher zone (San Pedro and Ugshapungo). However, not all elderly in lower lying communities agree with this:

> The oca gives a good harvest here in this community, but the custom of sowing it has been lost, now you cannot even find seeds. My parents also used to grow sweet potato, arracacha, mauka and ulluco, but nowadays people don't grow it anymore. And we say that those things only produce further up in the hill, but that is a lie. Yes, it does produce here, it is just that we have lost the custom.
>
> (Male farmer, 75, Quitugo)

On the other hand, an increase in cultivation of introduced vegetables and fruits among the surveyed farmers was found. In the canton as a whole, area devoted to commercial crops has increased

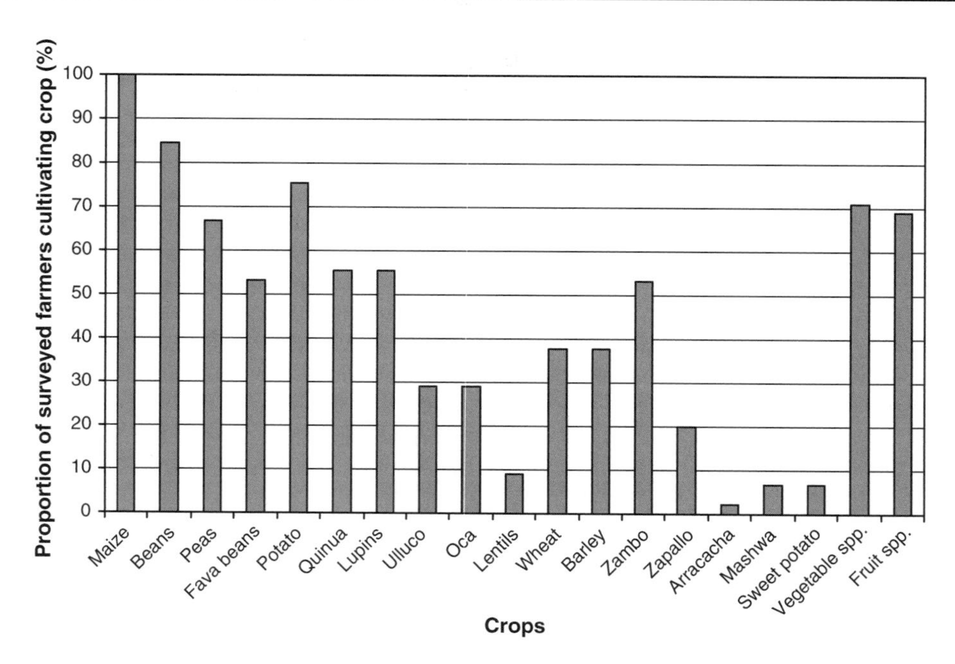

Fig. 9.1. Proportion of surveyed farmers growing the major crops of the area ($n = 45$).

due to the development of a greenhouse industry based on fruit, flower and asparagus production while the area devoted to native crops is declining (Table 9.2).

Diversity at the variety level

The results of this study indicate that Cotacachi still contains a wide diversity of crop varieties. Within 12 major crops, 139 varieties were recorded. When root crops and cucurbits are included, the aggregate variety number increases to 153. The majority of the varieties are landraces, although modern varieties (MVs) are also documented for potatoes, beans and maize. Most of the MVs are grown in Peribuela (maize, beans and potatoes) and Ugshapungo (potatoes). Eighty-nine per cent of all the varieties were not found in more than one or two of the five communities and 69% were grown by just one or two of the 45 interviewed farmers. The continued existence of these varieties depends upon the continued interest of these particular farmers in saving their seed and sowing it. Moreover, informants pointed to the total loss of 62 varieties (Table 9.3).

Losing the Inca Crops?

Genetic erosion is a much debated and contested concept. The Food and Agriculture Organization of the United Nations (FAO)'s report on the State of the World's Plant Genetic Resources says that 'recent losses of diversity have been large and that the process of 'erosion' continues' (FAO, 1996). However, recent studies suggest that landraces are more robust than predicted and, even if their area declines, many still survive in some fields (Brush, 1992; Perales *et al.*, 2003). Although we have seen evidence of a high aggregate diversity in Cotacachi, some crops and varieties are reported to be abandoned. Several crops and a high portion of varieties are only kept by a small portion of the farmers (Table 9.3). If the trend continues, these disappearing crops face an insecure future.

The lack of baseline data on genetic resources makes it impossible to measure

Table 9.2. Status of Cotacachi crops: lost, declining or rising.

Lost crops	Declining crops	Rising crops
Mauka (*Mirabilis expansa*)[a]	Winter squash (*Cucurbita maxima*)[a]	Vegetable spp.
Chickpea (*Cicer arietinum*)	Mashwa (*Tropaeolum tuberosum*)[a]	Fruit spp.[a]
Rye (*Secale cereale*)	Arracacha (*Arracacia xanthorrhiza*)[a]	Flowers, greenhouse production
	Sweet potato (*Ipomoea batatas*)[a]	
	Oca (*Oxalis tuberosa*)[a]	
	Ulluco (*Ullucus tuberosus*)[a]	
	Lentil (*Lens culinaris*)	
	Fava bean (*Vicia faba*)	

[a]Native crops.

Table 9.3. Trends in variety diversity within 12 important crops.

Crop	Total no. of varieties presently grown	Total no. of varieties reported lost	No. of varieties found in < 3 communities	No. of varieties found on < 3 farms
Maize	14	9	8	4
Beans	27	11	26	19
Peas	13	3	11	9
Fava bean	9	1	6	6
Potato	26	19	21	18
Quinua	9	3	8	6
Lupins	4	2	3	3
Ulluco	7	2	6	3
Oca	5	0	3	1
Lentils	3	6	3	3
Wheat	9	4	8	6
Barley	13	2	13	11
Total	139	62	116	89

accurately how agrobiodiversity has changed from past generations (Guarino, 1999). However, experienced farmers may help us by sharing their views:

> When I was young, I worked with my grandfather and with my dad. So I learned a lot about working with agriculture. With my grandfather I was ploughing the fields, caring for all the plants. We used to grow maize, beans, fava beans, all the *granos* there were; oca, ulluco, quínua, lupins, peas, lentils. Good lentils, coloured wheat, large barley; those three varieties are gone. Now we just grow a little wheat, but the seed is not from here, it's wheat from America or from Africa, something like that. Our own wheat is the coloured wheat, it had awns. Green lentils we also had, and *alverjano*; that one was similar to peas, but it was not the same. It was grey. But now it doesn't exist anymore. There also were black lentils. Now we only sow a small barley, I don't know what it is called. The barley we used to have is gone, the wheat from the old days is gone, so are the lentils and oca we don't sow. A few are still sowing ullucos, up on the hill. We sow a little of quínua, but earlier we grew a lot. (. . .) There are but a few potatoes, but we used to grow many.
>
> (Male farmer, 80, San Pedro)

Why? Reasons Behind Changes in Agrobiodiversity

Will the trend towards genetic erosion continue in Cotacachi? To answer this question, it is crucial to understand the reasons farmers give for changes in Cotacachi's agrobiodiversity. In the end, the farmers themselves choose what seed to sow. Therefore, I turned to them when seeking an answer to this question. In their answers, farmers talk about agricultural change, climate change, pests, lack of manure, currency change, lack of land, migration and cultural changes (Fig. 9.2). The rest of this chapter will present and discuss these factors and how they influence agrobiodiversity in Cotacachi.

Impacts of the 'Green Revolution': commercial farmers

> Earlier it was better, because then one didn't use chemicals. Today you have to use it for the crops to grow. It's like the soil is exhausted, we have to put more and more.
> (Female farmer, 27, Peribuela)

The Green Revolution with its modern varieties and technological and chemical inputs is often blamed for much of the genetic erosion taking place (Food and Agriculture Organization, 1996). In two of the communities, Peribuela and Ugshapungo, farmers also state that these changes have had major influences on their agriculture and agrobiodiversity. In Ecuador, as in other countries, urbanization has created an increased demand for agricultural products, and government policies have enhanced commercial production. Farmers who manage large fields have mechanized cultivation and harvesting in order to produce sufficient quantities for the market. Many look upon pesticides and fertilizers as a necessary evil. Even if they see negative environmental consequences, all apply them since to do otherwise would mean a poor harvest.

Impacts of the 'Green Revolution': subsistence farmers

> We prepare the land, ploughing, removing the weeds, applying manure, from cattle or sheep, whatever manure. We are not applying fertilizers, neither do we spray. This

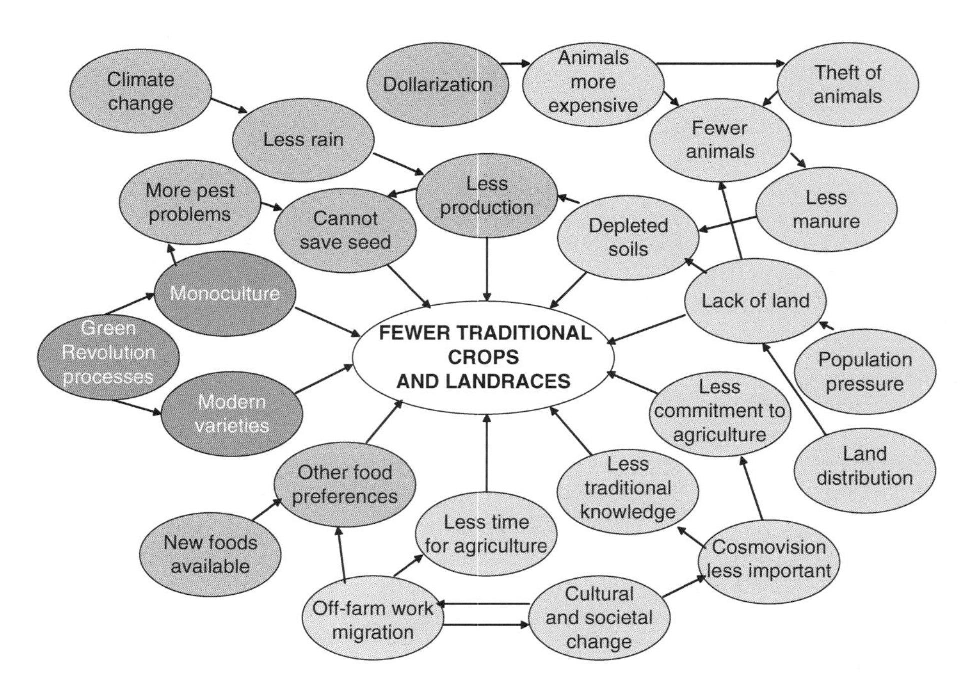

Fig. 9.2. Influence diagram: farmers' explanations of change in agrobiodiversity. The factors in dark grey ovals mostly relate to commercial farmers, those in light grey are particularly pointed to by subsistence farmers, while the rest relate to both groups.

> way we get more tasty products, whereas in the haciendas they put fertilizers and pesticides.
>
> (Male farmer, 64, El Batan)

Subsistence farmers face a different situation than commercial growers since their small fields are supported primarily by manual labour, and none apply fertilizers or pesticides. The absence of chemical inputs can be explained by the small-scale nature of their activities and lack of capital. Several farmers further expressed that they do not desire to apply such inputs:

> We have never sprayed. The chickens eat the worms and clean the peas. Here in Quitugo we are not using chemicals. For us it is very important to take care of the soil. There were some people who used chemicals, and it produced well for some years. But later the land was tired, and didn't want to give. For us the future is what is most important, and it isn't worth it if you have a good production for five or six years, and afterwards you have nothing. If we spray, we use water from eucalyptus, chilli pepper or lupins.
>
> (Female farmer, 33, Quitugo)

Impacts of the 'Green Revolution': consequences for diversity

> The seed changed, there came others, better ones, and so the old ones came to an end.
>
> (Male farmer, 90, Ugshapungo)

The simpler crop composition in commercial farmers' fields can be associated with 'Green Revolution' developments. Farmers choose crops which are profitable, can be produced efficiently and are demanded in the market. In Peribuela, this has resulted in monocropping of maize, beans, peas and potatoes, while in Ugshapungo farmers primarily grow potatoes as a cash crop. These same food crops have also been the focus of government and research programmes which have provided modern technology and varieties. Potatoes, beans and maize have suffered the most severe loss of landraces (Table 9.3). In commercial farmers' fields, landraces have mainly been replaced with MVs. As a 27-year-old female farmer from Peribuela comments: 'Of course you need improved seed, to get a better price, to have large ears.' In Peribuela, traditional crops other than maize, beans and potatoes have largely been abandoned, while in Ugshapungo roots and tubers such as ulluco, oca and mashwa are still grown. Except for potatoes, this high zone is too cold for crops where MVs are available, and landraces are thus maintained. Nevertheless, when demand for these crops diminishes, landraces fall out of use:

> Mashwa we only grow a few of, because mashwa they are not buying. Before we sowed a lot. Now we only sow the *chaucha*, but we used to have a reddish one, yellow ones, coloured ones. They produced so much. But since a long time we only sow the *chaucha*.
>
> (Male farmer, 90, Ugshapungo)

Whereas developments after the 'Green Revolution' can explain the abandonment of landraces by commercial farmers, many subsistence farmers also report dropping landraces even without having sown MV seed. Searching for further explanations for the changes in agrobiodiversity, we need to look beyond the wonder and power of agro-industrial inputs.

Economic pressures

> They want me to give it to them. In the last harvest the beans went for four dollars a bulk, it doesn't even cover a farm worker's payment.
>
> (Female farmer, 64, Peribuela)

Many commercial farmers today express concern about declining revenues resulting from expensive inputs and low product prices. The economic turbulence in the country during the 1990s, leading to reforms that peaked with the adoption of the US dollar as the monetary unit in 2000 in combination with free trade agreements with neighbouring countries, has put pressure on Cotacachi's farmers. Commercial farmers search for viable alternatives in production strategies that secure an income. In Peribuela, fruit cultivation has become increasingly popular, while farmers in Ugshapungo keep cattle for milk production and have established a microdairy with the support of the non-governmental organization (NGO), Ayuda en Acción. With these changes, fields serve new purposes and a new agrobiodiversity evolves.

Climate

> The weather is totally changed, it rains when it is not supposed to, and the summer comes in the months of winter. For this reason we have problems with disease and drought.
>
> (Male farmer, 80, San Pedro)

Climate change is recognized by farmers in all communities (Rhoades *et al.*, Chapter 5, this volume). Irregular rainfall patterns and a delayed onset of the rainy season in recent years have made farming more difficult and led to crop failure during prolonged periods of drought. Also, the glacier on volcano Cotacachi has disappeared in recent years, and with it a constant flow of melting water. This may have led to a decrease in stream flow and less water available for irrigation to offset drought stress. These altered conditions constrain crop production not only by drying out crops, but also by making them increasingly susceptible to pests. This again has implications for agrobiodiversity, as populations may be lost or varieties and crops that are not easily adaptable may lose priority.

Pests and diseases

> You cannot sow like before, we didn't use to have those pests.
>
> (Female farmer, 56, Ugshapungo)

In relation to changing climate, the increased incidence of late blight in potatoes (*Phytophthora infestans*) is often mentioned. The majority of farmers still cultivate potatoes, but among subsistence farmers many do not produce enough for a whole year's consumption, and instead supply their stores from the market. New planting material is selected from purchased sacks of food potatoes. This decreased degree of self-sufficiency is related to late blight problems. Further, attacks from insects originating in flower greenhouses have harmed crops. Many people also commented on weevils attacking stored maize, as described by a 33-year-old farmer from Peribuela:

> Now you cannot save it anymore, because the beetle enters, makes it into flour and you cannot use it for anything. Not even the pigs eat it – the taste is bitter and acidic like lemon. Earlier we didn't have this bug. Up to two years we used to keep the maize, but not anymore.

In Peribuela some farmers treat the maize with chemicals to avoid the weevils, while more traditional strategies include washing with water and sun drying. These storage problems may have severe effects on maize diversity because people who are forced to give up storing their own seed become dependent upon procuring planting material in the market, which most often offers non-native varieties.

Loss of soil fertility

> The fields are without manure and become weaker day by day.
>
> (Male farmer, 68, Quitugo)

Many farmers, particularly those who do not apply chemical fertilizers, point to loss of soil fertility as a major problem. The reduced fertility is explained by increased land pressure and lack of manure. As a result of increased land scarcity, farmers have tried to intensify their production, which has led not only to soil depletion but also to a priority on human food over animal feed. Fallows become rarer, resulting in fewer resources for livestock. One 34-year-old female farmer from El Batan explains: 'With just this narrow field, there is no way to keep many animals.' In Quitugo, El Batan and San Pedro, people say they used to have more livestock, but today keeping livestock is difficult because animals are more expensive now, referring to macroeconomic changes and inflation after dollarization. Animals are thus harder to obtain and are also more attractive to thieves: 'There are no animals anymore. We used to have many sheep, many pigs and many cows. But they finished them, robbing. They just took four cows from here, and they are still robbing. Now we cannot apply manure, this is the problem' (Male farmer, 90, San Pedro). Difficulties in keeping animals also relate to scarcity of pasture and feed. Prior to agrarian land reforms, hacienda owners allowed indigenous people to graze their animals on hacienda pastures. The larger amount of animal manure allowed intensive cropping without soil degradation (Bebbington, 1993). After agrarian reforms, these grazing rights were lost and indigenous people had less space to graze their herds and flocks. In addition, much former hacienda land is planted today with eucalyptus, or is intensively farmed (e.g. with greenhouse production). These constraints on livestock husbandry are mirrored in old people and children carefully guarding little herds of cows, sheep and pigs so they can be nurtured by the greens of roadsides.

Lower production

Ya no da. (It doesn't produce anymore.)

Drought and lower soil fertility restrain crop production. When asked about crop production level, 57% of respondents claim a reduction in recent years. In the subsistence-oriented communities of San Pedro and Quitugo, 80–90% reported a drop, while the number is lower in El Batan (40%) where farmers have access to irrigation. In commercial communities, 25% report reduction, while 42% say the level is kept steady, but only due to an increasing amount of applied chemical fertilizers. Lower production may thus be a major reason for losses in agrobiodiversity. '*Ya no da*' is a common saying when talking to farmers about the reasons why zapallo, lentils and faba beans disappear from the fields.

'Minifundio': declining farm size

There is just no space anymore.
(Female farmer, 23, Quitugo)

The population in most of Cotacachi's rural communities has increased since land reform started in the early 1960s, while average landholding by household has decreased. Based on the partible inheritance system, all children have the right to inherit land from their parents, implying a severe reduction in size of land holding for every generation. This, in combination with land reforms, has resulted in a significant decrease in average farm size in Cotacachi during the last four decades (the phenomenon is known as *minifundio*). Farm units < 3 ha cover the largest part of the cultivated area (Zapata *et al.*, Chapter 4, this volume). Farmers in Quitugo and El Batan have especially small fields, several not more than 0.1 ha. Left with a plot of land too small to produce enough food for the household, farmers must prioritize when choosing what to grow. Several farmers explain the abandonment of certain crops or landraces with the lack of land: 'Yes, we would like to sow the old crops, but we don't have land for that. Where my husband works they have seeds of sweet potato, but I tell him not to bring it, because we don't have land' (Female farmer, 52, Quitugo).

Changes in the Traditional System

In one's own house one has all granos, may it be dry or fresh. For example of the fresh ones first come the bush beans, and then we begin to have fresh peas. When the peas begin to dry, we have fresh maize, and when the fresh maize begin to end we have fresh climbing beans, we have fresh zambo, fresh zapallo, later we have quínua, the fava beans, the lupines. When we have finished all those fresh granos, we begin with the dry ones, and we begin storing them to have food during the whole year. The best seed we select and store to sow again. And this way our own seed will not be lost. Later we have barley, we have wheat, we have potatoes, arracacha, and we have sweet potato.

> (. . .) When one of the crops fails, one can harvest of any of the others. Everything will never fail, there is always some seed that will give me a good harvest.
>
> (Male farmer, 62, Quitugo)

Traditionally, every household has produced and harvested most of its own food. Harvests were staggered so that during the year food was always available. Elderly respondents from all communities tell about a time when they only purchased salt and nothing else. Since each altitudinal zone had its own specifically adapted crop, an interchange between zones took place. A 56-year-old woman from the high-altitude community Ugshapungo tells about how she has exchanged produce with colleagues at lower altitudes: 'Further down, there is for example fresh maize and zapallo. I went down there with lots of potatoes and faba beans.' Exchange of produce and labour has tied people together in reciprocal relationships. A woman from a lower lying community tells about how her parents-in-law cultivated some crops just for in-kind payment of labour and interchange with others:

> The rye and the lentils were only for the interchange. My grandmother came from a very poor family. As my parents-in-law had a lot of *granos*, they were rich, my grandmother washed clothes for them, and they sowed those crops only to give her when she washed. (. . .) They came with *panela*, bananas and bread and we gave them rye and lentils. For the interchange we also used zambo, zapallo and mauka.
>
> (Female farmer, 70, Quitugo)

Cultural changes – blending with the outside world

Although they once lived in a relatively isolated society, Cotacacheños today have access to a range of products from the outside world. They get international news and inputs through television screens and contacts with the numerous governmental organizations and NGOs that operate in the area. Cotacachi is also part of a region that is visited by tourists from far and near. In addition, the people themselves travel and migrate to nearby towns, to Quito and even to foreign countries on distant continents. This globalization has resulted in changing priorities, world views and ways of organizing livelihoods.

Today's system

> The young will have to work to buy food.
>
> (Female farmer, 46, Quitugo)

Today's market economy has, to some degree, replaced the traditional economic system of Cotacachi. However, people are still bound together by reciprocal non-monetary ties. They exchange produce, and hired labour may be paid in kind through a share of the harvest. Communal work, *minga*, is quite frequently arranged both privately in farmers' fields or when the community chapel needs restoration or some other public activity. Yet people have also become more independent of each other, and access to money has become crucial. Cash is needed for clothes, household equipment, children's education, transportation, medicines and food. 'The community is small. Today there are many small children, so in the future there are likely to be more houses, and there will be no space left for sowing. The young will have to work to buy food' (Female farmer, 46, Quitugo).

Food

> The food from before we have changed for the food of today, which is the one from the shop.
>
> (Male farmer, 75, Quitugo)

Agrobiodiversity is closely related to food preferences. The fields of subsistence farmers are planted with crops the household prefers, while commercial farmers opt for crops which are demanded by the market. Food is fundamental, not only for its nutritional value, but also as an expression of identity and culture. Food traditionally prepared in Cotacachi is hearty, based on the *granos* produced in the area (Camacho,

Chapter 11, this volume). However, local food habits are changing; as new foods become more popular, the old ones leave the rural kitchen.

Cotacacheños like to vary their food: 'We change every day, one day this, the other day that' (Male farmer, 64, El Batan). The relatively high diversity of area crops reflects this, but now the variation includes purchased foods. In the local market, an array of imported fruits and vegetables are sold. NGOs and government agencies provide seeds and planting material in order to increase production in Cotacachi: 'Before they only had salt and onion to season the food, but now we also have vegetables' (Female farmer, 23, Quitugo). Fruit production is attractive for commercial farmers, but subsistence farmers also see it as a good source of extra income, as a 33-year-old El Batan farmer explains: 'We should plant some trees, like avocados, for the money it gives.' Exotic vegetables and fruits may contribute to better nutrition and vitamins but it should not be forgotten that traditional Andean food is quite healthy.

According to those surveyed, an even more important change in local cuisine has been the incorporation of rice and noodles. Eating rice has been associated with mestizo food habits. The survey for this study showed the consumption of rice, noodles and meat was higher in Peribuela, Ugshapungo and El Batan than in Quitugo and San Pedro. Rice is grown on the Ecuadorian coast and imported from the USA, and is now cheaply available in the little stores in every community. In the survey, 60% state that they purchase more food products and 55% emphasize the decreased role *granos* now play compared with rice and noodles. 'Sometimes we buy rice and noodles to change with the potatoes. Yes, we like it. Just from potatoes, potatoes, potatoes, beans, beans, ocas, ocas, ullucos, ullucos, you become bored' (Female farmer, 56, Ugshapungo). Nevertheless, many regret this increased reliance on *comida liviana* or 'light food.' Thirty-one per cent of my informants say the food was healthier before. 'The food of my parents was better, and didn't leave you hungry. One could make it just having one meal a day, but if I eat a soup of rice, I am hungry again after five minutes. It doesn't sustain the stomach and the children become sick and weak, they don't have strength' (Female farmer, 58, Quitugo). A 24-year-old female farmer from Peribuela elaborates:

> Earlier there was quínua, *chuchuca* [a kind of dried maize] and zapallo. There still is zapallo, but they don't prepare it like before. It was better before, more nutritious. We have searched the easiest way of production. (. . .) Now the food is contaminated, the fruits and everything. Earlier one ate healthier. I would like to eat healthier, but I don't know about going back to eating that way . . . only sometimes we do.

Some explain that they prepare rice because the children appreciate it so much: 'We don't make that [traditional dish] anymore, because the children learned to eat 'fine food', as we call it' (Male farmer, 49, El Batan); 'They don't want to eat food from what we harvest here, the *chuchuca*, barley, they say they are not hungry, but if you cook rice, there will not be enough. I think that sometimes it is the parents' fault as well, we let them win, and don't make them accustomed to eating *granos*' (Male farmer, 33, El Batan). In the survey conducted among children, 42% out of 50 children in the different communities declared a rice dish as their favourite.

Less time

> I would like to sow sweet potato, oca, mashwa, arracacha and mauka, I already have the seeds, but I don't have the time.
> (Female farmer, 33, Quitugo)

Migration and off-farm work also imply a severe reduction in the time available for cooking and working the household's own land. Many traditional dishes take a long time to prepare. Since mothers today have less time to spend at the stove, it is more convenient to prepare a soup with macaroni than a quínua soup. Labour-demanding crops also receive less priority. For example, people who have given up farming quínua say it demands too much work and it is easily obtained in the market. However, most households do not have access to sufficient money, and in the end cheaper foods are prioritized. A 33-year-old male from El Batan talks about all the *granos* his grandparents used to grow, many of which he does not have in his field: 'When we have the money, we buy it, when we don't, we are left with a bag of noodles.'

The traditional way and dissolution of beliefs

> It is not only a question of working, weeding or looking at the moon. If a person is weeding without believing in God, he will not have a good harvest. My mother said that when a person goes to sow, he first has to have a dialogue with mother earth. Because to us, the indigenous, the mother earth has life, the plants have life, the stars have life, the stones have life. All those things have a relation with man (. . .). My mother said that you always have to believe in the mother earth, you have to believe in God, you have to believe in the hurricane, in the blow of the wind. You have to respect them, so they also respect our crops, she said. When a harvest failed, she asked 'who of you did not respect the mother earth? You have to regret and apologize.' It was prohibited to cut trees without looking at the moon, sow without watching the moon. She said that one always has to be grateful to the mother earth, to be grateful to the crop. Because the crop one does not cultivate just by cultivating, but to sow you have to sow with love, thinking of how it is going to produce.
>
> (Male farmer, 89, Quitugo)

> The people have changed their beliefs. They are not calling for mother rain anymore.
>
> (Woman, 58, Quitugo)

In the Andean cosmovision, not only do human beings have a soul, but the environment is filled with spirits. Even after the introduction of Christianity, these animistic spirits lived on in the Andean world. To ensure a good harvest, a well-attended relationship to the spirits and Mother Earth (*Pachamama*) has been deemed just as necessary as preparing and sowing in the right way at the right time. Whereas other rituals now are performed in the name of Catholicism, agricultural rituals have maintained the relationship to Mother Earth and the hill spirits (Custred, 1980). Interacting with the modern world, those growing up today do not have an intimate relationship with the soil and its fruits. Forty-three per cent of respondents never received any formal education. They grew up before schools were introduced to the area, or were kept away from schools by parents who kept them at home to work. This was the last generation to grow up in relative isolation from the outside world. In contrast, today's young spend considerable time outside their homes and fields. Given that their world views are shaped by forces different from those of their grandparents, a belief in Mother Earth seems irrational. As the beliefs and relationship with the land dissolve, incentives for growing the traditional crops also wither. An old woman tells about how they used to bring both soil and seeds to be blessed in the church before they sowed, but: 'Today we don't believe in those customs anymore, even I have given up the beliefs. It must be because my children say that I look like I'm crazy doing those things, that everything is a lie, that one should not believe. Now I am only sowing what one uses most: peas, maize and beans' (Female farmer, 70, Quitugo). The 89-year-old farmer from Quitugo who is previously quoted describing the importance of cultivating with love continues: 'Today they are sowing with sloth, they run away, and there they leave it. And everything fails, and they blame it all on God or on the climate change. It is not that the climate has changed, the one who changes is the person – he goes losing the beliefs and the respect.'

Lack of commitment and knowledge

> To keep all those crops, you need to have strength and know how to work in the field.
>
> (Male farmer, 68, Quitugo)

Many old people, in complaining about the lack of interest and commitment among the young, say they lack the knowledge necessary to produce a good harvest. Younger people acknowledge this: 'We are not used to work so much anymore' (Female farmer, 34, El Batan). Children still help their parents in the field, receiving their first chores at a young age. Nevertheless, they do not learn farming the way their grandparents did since they spend a larger part of their time in other arenas. Moreover, many farming practices of yesterday have disappeared and been replaced by new ones. For instance, one informant points to the importance of letting at least 8 days pass between sowing every parcel. In this way you will be better secured in case of a failure, and you can harvest from the crop over a longer time period. This is practical when all the work is done manually. However, as soon as you rent a tractor to prepare your fields, you will have to do the operations all at once to be more efficient. Efficiency is also a must, both for commercial farmers and for those that have to take care of off-farm jobs.

People complain that nowadays a good harvest from traditional crops is difficult to achieve. When asked whether they would sow quínua or zapallo if someone brought the seeds, they reply that it has been tried but with poor results. This experience has led them not to wish for seed. The poor result is probably grounded in depleted soils and drier conditions, but on the other hand it may also be that the knowledge their parents had about the cultivation was not transmitted or, if it was, time for and interest in the practices have been lost. A woman from Quitugo talks about her nephew with whom she shares some land: 'He says that it's a land that is worth nothing, that it doesn't produce much. But that is a lie, it's because he doesn't know how to work in the agriculture. When a person knows how to work with the crops, it produces very well. When it is raining, and one works well, one will have a good harvest' (Female farmer, 52, Quitugo). A young female farmer from the same community explains: 'For example my sister, she doesn't like the agriculture, because of lack of understanding. A little bit of motivation is lacking, there are no people who say that the agriculture is important.'

What Will be Grown? Blending in or Wiping Out?

Finally, what will Cotacachi children of the future decide to grow? The current trend of loss of biodiversity will continue until it is countered by some force. *Culture* is one of the crucial factors for the continued existence of agrobiodiversity in Cotacachi (Rhoades and Nazarea, 1998). Traditional crops are associated with indigenous culture, and movements towards a more 'modern' kitchen result in distancing from the past; a step towards a 'mestizo' lifestyle. However, recent countertrends indicate that indigenous movements are flourishing in Ecuador. During the latter half of the 20th century, peasant federations grew large and gained political influence. Cotacachi's first indigenous Mayor is now in office, and the same is the case in the nearby town Otavalo, which also hosts one of the most famous and visited Indian handicraft markets in South America. Indigenous people are proud of their ethnicity, and this reflects identification with traditions and history that is a crucial factor for the continuum of traditional foods as well. It might not be rational to continue growing *only* traditional crops in their small land plots when cheap substitutes can be bought in the market, but if traditional foods are important to them, then they have an incentive to continue growing them. There are prospects for increasing urban demand for *granos*. The International Potato Center has several on-going projects to develop and promote entirely new products based on various roots and tubers (CIP, 2002). In Brazil, for example, the *arracacha* has become popular among urban residents (Hermann, 1997). In the USA, 'Lost Crops of the Americas' has become a patented brand name (Caplan, 1996). As a woman from Quitugo puts it: 'Earlier the mestizos didn't eat *granos*, only rice, but now it is different. Now they eat *granos* and buy *nabo* [*Brassica rapa* ssp.

rapa – a weedy plant gathered for food] too, the mestizos. Now they are eating cooked barley, earlier they said it was food for the Indians.'

If a large market for traditional crops develops, commercial farmers will sow them. If the family wants food based on traditional crops, the subsistence farmer will continue to sow roots, tubers, cucurbits and grains. If landraces of maize and potatoes are preferred on some occasions, their chances of continued existence will increase. The knowledge is still there, and many seeds can still be found although they may be hard to locate. Many factors have contributed to this obscurity, but those who search will find answers in the communities of Cotacachi. They will come if society asks for the roots from the past, if a voice inside tells them that this is important.

References

Bebbington, A. (1993) Modernization from below: an alternative indigenous development? *Economic Geography* 69, 274–292.

Brush, S.B. (1992) Ethnoecology, biodiversity, and modernization in Andean potato agriculture. *Journal of Ethnobiology* 12, 161–185.

Caplan, F. (1996) Marketing Lost Crops™ of the Americas. In: Janick, J. (ed.) *Progress in New Crops. Proceedings of the Third National Symposium New Crops, New Opportunities, New Technologies.* ASHS Press, Alexandria, Virginia, pp. 127–129.

CIP (2002) Tapping into biodiversity: research on neglected Andean crops set to pay dividends. In: *Broadening Boundaries in Agriculture: Impact on Health, Habitat and Hunger. International Potato Center-Annual Report 2001.* International Potato Center, Lima, pp. 55–64.

Custred, G. (1980) The place of ritual in Andean rural society. In: Orlove, B.S. and Custred, G. (eds) *Land and Power in Latin America.* Holmes and Meier Publishers, Inc., New York, pp. 195–209.

Food and Agriculture Organization (1996) *Report on the State of the World's Plant Genetic Resources.* Food and Agriculture Organization of the United Nations, Leipzig, Germany.

Frankel, O.H. (ed.) (1973) *Survey of Crop Genetic Resources in their Centres of Diversity.* First Report. Food and Agriculture Organization of the United Nations, Rome.

Guarino, L. (1999) Approaches to measuring genetic erosion. *Proceedings of Technical Meeting on the Methodology of the FAO World Information and Early Warning System on Plant Genetic Resources.* Food and Agriculture Organization of the United Nations, Prague.

Hermann, M. (1997) Utilization of arracacha. In: Hermann, M. and Heller, J. (eds) *Andean Roots and Tubers: Ahipa, Arracacha, Maca, Yacon. Promoting the Conservation and Use of Underutilized and Neglected Crops.* Institute of Plant Genetics and Crop Plant Research, Gatersleben, Germany and International Plant Genetic Resources Institute, Rome, pp. 118–126.

Hernández Bermejo, J.E. and León, J. (1994) *Neglected Crops: 1492 from a Different Perspective.* Food and Agriculture Organization of the United Nations, Rome.

Izquierdo, J. and Roca, W. (1998) Under-utilized Andean crops: status and prospects of plant biotechnology for the conservation and sustainable agriculture use of genetic resources. In: Scannerini, S., Baker, A., Charlwood, B.V., Damiano, C. and Franz, C. (eds) *ISHS Acta Horticulturae 457: Symposium on Plant Biotechnology as a Tool for the Exploitation of Mountain Lands.* Fondazione per le Biotechnologie, Turin, Italy.

National Research Council (1989) *Lost Crops of the Incas.* National Academy Press, Washington, DC.

Perales, H.R., Brush, S.B. and Qualset, C.O. (2003) Landraces of maize in central Mexico: an altitudinal transect. *Economic Botany* 57, 7–20.

Rehm, S. and Espig, G. (1991) *The Cultivated Plants of the Tropics and Subtropics.* Verlag Josef Margraf, Weikersheim, Germany.

Rhoades, R.E. and Nazarea, V. (1998) Local management of biodiversity in traditional agroecosystems: a neglected resource. In: Collins, W. and Qualset, C.O. (eds) *Importance of Biodiversity in Agroecosystems.* Lewis Publishers, CRC Press, Boca Raton, Florida, pp. 215–236.

Rodriguez Cuenca, J.V. (1999) *Los Chibchas: Pobladores Antiguos de los Andes Orientales. Aspectos bioantropológicos.* Colciencias Universidad Nacional de Colombia, Santafé de Bogotá, Colombia.

Scheldeman, X., van Damme, P., Ureña Alvares, J.V. and Romero Motoche, J.P. (2003) Horticultural potential of Andean fruit crops exploring their centre of origin. In: Düzyaman, E. and Tüzel, Y. (eds) *International Symposium on Sustainable Use of Plant Biodiversity to Promote New Opportunities for Horticultural Production Development.* International Society for Horticultural Science, Anatalya, Turkey.

Scott, G.J., Rosegrant, M.W. and Ringler, C. (2000) *Roots and Tubers for the 21st Century: Trends, Projections, and Policy Options.* International Food Policy Research Institute and International Potato Centre, Washington, DC and Lima.

Smartt, J. and Simmonds, N.W. (1995) *Evolution of Crop Plants,* 2nd edn. Longman Group UK Ltd, London.

Sperling, C.R. and King, S.R. (1990) Andean tuber crops: worldwide potential. In: Janick, J. and Simon, J.E. (eds) *Advances in New Crops.* Timber Press, Portland, Oregon, pp. 428–435.

Tapia, C. (2001) Conservation of the Andean tuber diversity on farms of farmers of the Huaconas, Chimborazo-Ecuador. *Proceedings of the In Situ 2001 E-Conference.* Consortium for the Sustainable Development of the Andean Ecoregion (Condesan) and International Potato Center (CIP). Available at: http://www.condesan.org/infoandina/Foros/insitu2001/

Weismantel, M.J. (1988) *Food, Gender and Poverty in the Ecuadorian Andes.* Waveland Press Inc., Long Grove, Illinois.

Zimmerer, K. (1996) *Changing Fortunes. Biodiversity and Peasant Livelihood in the Peruvian Andes.* University of California Press, Berkeley, California.

10 Women and Homegardens of Cotacachi

Maricel C. Piniero
CATIE/NORAD, Casa No. 7, Avenida Libertad, Ciudad Flores, Peten, Guatemala

Introduction

Homegardens, according to Fernandes and Nair (1986, p. 279), 'represent land use systems involving deliberate management of multipurpose trees and shrubs in intimate association with annual and perennial agricultural crops and, invariably, livestock, within the compounds of individual houses.' A homegarden is a domestic area cultivated by all members of the household where the main purpose of the produce is for home utilization. Homegardens are omnipresent in rural areas and also are often found in urban and peri-urban regions, especially in countries of the developing world. Although this household strategy has existed for centuries and was the earliest form of food production, only recently have homegardens been studied in terms of their contribution to household nutrition and, to a certain extent, the family's income (Niñez, 1989). A majority of these studies deal with the structures and functions of homegardens or the types of plants cultivated in the area (for example, see Fernandes and Nair, 1986; Rico-Gray, 1990). While these kinds of studies are important in evaluating the subsistence function of homegardens, it is just as necessary to understand homegardens as areas where gardeners, usually women, experiment with and nurture diversified species of crops (Padoch and Jong, 1991). In this regard, issues concerning changes of crop diversity over time and what influenced their transformation – including opportunities and constraints gardeners encountered in maintaining and managing gardens – are important aspects in the study of homegardens. This chapter discusses women's homegardens in Cotacachi, Ecuador, focusing on women's plant classification system as well as a comparative analysis of women's garden maps using the variables of age and socioeconomic class. These are contextualized using excerpts from women informants' life histories.

Description of Research Site

The research was conducted in the community of La Calera located 2 km from the Cotacachi town centre in the parish of San Francisco. La Calera's approximately 250 households include 1250 inhabitants (F. Yepez, 1999, personal communication). The population is predominantly indigenous, with very few mestizos who have blended in well with indigenous culture. The common crops cultivated in the area are maize, potato, peas, beans and some vegetables. Some families raise cows, pigs and chickens. The majority of the people are subsistence farmers, but there are a few families who sell their chickens and pigs for cash or donate them for special occasions such as weddings and baptism ceremonies. There are also some individuals, mostly men, who engage in selling handicrafts outside the country. The quality of housing and other properties, such as cars and stores, tend to be the criteria being used to distinguish rich from poor families. Families that are considered wealthy usually have two-storey concrete houses with appliances and furniture and own a car or two. Middle-income families have smaller concrete houses and own fewer less expensive appliances and furniture. Poor families may or may not have concrete houses with fewer rooms, furnished only with basic items such as beds and chairs. The very poor families have one-room houses made of adobe and pieces of wood or bamboo, with little or no furniture.

Research Methods

A total of 24 women representing different generations and socioeconomic levels participated in the investigation. To evaluate the changes taking place in the homegardens and to incorporate the gardeners' perceptions of homegardens and their plants, I employed different research techniques. These included objective mapping and listing of all plants and their varieties, measuring land area through pacing, and drawing by informants of cognitive maps of their gardens at three different time intervals – 10 years ago, the present time and 10 years from now. This information was elicited to assess women's evaluation of changes over time and to get an idea of what development direction these women want to take in terms of crop diversity. The general trend of changes in the garden is analysed using the three temporal maps, while only the 'present' map is used in comparing cognitive maps with my researcher-drawn scientific maps to identify culturally significant crops.

Age and economic groups are the variables used for comparative analysis. In this study, I categorized age into three groups: grandmother or the first generation, mother or the second generation and daughter or the third generation. Economic groups also have three categories: high (considered to belong to a more affluent family and owns some properties), middle (families who do not own big properties but own more than what poor families have) and poor (characterized as having no or the least property). The crops drawn on the maps were coded and the results were tallied and quantitatively analysed. These data were also complemented with excerpts from life histories to emphasize various issues.

Diversity and Integration: Indigenous Homegardens

The enduring and intimate relationship that exists between women and the environment can be understood from various perspectives. One explanation deals with the symbolic relationship of women with nature, emphasizing how Mother Earth is associated with women (Sachs, 1996). According to this theory, a woman's ability to reproduce other human beings and to produce materials (e.g. milk) needed for survival can be equated with the ability of nature to provide the needed materials for human survival. In other words, all activities related to household and childcare are simply an extension of a physiological make-up that nature intended women to have (Mies *et al.*, 1988; Merchant, 1995). Other feminists argue that this relationship parallels women's subordination to men in the same manner that nature is dominated by human technological advances (Werhof, 1988). Some feminist scholars (e.g. Shiva, 1989; Agarwal, 1992) argue that women are closer to nature than men because of the material and utilitarian value of the environment. Since the beginning of time, women's ability to utilize the environment, particularly plants, is well documented in the literature (Fowler and Mooney, 1990; Heiser, 1990). This can be traced back to the division of labour by gender or the man-the-hunter and woman-the-gatherer dichotomy where women have the major responsibility of feeding the family through the plants they collected and propagated (Southeast Asia Regional Institute for Community Education, 1995a; Sachs, 1996).

In rural agriculture-based households such as those found in La Calera, women's major responsibilities include providing food for the family and making sure that their children's general health and well being are addressed. Women usually spend a significant amount of time in activities that involve food preservation, such as drying beans grown in their gardens, and in cultivating plants that heal common diseases. Because of their responsibilities, women observed and learned more about various crops that are adapted for medicine, fodder and food (Consultative Group on International Agricultural Research, 1997; Sachs *et al.*, 1997). Therefore, women often decide what crops are good for storage, which variety of maize is best to cook with beans, and what food items are excellent sources of nutrients. These women keep, maintain, propagate and conserve various crops grown in their homegardens because they are aware that these materials are necessary for their survival.

Although women have an active role in the utilization and propagation of plant diversity to ensure food availability, this role remains socially invisible and, therefore uncompensated. Aidoo (1988) points out that development has historically been planned and controlled mainly by men, while the work of women has remained 'invisible.' Even activities involving food production – the very basis of human existence – are considered secondary by development researchers who assume that all farmers are male (Consultative Group on International Agricultural Research, 1997). Women's knowledge about crops and their management is eroding as modern seeds and technologies are introduced in rural regions. This trend negates the fact that women's homegardens, as well as women's knowledge about the crops being cultivated in Cotacachi, are crucial in sustaining human lives and the integrity of the environment.

Considerable effort and money are devoted to research on collection and conservation of plant genetic resources in formal institutions such as international gene banks or *ex situ* conservation (Rhoades and Nazarea, 1999). Recently, similar efforts are being undertaken in informal settings in the form of *in situ* gene banks and on-farm conservation. *Ex situ* and *in situ* conservation are now perceived as complementary approaches rather than as mutually exclusive alternatives in conserving plant genetic resources. According to Brush (1999), *in situ* conservation is critical for the maintenance of key elements of genetic resources. In some cases, maintenance cannot be done off-site primarily because of its dynamic nature. The natural habitat continues to generate variation through various mechanisms such as mutation, recombination and gene flow (Jana, 1999). *In situ* gardens (or on on-farm conservation areas) can be treated as laboratories or experimental stations where these different processes can occur. Moreover, since deterioration and human error can adversely affect off-site conservation facilities, on-site conservation can serve as back up for the existing off-site gene banks.

Despite the increasing legitimacy of *in situ* conservation, however, the significance of women's homegardens continues to be seriously under-rated in the conservation agenda. Because women remain invisible in the agricultural process, their effort concerning the conservation of crop genetic diversity is overlooked as well (Aidoo, 1988; Henderson, 1995; Martelo, 1996). In fact, it is often forgotten that *in situ* conservation is critical not only for preserving plant genetic resources but also for safeguarding cultural knowledge important for the conservation of plant genetic resources (Nazarea, 1998).

Padoch and De Jong (1991) point out that house gardens not only provide subsistence and cash resources but also serve as repository and testing sites for uncommon species and varieties of plants. Bittenbender (1983) characterized gardens in Africa as areas where vegetables, staple crops and tree crops are found. In Puerto Rico, the gardens provide various functions in the household including food, fibre, medicine, construction materials and places to undertake recreational activities. In Yucatec Maya, the garden remains a productive subsistence production strategy, open to poor peasants since it requires limited land, water and capital (Kints and Ritchie, 1997). Even in the USA, many families engage in gardening for economic reasons (Gladwin and Butler, 1984). In the Andean region, particularly on the high plateau of Peruvian Puno, women cultivate and maintain bitter potato varieties in their gardens to use for a particular drink. In Cajamarca, Peru, a potato grower named Rosa prefers to grow different varieties of potatoes, particularly the native kind, because, in her opinion, native varieties never fail to produce (Tapia and de la Torre, 1998). Numerous landraces of different crops are typically found in the garden, mainly because of their multiple functions in the household.

Midmore *et al.* (1991) describe homegardens as areas usually located close to the house and planted with various crops. This description holds true in La Calera where women also include plots planted with a more uniform crop situated adjacent to the house. A typical garden for indigenous Cotacacheños is surrounded by flowering plants and several trees (Fig. 10.1). Vegetables and other food crops are usually planted 3–5 m away from the house, either

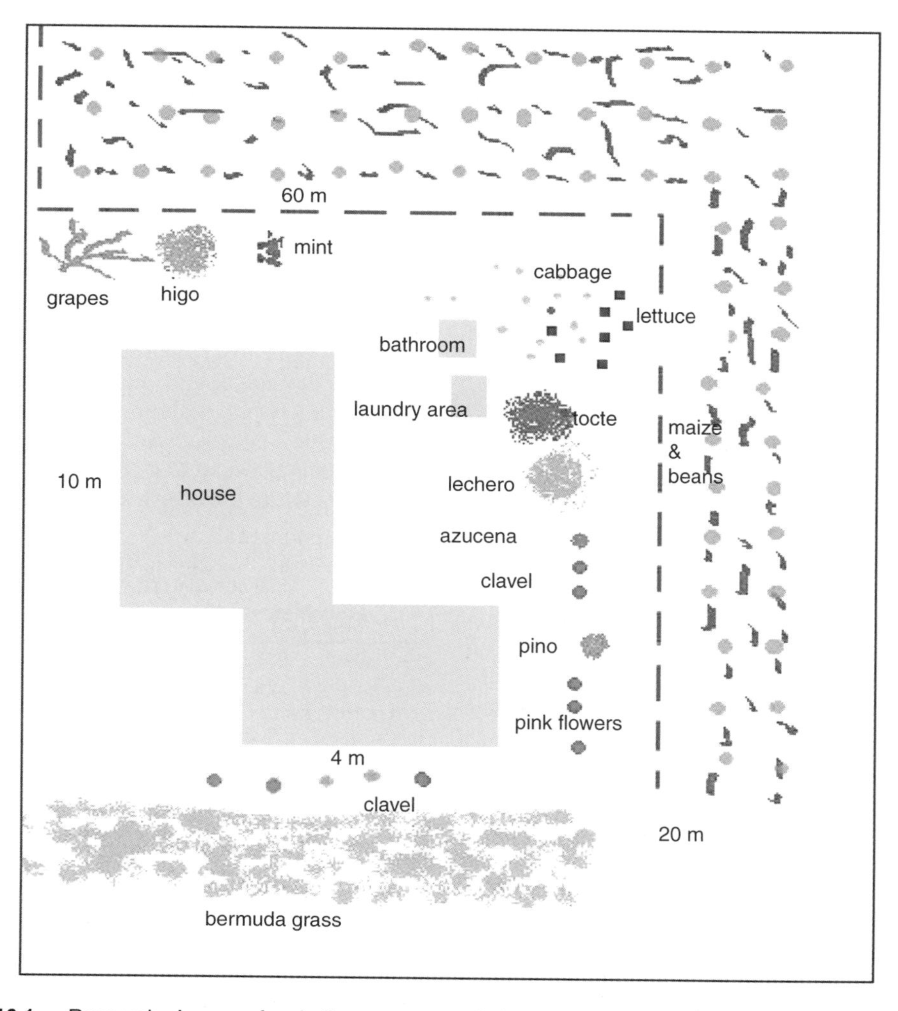

Fig. 10.1. Researcher's map of an indigenous woman's homegarden showing crop diversity.

at the back or at the side, while a single crop (e.g. potato, beans or peas) is cultivated adjacent to the house. This area is still included in the garden because the produce is usually for home consumption. In some cases, very few crops are cultivated close to the house, and the rest are planted several metres away. The land area of homegardens ranges from 20 to 300 m^2, depending on the economic status of the household. However, land area alone is not an indication of crop diversity. I found that low-income families with their smaller gardens tend to have more diversified crops than the high-income families because the former do not have enough money to buy all their food needs and therefore depend on their gardens for some of these necessities.

The method of land preparation for homegardens varies according to size. A wooden plough is used when the land area is large enough for cows to move around, or approximately 300 m^2, while shovels and machetes are used when the area is small (e.g. 2 × 4 m). Males in the family prepare the land. No particular garden layout is used since the plants cultivated are generally mixed and diversified. Most plants are randomly scattered in the area, except for flowering plants, which women tend to cultivate in front or in areas that are highly visible to outsiders. If a larger area is planted to a single crop such as maize, potatoes, beans or peas, the person who prepares the land makes ridges or furrows in the ground, 31 cm wide and 186–310 cm long. Women weed regularly where staple crops are growing and occasionally around plants cultivated closer to the house. Watering generally is done only during drought conditions. Pests and plant diseases are not considered as problems deserving any management effort.

The variation in crop diversity of homegardens in La Calera community can be explained from different angles. As mentioned earlier, indigenous high-income families, although viewed as rich families within La Calera, fall under the middle- or low-income category compared with other mestizo communities in Ecuador. Hence, the majority of families in La Calera are still economically marginalized. Nazarea (1998) points out that farmers at the margins tend to retain a variety of landraces to sustain a family's multiple needs. This explains why indigenous women, regardless of whether they are 'rich' or 'poor' (in local people's definition of wealth), maintain a higher diversity of crops in their gardens.

The degree of market integration can also influence the manner by which homegardens are maintained. Compared with other women who live in areas where commercialization is the driving force for production, the majority of women in La Calera do not have outside jobs that demand full-time attention. High-income women still engage in subsistence farming and depend mainly on their garden for food. Women who are engaged in store keeping, dressmaking or handicraft do not need to attend to their work-related responsibilities on a specific time schedule. Hence, they have more time to work in their gardens. The most common crops cultivated in the homegardens of the 24 indigenous women interviewed are presented in Table 10.1.

For Ecuadorians, particularly indigenous people, *aji* is as indispensable as salt in the households. A typical plate comprised mainly of boiled potato, maize and beans will not be served without *aji*. Lechero, a type of tree, is generally maintained in homegardens because it serves as a post for tying cows while they are being milked.

Table 10.1. Most common plants cultivated in La Calera homegardens.

Common names	Scientific names	Cultivation (%)
Potatoes	*Solanum tuberosum*	50
Cabbage	*Brassica oleracea*	50
Maize	*Zea mays*	40
Onions	*Allium cepa*	40
Beans	*Phaseolus vulgaris*	30
Aji	*Capsicum annuum*	40
Oranges	*Citrus sinensis*	20
Tocte	*Juglans neotropica*	40
Lechero	*Euphoria laurifolia*	40
Peas	*Pisum sativum*	30

Potatoes, maize and beans are staple crops in La Calera and are used in making soup and as a main dish. These crops are considered 'staple', because they anchor each meal. Except for these staple crops, harvesting is done on a staggered basis and only when the produce is needed in the household. Maize and beans are usually planted during September and October and harvested 5 months later, while potatoes and peas are cultivated and harvested during the months when maize and beans are not planted. For these types of crops, relay cropping is practised, with maize being followed by potatoes. Maize is always mixed with beans, while potatoes are intercropped with peas. The mixing of maize and beans is a practical agricultural strategy. Maize stalks serve as supports for climbing beans, therefore saving the gardeners time and energy in putting up poles. Maize stalks are also fed to guinea pigs and eaten like sugarcane as a sweet snack. Beans help replenish the nutrients of the soil by fixing nitrogen.

Homegardens also serve as a venue for women of La Calera to socialize with other women, as neighbours and friends provide help during planting and harvesting. Five women who were observed, each planting maize, moved down one furrow in alternate directions so the adjacent women would meet at the centre before moving on to opposite directions. Planting was done using a wooden pole (digging stick) to poke small holes in the ground in which to drop kernels. They cover the holes using one foot. While putting 2–3 kernels in each small hole, the women chatted about what was happening in their village. When a conversation topic revolved around interesting events, women giggled while discussing it. Issues such as wife battering were also included in the discussion, accompanied by sighs of frustration. Before lunch, the owner of the garden went home to prepare the food, accompanied by one of the women who helped her in the preparation. These women continued to chat freely about their private lives over lunch.

Staple crops such as potatoes and maize constitute a large portion of the families' diets and contribute a fair amount to their nutritional demands. Maize and potatoes, for example, supply carbohydrates and vitamin A, while leguminous plants such as beans and peas provide protein. The people in this area also eat insects such as white grub, which is found in homegarden soil. In one instance, a doctor recommended that a mother feed her son this insect to prevent an asthma attack. Hence, while many families are poor and cannot afford to buy meat, protein deficiency is not a significant problem in this community. Cultivation of vegetables also supplies other vitamins needed by the family.

Homegarden Maps: Local Categories

A map is a visual way to connect physical space and social relationships through people's memory and representations, reflecting the interests and priorities of its makers (Jescavage-Bernard and Crofoot, 1993). Maps are also used to evaluate the relationships of humans with their environment. Cognitive mapping of local people's point of view about their interaction with and use of the environment has become a popular research method in recent years (Garling *et al.*, 1984; Nazarea, 1995; Rocheleau *et al.*, 1995). A majority of these mapping exercises has focused on macroenvironments such as large watersheds or urban cities. In the process, the immediate and everyday environment, such as around individual houses or farms, is generally taken for granted. Utilization of natural resources including rivers, animals and plants, for example, is studied by asking people to map a whole watershed while, in fact, similar kinds of information can be generated on microscale environments as well. Mapping homegardens, for instance, can provide pertinent information about plants, varieties of plants and allocation of resources – land and time in particular – that can be significant in the overall use and management of the natural world. The data generated may not be as encompassing in representing a whole watershed or a province, but the information may provide in-depth understanding of people's interaction with nature.

The cognitive mapping technique was used to elicit information from women about crops grown in homegardens. The women were asked to represent their homegardens for three time periods – the present, the past (10 years ago) and the future (10 years from now) – in order to evaluate the perceived changes over time. Unfortunately, the activity was done all at one time, making it difficult for the women to complete the three maps. Instead of drawing the crops on their 'future' maps, the women opted to just list the names of crops that they want to grow. Having the three time period maps can be a good means to elicit information about what crops were lost or added in the homegardens and what type of homegardens these women will want to maintain in the future. Ideally, however, the mapping session should not be confined to one interview setting but at least two, one for the 'past' and one just for the 'future.'

The maps of three time periods were analysed to understand the changes taking place in the homegardens through women's eyes. Crops drawn or written on homegarden maps were organized according to categories of local classification. The data generated were coded and the results were analysed using the SPSS program (Statistical Program for Social Sciences). The results were disaggregated based on economic groups and the ages of women informants. Based on relative frequency, the women's homegarden maps were also analysed to reveal the relative salience of crops and different categories of women.

The most important categories for these women are food crops that include 'cereals/grains', 'fruits' and 'vegetables' (Fig. 10.2). This pattern is particularly true in the maps depicting the present and the future. This implies that the women will maintain the cultivation of these crops if they are already being grown, and, if not, will start planting them in the future. This trend can be explained by the contemporary economic hardship that everyone in Ecuador is experiencing. In 2000, the currency of Ecuador was changed from *sucres*, a national currency, to the US dollar to stabilize the country's economy. The conversion process, however, caused significant difficulty for people in rural areas. As a consequence of 'dollarization', prices of commodities rose and even basic goods such as food items became harder to purchase. Therefore, the importance of the contribution of homegardens to household food security has increased.

Although the temporal representation of 'cereals' appears to decrease overall, this

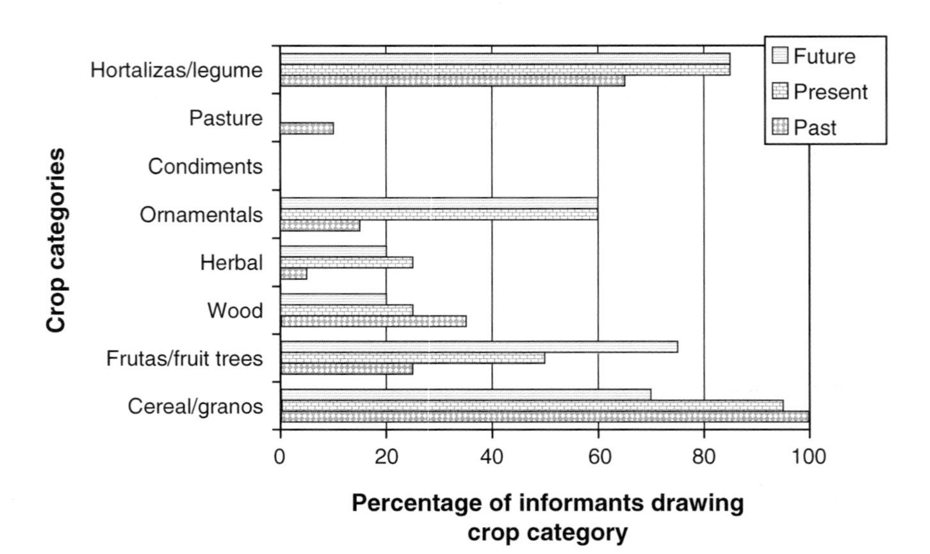

Fig. 10.2. Crop representation by indigenous women.

crop group remains salient over the three time periods. The decline in the planting of the cereal group could be due to the shrinking land area allocated for homegardens as a result of women expanding their fruit tree production. The convergence of strategies being employed by these women probably emanates from problems of poverty and the continuing depletion of natural resources. Vegetables and other food items are becoming more expensive because of the weakened economy. To alleviate this situation, women plan to increase the production of basic food items in their future gardens.

Vegetables have always played a dual role in the household economy as food for subsistence and as cash crops for the market. Hence, they have been central to the lives of these women. 'Vegetables are for the nutritional needs of my family and are also good for the market, hence it is good to cultivate them', a middle-income mother explained. This example underscores the adaptiveness of homegardens for these women. They are coping with the economic situation of their country by planning to cultivate more food crops in the future.

Although it seems that some plant categories tend to decline across the three time periods (e.g. wood and cereals), the majority of them have the tendency to remain the same or to increase (e.g. fruits, ornamentals and vegetables) in the future. This increasing diversity can be explained by the fact that indigenous peoples in different parts of the world always engage in plant conservation, gaining important knowledge through their long-term experience with plants that is crucial in the continuation and proliferation of biodiversity (Shiva, 1996; Cleveland and Murray, 1997; Brush, 1999). Aside from considering 'Mother Earth', or *Pachamama* in Quichua, as sacred and certain plants as worthy of being respected and conserved, indigenous people also engage in the conservation of biodiversity because of various practical reasons. Nazarea (1998) emphasizes this issue when she explains why local farmers in the Philippines cultivate different varieties of sweet potato. According to Nazarea (1998, p. 68), for marginal populations, 'diversity is actually a natural state of things' due to the multiplicity of criteria for choosing varieties of sweet potatoes for different occasions and seasons, individual tastes and even for different kinds of farmers.

Differences in homegardens according to age

Based on the map analysis, 'cereals', 'fruits' and 'vegetables' are the three main crop categories represented most frequently by women across the three time periods. This consistent pattern is driven by women's role in the households, where they are responsible for making sure that food is available on a day to day basis. As stated earlier, women are often responsible for all their families' agricultural production, especially when husbands and children are engaged in non-farm activities, as illustrated by this example:

> I am the one who attends to the farm since my husband is always out of the country. He sells handicrafts in Colombia.
> I am responsible for everything: land preparation, renting a tractor or hiring laborers, weeding, and harvesting . . . There used to be 6 people working with me on the farm – my kids – but since all of them are working outside now, I was left to tend the garden by myself. However, I prefer to work on the farm than to go to other places to work because it is better to cultivate different crops since we do not have money to buy food. We can simply harvest these crops that we planted and use them at home.
> (Indigenous mother, 57 years old)

In addition to a high percentage of representation of the three crop categories, particularly cereals and fruits, the oldest or the first generation, the 'grandmothers', also have the highest percentage in medicinal categories compared with the other two generations. These older women have lived in the area for a long period of time, and they have seen a number of changes in the community. They emphasized the undesirable results of these changes, as illustrated by these quotes from their life histories:

> Now agriculture has changed. Now agriculture is bad. Crops do not grow anymore. Now people are using fertilizers.

> These chemicals force the plants to mature early. Some plants are irrigated using the river while others are dependent on the rain for irrigation. We have to irrigate the crops so that they will give a better yield. During summer, crops do not produce a lot, but on rainy days, grasses are the ones growing. When I was a little girl, I liked cultivating flowering plants. Actually, there were more varieties of crops in the past. Before, there were a lot of leguminous plants and vegetables such as cauliflower, onions, and flowering plants including dahlias, and clavel. Now there are fewer varieties.
> (Indigenous grandmother, 76 years old)

Although all age groups underscored the categories of 'fruit', 'cereal' and 'vegetables' on their maps, the third or youngest generation's future maps shows a dramatic increase in the fruit tree and ornamental categories. While it is true that all generations are being threatened by the impact of development that encroaches on even the most rural areas in Ecuador, the youngest generation is more aware of the environment and its problems because of their exposure to these issues in school. A 19-year-old daughter recognizes the value of trees as shown by this comment:

> I want to cultivate fruit trees in the future because trees purify the air. Now we have a drought problem. All plants are dying because there is no water. There is no water because there are no trees. Before, rivers were abundant with water, now all rivers here in La Calera are drying up.

This level of exposure helps explain why 'daughters' give priority to fruit trees on their future maps. The economic issue also surfaced as an important factor that the younger generation has taken into consideration in deciding on what crops to cultivate in the future.

Differences in homegardens according to economic status

Economic class is another variable that is important to consider in analysing the garden maps. It should be noted that 'affluence' or the high-income group as defined by local people, particularly in indigenous communities, does not reflect large differences in economic conditions. Thus, women in indigenous communities all belong to relatively low-income groups. As with the age variable, crop categories that are salient for all economic groups are cereal, fruits and vegetables. As shown by their homegarden maps, the women belonging to the low-income group appear to have cultivated fewer vegetables in the past and are planning to cultivate more vegetable crops in the future. These trends can be attributed to the country's current poor economic climate. Foodcrops cultivated in the garden are valued not only for food but for their commercial uses as well. Instead of buying food, these women need only to harvest from the homegarden and, if the produce exceeds their household needs, it can always be sold.

The percentage of women cultivating wood trees has declined according to the homegarden maps of all three economic groups. However, a more drastic decrease is evident with the high-income group (from 40 to 10%). As modernization has swept Cotacachi, new technologies have been introduced, including modern amenities for the home. The more affluent families purchased appliances such as gas ranges or ovens, eventually leading to lower utilization of firewood. Families who could not afford to buy appliances continued to maintain trees for firewood. Native trees for fuelwood have also decreased tremendously in these two areas because of deforestation associated with intensive agriculture. Marcelo Cruz (1999) similarly pointed out that in Colta, Chimborazo, Ecuador, farmers are changing their lifestyles because of modernization that arrived in rural areas in the form of fertilizer, radios, new clothes and vehicles.

Relative salience: culturally significant crops

The researcher's maps of women's gardens are used as a point of comparison to evaluate how perceptions of realities differ from the actual physical locations of plants in the garden. These maps were compared with the set of 'present' maps drawn by the informants from the community.

According to Gould and White (1986, p. 28), people's action is 'affected by that portion of the environment that is actually perceived.' Hence, environmental features that are perceived as important will always be highlighted or exaggerated, whether on road maps, building maps or garden maps. Cognitive maps, Diane Austin (1998) explains, not only represent how people perceive space but also include physical attributes of the place and stories about them. Maps not only represent the physical features of an environment but also epitomize people's interpretation and perception of realities. After all, all maps reflect the interests of their creators and those who use them (Jescavage-Bernard and Crofoot, 1993). The maps are drawn to different scales and with different objectives. In cognitive maps, people emphasize features that are significant to them.

Garden maps drawn by women show salient features that can be attributed to different factors affecting women's perception of reality. The feature that is common in all maps is the presence of the houses. Although instructed to draw their gardens, women included their houses on their maps. It can be inferred that women always associate their homes with their gardens. This is related to Agarwal's (1992) explanation regarding the materialist basis for why women are closer to the natural environment and Sachs' (1996) explanation of rural women's work. Because of both their traditional roles in the household and the changing global economy, cultivating plants closer to the house provides the needed food without necessarily engaging in commercial production (Sachs, 1996). Furthermore, women's other family responsibilities such as childcare and food preparation require them always to be close to the house. If homegardens were situated away from the homes, it would be difficult for women to perform all the various tasks associated with the maintenance of home and garden.

Although not present in all informants' maps, another important feature that is included in a few maps is animals (Fig. 10.3). Domestic animals have always played an integral part in the household agricultural system, especially in rural areas in the developing world (Stephens, 1990a;

Fig. 10.3. Sample of a woman's homegarden map showing the incorporation of animals in the garden.

Southeast Asia Regional Institute for Community Education, 1995b). Women are responsible for collecting fodder for animals. Some plants collected for fodder are from homegardens; hence, animals are always associated with the homegarden. Furthermore, the basic function of a garden in the household is to provide food and/or additional income for the family. Livestock also serve multiple roles in the household. They serve as assets that families can sell in times of financial difficulty and can be slaughtered for special occasions including weddings and birthdays (Stephens, 1990b).

In comparing the 'objective' (researcher) maps with the maps that local women drew of their homegardens, it is evident that women do not emphasize the actual physical location and number of plants cultivated in the garden. Usually, what is emphasized on maps are the major crops utilized at home. For example, potatoes, maize and beans are always included on maps of indigenous women.

Other crops such as flowering plants, which are included on my maps, were only incorporated by a few informants into their drawings. Local names for the majority of ornamental and flowering plants are not known. Also, crops are represented as disproportionately large compared with the house. Some women drew dying flowers to illustrate a lack of water brought about by the drying of the stream located close to the house (Fig. 10.4). This stresses the point that a cognitive map represents not just the physical elements of the natural world but also epitomizes processes and relationships

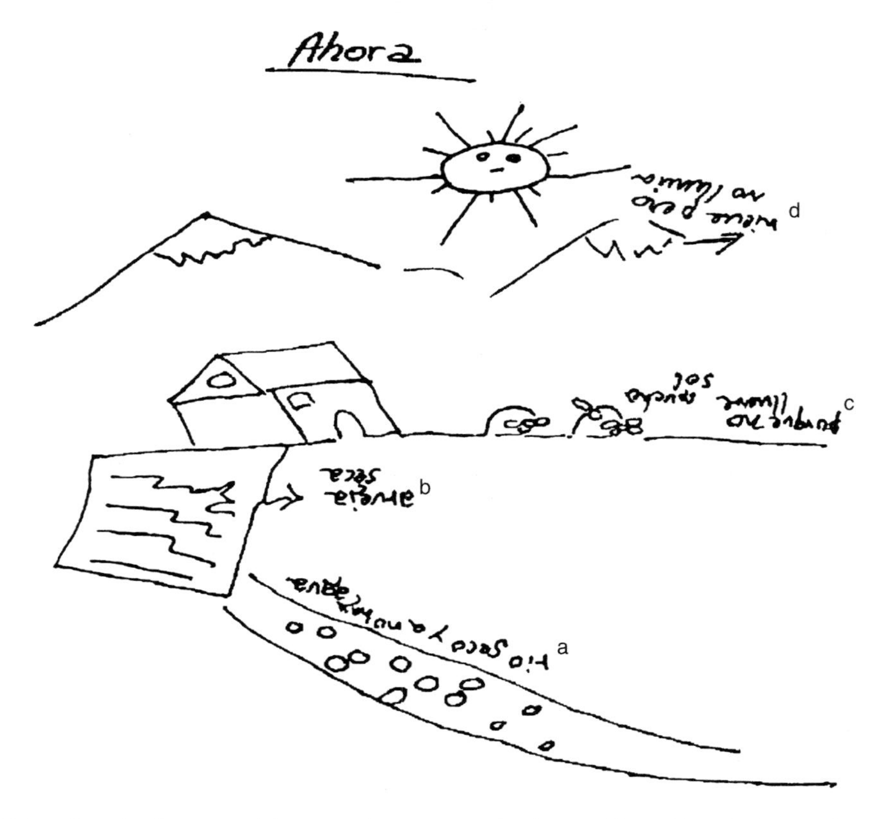

Fig. 10.4. Sample of a woman's homegarden map showing the condition of the plants and climate conditions.
The Spanish notes from bottom to top are: [a]'dry river, no water'; [b]'dry peas'; [c]'because there is no rain, much sun'; [d]'snow but no rain'. These observations demonstrate the woman's environmental perceptions in relation to her plants.

among various elements drawn on the maps, for example the plants and the stream.

Among indigenous women, other structures such as laundry area, oven and animal pens were included in their homegarden maps. According to Soemarwoto and Conway (1991), a homegarden is considered one of the oldest forms of agroecosystems. For this reason, people regard it as a sustainable unit where various activities in the households are seen as interconnected. The inclusion of features such as animal pens and a laundry area suggests that these structures have a role in the maintenance of homegardens. An animal pen, for example, implies that animals are being raised, creating manure that the women in La Calera use as fertilizer. Women who have pigs collect the droppings from their pen and pile them in one corner of the garden where they also place other waste materials from the kitchen. This compost pile is then used as garden fertilizer as needed.

Rappaport (1979) points out that humans act on the environment based on both the actual world, or the operational reality, and how they perceive the world, or their cognized models. Nazarea-Sandoval (1994) investigated this interaction among different categories of rural cultivators through cognitive mapping. She demonstrated that humans tend to filter the various realities with their own models that are affected by various factors including economic condition, social status, ethnicity and age. Although homegardens exist as a physical entity for the indigenous and mestizo women, they tend to represent actual species of plants and their numbers from different perspectives, based on their association with these biological entities or their functions in women's lives. The objective maps might be more accurate in terms of representing what crops are planted in specific locations, what land area is allocated for homegardens or how many varieties of a certain plant are being cultivated, but they do not represent various interactions that exist in the household. The objective maps also fall short in showing what crops are being prioritized by women of various ages and economic backgrounds. As this study shows, women from La Calera see their homegardens differently because of differences in age and economic circumstances.

Changes in inter- and intracrop diversity

The three period homegarden maps indicate an increase in plant species between the past homegardens and the present ones. More vegetables, fruit trees and ornamental plants have actually been added to the garden and, based on women's projections, this trend will probably continue in the future. Various factors that could have increased the interspecific crop diversity include the exchange of planting materials, either seeds or actual plants, among women in this area. It is very common in rural areas to exchange resources and commodities, as this woman explained:

> We give or exchange our produce with our neighbors, particularly vegetables such as cauliflower and carrots. We also give medicinal plants when our neighbor's child is sick. For example the flower clavel. When someone is sick with fever in the family, they come here to ask for clavel. In terms of the amount, it depends on how much we have. We also give some portion of our produce such as corn to people who helped us during planting and harvesting.
>
> (56-year-old mother)

Bourque and Warren (1981) in their study of two villages in Peru show that women are experts in maintaining their kinship and trading networks. They always give gifts or favours to other women because they know that when the time comes that they also need help, the women to whom they extended some form of favour would reciprocate. In the same spirit, planting materials are exchanged among neighbours and friends in this Ecuadorian village, thus increasing the diversity of plants in homegardens. Unfortunately, this kind of exchange might generally wane in the future for major subsistence crops because production is also declining every year. One informant said 'We are not exchanging seeds with our neighbors. It is hard for us to store the seeds, or to give them away. They are all consumed at home because of small harvests.'

Another possible explanation of the change in diversity is the presence of various national and international projects that support agricultural production. In this village, a women's group met to discuss how vegetable seeds, donated by a doctor who worked for an international health project, should be distributed among the group members. This type of approach is commonly implemented in countries of the developing world where seeds of crops said to be rich in vitamins and minerals are distributed to rural areas where malnutrition is prevalent. Although the objective is worthwhile, this development strategy does not address the real issues in rural areas. Sometimes, the introduced plants are not even adapted to the local environments or local people do not know how to utilize these plants. Although the introduced seeds augment the plant repertoire, introducing plants that are not appropriate to certain areas is like introducing technology packages without first understanding what local people want and/or need.

Based on my interviews, there is a decline in intraspecific variation of grains and/or cereals and Andean tubers. According to the women interviewed, there is a decrease both in the quantity of production and in the number of varieties of these traditional crops. The women stated that tubers such as oca and melloco are no longer cultivated in the garden because more and more people are cultivating high-yielding varieties of commercial crops. Although potatoes remain a popular crop, older heirloom varieties have been replaced by modern ones. One woman remembered that her parents used to cultivate five varieties of potatoes with different shapes and colours. Now, she cannot even recall the names of these varieties. Hence, while interspecific diversity is increasing through plant exchange and introduction, intraspecies diversity is in fact decreasing. Cruz (1999) pointed out that the increased market orientation has favoured cash crops over the traditional Andean crops. This pattern reflects two things: the impact of modern agriculture on the two areas, resulting in people tending to allocate more land for crops with high commercial value; and the narrowing options development workers provide to local people, particularly women who are the caretakers of homegardens, in terms of the kinds of crops that they can cultivate.

Conclusion

This chapter highlights the importance of homegardens in the lives of women in La Calera, particularly in terms of how they are coping with the development process in many developing countries. I used two methodologies to elicit information that are crucial in my analysis and singled out patterns or trends that are distinctive among the different variables used in my interpretation of the data. The use of the cognitive mapping technique, for example, reveals important issues that are sometimes ignored in identifying development projects suited for rural regions. To illustrate, both the researcher's maps and the women's maps depict the same geographical space; however, the women's homegarden maps show features that emphasize what they value in their gardens, including crops they find useful in the households. The researcher's homegarden maps can be very objective and accurate in terms of the number of plants and their varieties, land area and distribution of crops, but they do not represent other important features that women highlighted on their maps. Giving importance to local people's perception of their environment is crucial in identifying the kind of intervention programme that development workers will implement.

To understand the constraints and opportunities that indigenous women face, I used variables such as age and economic status in my analysis. Various age groups, for example, illustrate the fact that regardless of their age or generations, women are planning to maintain diversified crops in their gardens because of their social role of being the food provider. None the less, there are also differences across age groups as influenced by exposure to education, as illustrated by the younger generation's concerns for the environment. It seems that, regardless of what economic group women belong to, the maps illustrate that the three class groups are planning to increase the cultivation of plants that are consumed at home mainly because of economic problems all of them are facing at the present time. There might be discrepancies in some categories, such as the tendency to decrease trees for high income families, but this can be explained by the introduction of technologies (use of firewood is no longer necessary) that some of them can afford to purchase. The other divergence that is prominent is the fact that low-income families want to increase the production of vegetables. Again, this is a reflection of how the declining economy influences the women's plan to continue planting crops that can both be used in the household and be sold in the market.

Women want to continue maintaining homegardens mainly for home consumption and sometimes to augment the household income. Factors that women have no control over, including limited area, lack of manual labour and the worsening economic condition, force them sometimes to neglect their homegardens. Understanding first what these women prefer to have or maintain in their gardens will not only help in food security but will also prevent the further erosion of biodiversity.

Recognizing the importance of informal gene banks such as homegardens and women's crucial role in their maintenance will make biodiversity conservation more sustainable in the long run. Women's close involvement with plants and food production make them more knowledgeable about environmental problems. For example, through their long-term interaction with nature, women learned which plants are best to mix with other plants or which ones are best to cultivate in certain months. Hence, this information will also be conserved if homegardens continue to be maintained.

In summary, maintenance or abandonment of homegardens is a reflection of the kinds of decisions women have to make in their everyday lives. Because of the potential of homegardens as a good source of food and nutrients for the family and as an additional source of income in the household, some women continue to cultivate and maintain various crops in the garden. Not only does the propagation of homegarden crops help in the conservation of food crop diversity, it also provides flexibility for women who rely on different species of crops with various characteristics. These attributes meet the household's specifications for different uses and play an important role in the maintenance of the resilience of the ecosystem.

References

Aidoo, A.A. (1988) Women and food security: the opportunity of Africa. *Development: Journal of the Society for International Development* 2 (3).

Agarwal, B. (1992) The gender and environment debate: lessons from India. *Feminist Studies* 18.

Austin, D. (1998) Cultural knowledge and the cognitive map. *Practicing Anthropology* 20, 21–24.

Bittenbender, H.C. (1983) *The Role of Home Gardens in Rural and Suburban Family Nutrition in the Third World.* Paper presented at the association of Women in Development Conference, Women in Development – A Decade Experience, October 1982, Washington, DC.

Bourque, S. and Warren, K.B. (1981) *Women of the Andes: Patriarchy and Social Change in Two Peruvian Towns.* The University of Michigan Press, Ann Arbor, Michigan.

Brush, S. (1999) The issues in in-situ conservation of crop genetic resources. In: Brush, S. (ed.) *Genes in the Field: On-Farm Conservation of Crop Diversity.* Italy: IPGRI; Canada: IDRC. Lewis Publishers, Washington, DC.

Cleveland, D. and Murray, S. (1997) The world's crop genetic resources and the rights of indigenous farmers. *Current Anthropology* 38, 477–515.

Consultative Group on International Agricultural Research (1997) *The Forgotten Farmers: Plant Genetic Resources, Women and the CGIAR.* CGIAR, Rome.

Cruz, M. (1999) Competing strategies for modernization in the Ecuadorian Andes. *Current Anthropology.* 40, 377–383.
Fernandez, E.C. and Nair, P.K.R. (1986) An evaluation of the structure and function of tropical homegardens. *Agricultural Systems* 21, 279–310.
Fowler, C. and Mooney, R. (1990) Origin of agriculture. *Shattering: Food Politics, and the Loss of Genetic Diversity.* The University of Arizona Press, Tucson, Arizona.
Garling, T., Book, A. and Lindberg, E. (1984) Cognitive mapping of large-scale environments: the interrelationship of action plans, acquisition, and orientation. *Environment and Behavior* 16, 3–34.
Gladwin, C. and Butler, J. (1984) Is gardening an adaptive strategy for florida family farmers? *Human Organization* 43, 208–215.
Gould, P. and White, R. (1986) *Mental Maps.* Allen and Unwin, Boston, Massachusetts.
Heiser, C. Jr (1990) In the beginning, seeds, sex, and sacrifice. *Seed to Civilization: the Story of Food.* Harvard University Press, Cambridge, Massachusetts.
Henderson, H.K. (1995) The gender division of labor. In: Henderson, H.K. and Hansen, E. (eds) *Gender and Agricultural Development: Surveying the Field.* The University of Arizona Press, Tucson, Arizona.
Jana, S. (1999) Some recent issue on the conservation of crop genetic resources in developing countries. *Genome* 42, 5562–5569.
Jescavage-Bernard, K. and Crofoot, A. (1993) Mapping to preserve a watershed. *Scientific American* May, 134.
Kints, E. and Ritchie, A. (1997) *Food Production in a Yucatec Maya Community: Rethinking a Sustainable System for Micro-level Household Kitchen Gardens.* Paper presented during the Applied American Anthropology Meeting, Tuczon, Arizona.
Martelo, E.Z. (1996) Modernization, adjustment, and peasant production. *Latin American Perspective* 23, 118–130.
Merchant, C. (1995) *Earthcare: Women and the Environment.* Routledge, New York.
Midmore, D., Ninez, V. and Venkataraman, R. (1991) *Household Gardening Projects in Asia: Past Experience and Future Directions.* Asian Vegetable Research and Development Center, Technical Bulletin 19.
Mies, M., Bennholdt-Thomsen, V. and von Werlhof, C. (1988) *Women: the Last Colony.* Zed Books, London.
Nazarea, V. (1995) *Local Knowledge and Agricultural Decision Making in the Philippines: Class, Gender, and Resistance.* Cornell University Press, Ithaca, New York.
Nazarea, V. (1998) *Cultural Memory and Biodiversity.* University of Arizona Press, Tucson, Arizona.
Nazarea-Sandoval, V. (1994) Memory banking: the conservation of cultural and genetic diversity in sweet potato production. In: Prain, G.P. and Bagalanon, C.P. (eds) *Local Knowledge, Global Science, and Plant Genetic Resources.* UPWARD, Los Baños, Philippines.
Niñez, V. (1989) Garden production in tropical America. In: Landauer, K. and Brazil, M. (eds) *Tropical Homegardens.* United Nations Press, New York.
Padoch, C. and De Jong, W. (1991) The house garden of Santa Rosa: diversity and variability in an Amazonian agricultural system. *Economic Botany* 45, 166–175.
Rappaport, R. (1979) On cognized models. *Ecology, Meaning and Religion.* North Atlantic Books, Richmond, California.
Rhoades, R. and Nazarea, V. (1999) Local management of biodiversity in traditional agroecosystems. In: Collins, W. and Qualset, C. (eds) *Biodiversity in Agroecosystems.* CRC Press, Boca Raton, Florida.
Rico-Gray, V. *et al.* (1990) Species composition, similarity, and structure of Mayan homegardens in Tixpeual and Tixcacaltuyub, Yucatan, Mexico. *Economic Botany* 44, 470–487.
Rocheleau, D., Thomas-Slayter, B. and Edmunds, S. (1995) Gendered resources mapping: focusing on women's spaces in the landscape. *Cultural Survival* 18, 62–68.
Sachs, C. (1996) *Gendered Fields: Rural Women, Agriculture, and Environment.* Westview Press, Boulder, Colorado.
Sachs, C., Gajurel, K. and Bianco, M. (1997) Gender, seeds, and biodiversity. In: Sachs, C. (ed.) *Women Working in the Environment.* Taylor and Francis, Washington, DC.
Southeast Asia Regional Institute for Community Education Review (1995a) *Women and Plant Genetic Resources* 2(1), Occasional Papers published by the Southeast Asia Regional Institute for Community Education.
Southeast Asia Regional Institute for Community Education Newsletter (1995b) Vol. 2, No. 2.
Shiva, V. (1996) Agricultural biodiversity, intellectual property rights and farmers' rights. *Economic and Political Weekly* June 22, 1621–1631.

Shiva, V. (1989) *Staying Alive: Women, Ecology, and Development*. Zed Books, London.
Soemarwotto, O. and Conway, G.R. (1991) The Javanese homegarden. *Journal for Farming Systems-Extension* 2, 95–117.
Stephens, A. (1990b) *Women and Livestock Production in Asia and the South Pacific*. Regional Office for Asia and the Pacific (RAPA). Food and Agriculture Organization of the United Nations, Rome.
Stephens, A. (1990b) Participatory methods development. In: *Proceedings of the Inaugural Planning Workshop on the Users' Perspectives with Agricultural Research and Development*. UPWARD, Los Baños, Philippines.
Tapia, M. and dela Torre, A. (1998) *Women and Farmers and Andean Seeds*. IPGRI and FAO, Rome.
Werlhoif, C. (1988) Women's work: the blind spot in the critique of political economy. In: Mies, M., Bennholdt-Thomsen, V. and Von Werlhof, C. (eds) *Women: the Last Colony*. Zed Books, London.

11 Good to Eat, Good to Think: Food, Culture and Biodiversity in Cotacachi

Juana Camacho
University of Georgia, Anthropology Department, 250, Baldwin Hall, Athens, GA 30601, USA

Introduction

Food and anthropology

The anthropological study of food is tied to the evolution and variability of the human diet, the cultural and historical diversity of food habits and cuisines, the intricacies of kitchen politics and the relationship between food, power and identity (Appadurai, 1981; Mintz, 1985; Harris and Ross, 1987; Ohnuki-Tierney, 1993; Counihan and van Esterik, 1997; Goodman *et al.*, 2000). Diet is commonly defined as what people eat, in what proportions, and nutritional value (Weismantel, 1988), but diet and nutrition are mediated by ecological and socioeconomic factors and cultural choices. According to local environmental conditions and economic possibilities, human groups select and classify what is edible and desirable, and rank what is most valuable nutritionally and socially (Messer, 1989). Cuisine, on the other hand, refers to the cultural rules, representations, beliefs and practices that govern cooking and eating in different societies and influence people's behaviours and identities (Fischler, 1995). Identity and sense of self are very much about what, where and with whom we eat; the common expression 'you are what you eat' refers to not just the very material and physiological need for nourishment and the psychological comfort provided by food, but to the social and symbolic dimensions of a people's food habits.

This chapter presents the results of anthropological research on food, culture and biodiversity among indigenous communities in Cotacachi, Ecuador. The main objectives of this work are the study of the principal food production and consumption activities and practices, the local perceptions and classifications of food and the effects of recent transformations in the local food system on local identities and social relationships. Information has been obtained by means of participant observation in indigenous households, lodgings, markets, restaurants and stores. Informal interviews were conducted with men, women and children from Morochos, Iltaqui, Turuco and La Calera, as well as with local leaders, officials and researchers. Fieldwork took place during various trips between 2003 and 2004.

Food and Culture in Time and Space

Historical and geographical context: Lo Andino

Cotacachi is part of what is known as the Andean world (*lo andino*), originally a geographical concept but one that has acquired other connotations when referring to a history and a culture with common features. Geographically, *lo andino* is characterized by the mountainous chain that crosses South America and whose snow-capped heights give origin to the rivers of the Pacific and Amazon basins. The rugged topography and the altitudinal variations produce a great heterogeneity of habitats and a notorious ecological diversity. This heterogeneity has been decisive in the adaptive processes of the native populations who historically made a vertical use of the various altitudes by means of the colonization of different territories in different climatic zones. Culturally, *lo andino* refers to the indigenous peoples that were under the influence of the Inca Empire, who used Quechua as the lingua franca, who shared elaborated technical and agricultural practices for erosion control and water conservation through agricultural terraces and irrigation systems, and who established significant exchange and reciprocity relationships. In spite of the apparent homogeneity, the Andean pre-Hispanic scenario was ethnically and sociopolitically diverse (Stanich, 2001). Ecuador was characterized by autonomous chiefdoms – some highly specialized, connected by exchange networks (Salomon, 1986). At present, as the result of rural–urban migratory processes, integration to the market economy, the influence of formal education, the media and tourism, the Andean world is characterized by plural and flexible livelihoods that extend beyond the Andean heights to national and international urban centres (Salman and Zommers, 2003).

The Andean food system has been moulded by the horizontal and vertical land use, through the utilization of various climatic zones, the establishment of exchange networks, the domestication of wild biodiversity and the transformation of the landscape for agriculture. The Andes are one of the centres of diversity and domestication of plants and animals of local and regional importance such as camelids (llama and alpaca) and guinea pigs (Archetti, 1997), and globally consumed crops such as potatoes and tomatoes. Many of today's basic products have been regularly consumed in the Andes since pre-Hispanic times, reaching ritual and sacred relevance, such as potatoes and maize (Brush, 1980; Murra, 1980; Harrison, 1989; Hastorf and Johannessen, 1993; Coe, 1994). Maize, quinua and tubers remain central crops for inhabitants of the sierra (Orlove 1987). The pre-Hispanic diet was based primarily on consumption of different varieties of maize, a crop that continues to have great ritual value, as well as tubers, legumes (beans and chochos or *Lupinus mutabilis*), squashes, quinua and wild plants and fruits (Piperno and Pearsal, 1998). Historically, nutritional deficiencies and hunger periods especially affected those sedentary communities of farmers dependent upon a few crops whose diets were less diverse than those of hunters and gatherers (Alchon, 1997).

As a result of Spanish colonization, the Andean diet was modified by the adoption of cereals such as wheat and barley, legumes (fava beans), vegetables (onions and cabbage), sugarcane and fruits, as well as domestic animals such as sheep, pigs, cows and chickens, which expanded animal protein sources (Patiño, 1984). Diversification of food products did not always mean improvements in the diet and health of indigenous people since they did not always have access to the new products. On the contrary, colonial socioeconomic institutions such as the *encomienda*, the hacienda, the mining enclaves and the textile workshops (*obrajes*) were based on the transformation of local landscapes and the exploitation of native resources and labour.

The agrarian reforms of 1964 and 1973 ended the majority of haciendas and the *huasipunguero* (hacienda tenant farmers under a system of sharecropping), and transformed the productive systems, but the social, economic and nutritional conditions of the native population did not improve substantially. In recent years, the presence of state institutions and market-oriented rural development policies, the introduction of Green Revolution technological packages and the prioritization of a few Andean products have propitiated changes in the agroecological systems and traditional knowledges and practices (Bebbington, 1993). Ecuador has one of the most uneven income distributions, and poverty is the common denominator of 69% of the country's households (Oficina de Planificación de la Presidencia de la República–Food and Agriculture Organization, 2001). According to the World Bank, 40% of the Ecuadorian population lives in rural areas and 60% of them are poor (World Bank, 2004).

The high poverty levels in the rural sierra are due in part to land concentration, property fractioning as the result of demographic growth, soil erosion in *minifundio*, and underemployment. Compared with the rest of the population, the sierra lags behind in nutritional and health indicators (Oficina de Planificación de la Presidencia de la República–Food and Agriculture Organization, 2001; Word Bank, 2004). This situation affects primarily indigenous populations and women who have a larger responsibility for the food security and well being of their families. At present, the highest chronic malnutrition and food insecurity index in the Andean region are among the indigenous rural areas, a trend that Ecuador shares with Peru and Bolivia (Larrea and Freire, 2002).

In response to deteriorating livelihoods and access to food, families have diversified their productive strategies with wage labour and temporal migration; with more access to money and the market, however, new products with low nutritional value are integrated into the food system. Information on changes in consumption patterns in the last decades is fragmentary but, according to nutritional diagnostic surveys from the Ministry of Health (Freire, 1988 cited in Oficina de Planificación de la Presidencia de la República–Food and Agriculture Organization, 2001), the main foods in Ecuador are rice and oats, wheat-derived products, potatoes and manioc, sugar, lard and oil, and in the rural areas cereals and noodles. According to the Integrated Social Indicators System (2003) rice, potatoes, plantains, banana, bread and sugar are important in the Ecuadorian diet. Some of these trends, with local variations, are observed among indigenous people in Cotacachi, whose food system is described next.

Food habits, diet and cuisine in Cotacachi

Food is a very important aspect of the ethnic identity and agricultural lifestyle of indigenous peoples. In the Ecuadorian sierra, the persistence of a diet largely based on native crops, slowly cooked and subtly seasoned in diverse porridges and soups, reveals the depth of its pre-Hispanic roots and its significance for Quechua identity. In Cotacachi, the diet and cuisine of indigenous farming communities are governed not only by ecology and habit but also by sensory preferences sedimented over time. As anthropologist Nadia Seremetakis (1994) has pointed out, memory, knowledge and experience are stored in specific everyday foods and crops whose tastes, aromas and textures are part of peoples' shared sensory landscapes and local histories.

As in much of rural Latin America, food habits in Cotacachi are still based on diversified agricultural cultivation of *chagras* (agricultural fields) and *huertos* (homegardens), mainly for household consumption, and marginal to the market. The agricultural calendar (Ramirez and Williams, 2003) governs the social and ritual life, and food is central to ceremonies such as weddings, baptisms and communal feasts (Moates and Campbell, Chapter 3, this volume). The traditional peasant diet consists of cereals, tubers, legumes, and wild plants and fruits. Consumption of animal protein is sporadic and associated with ritual events. The economic possibilities of the family determine the consumption of meat, eggs or cheese. Indigenous Cotacacheños define food as that which is grown in the field and provides strength, energy and health. Staples such as maize, quinua, *chochos*, beans, potatoes, various squashes and Andean tubers such as *oca* (*Oxalis tuberose*), *mashua (Tropaeolum tuberosum)*, *melloco* (*Ullucus tuberosus*) and white carrot (*Arracacia xanthorrhiza*) are classified as *granos* (grains). Grains are synonymous with food because they are nutritious and give a feeling of satiety and strength. The importance of grains is evidenced in the belief that to dream of maize, beans and peas is a sign that one will receive money.

Nutritious food is needed for everyday survival and hard work, and is especially recommended by midwives for women after childbirth. Food is also associated with that which is cooked and requires considerable time and labour by women in processing and preparation activities. The main preparation categories are wet/liquid and dry. Wet foods include thick porridges (*mazamorras* and *coladas*) made with ground grains, quinua or barley, as well as non-flour-based soups (*locros*), and broths (*caldos*). Dry foods range from toasted and crunchy meals to finely ground grain flours, easily transportable to the fields, and specially fit for toothless elders. Sweet foods are also prepared in wet and dry preparations such as sweet thick porridges or drinks called *mazamorras* and *coladas*, *chichas* (drinks) and *dulces* (sweet dishes). Culinary preferences are characterized by bland or subtly seasoned meals, mostly with wild or cultivated aromatic herbs such as turnips, *paico* (wormseed), spearmint, cabbage and cilantro. Foods are accompanied by ají (chilli), groundnut or *zambo* (*Cucurbita ficifolia*) seed sauces.

Productive processes, dietary repertoires and people's relationships to food, however, are being modified as the socioeconomic and cultural landscapes of the rural Andes experience the impact of national development policies (dollarization, economic adjustment, elimination of agricultural subsidies and liberalization of land markets), and globalized relationships of production and consumption (trade liberalization and food imports). No longer self-sufficient, household economies are increasingly market dependent and rely upon the wage labour of family members. In addition to agricultural production for family consumption, domestic economies are complemented by income generated by remittances of urban and international migrants. This situation is similar throughout the Ecuadorian sierra (Bebbington, 1993).

The scenario is similar to what Seremetakis (1994) describes for her native Greece: with the expansion of free market rationalities, foreign products and new tastes replace everyday common referents, transforming local material cultures of production and consumption. Together with the disappearance and replacement of traditional crops and foods, she argues, is the erosion of sensory–perceptual experiences, local epistemologies and social identities. With more market dependency and exposure to *mestizo* culture, new foods penetrate the local diets and social differentiation is measured by what people eat (Weismantel, 1988). When youth seek employment in urban areas or migrate abroad, they are further exposed to new values and tastes and they modify their living and eating expectations (see Flora, Chapter 18, this volume). Also, when women join the wage force, they have less time to cook the traditional slow meals and resort to fast-cooking store-bought foods. Rice and noodles, for instance, which were once considered famine foods, are now part of daily meals.

In Cotacachi, people often express their concerns and ambiguities with respect to their changing identities and life styles in the language of food. As fast foods enter into competition with the grain-based meals, children prefer 'dry', spicy and high status *mestizo* foods, and the desire for home-grown crops and the bland, but naturally flavoured, foods declines. Parents and elders blame these new tastes and foods for the weakening of the children, who no longer engage in the hard chores of rural life, but rather seek other economic opportunities. The consumption of new foods is also seen as the desire to become *mestizo* and to assimilate into the master culture.

However, indigenous communities have not remained passive to transformations in rural livelihoods, culture and diet resulting from neoliberal development policies and free trade agreements. Resistance is expressed domestically and locally in the cultivation of native crops and the maintenance of traditional culinary practices for daily life and special occasions. Food security initiatives that include agroecological and economic diversification, and food aid programmes have been sponsored by local indigenous organizations, non-governmental organizations (NGOs) and state institutions. At a higher level, local communities and UNORCAC have joined larger social movements that advocate food sovereignty and social justice, especially in light of growing food imports. Implicit in the challenge to 'deglobalize the belly', as expressed by some indigenous leaders, is the need to understand and value local food systems and people's cultural relationship to food and productive systems. In the larger trend of cultural revitalization, recuperation of native agricultural varieties, emphasis on the consumption of traditional Andean crops and the valorization of native agricultural practices through agrotourism initiatives have been important sources of ethnic renewal.

Given the centrality of food to everyday life, much understanding of social life can be gained through the study of a people's food system. Indigenous food habits in Ecuador have been studied for different historical periods and from different disciplinary perspectives. Contemporary anthropological studies include Casagrande (1981), Weismantel's (1988) descriptions of Zumbagan communities, Bourque (2001), and Corr's (2002) analysis of food rituals in the Tungurahua province. The publication of the Agroculinary Guide of Cotacachi (Ramirez and Williams, 2003), elaborated under the auspices of UNORCAC and *Runa Tupari*, is an important collaboration effort among researchers, communities and institutions. In this chapter, my goal is to present some elements about the indigenous food system in Cotacachi in relation to the use of local biodiversity and Quichua cultural identity (Fig. 11.1).

Fig. 11.1. Women of Cotacachi enjoy a traditional meal of Andean foods. Magdalena Fueres (right, foreground) was SANREM's primary collaborator in Cotacachi (Photo: Robert E. Rhoades).

Food System and Diet

Agricultural production

In Cotacachi, land tenure, topography and climate influence the agricultural system and the management practices employed by farmers. Land tenure is characterized by *minifundio*, small properties ranging between 0 and 5 ha, which in most cases are not enough to cover the subsistence needs of families (for a more detailed characterization, see Zapata Ríos *et al.*, Chapter 4, this volume). Land continues to be the most valued possession and the most important measure of wealth, but with population growth, holdings have been increasingly fragmented as they are divided among family members as part of inheritance rights. The purchase of additional land has been a common practice to diversify domestic production. Like in other agricultural societies, productive diversification in time and space has been a strategy employed to spread out risk and prevent famine.

In addition to land scarcity, water and soils are limiting factors for agricultural production in Cotacachi. Regarding edaphic conditions, a unity in parental soils of volcanic origin characterizes the area, but soil type and quality vary according to altitude (see Zehetner and Miller, Chapter 2, this volume). The most fertile soils, with better structure and less erosion, are in the higher elevations given the slower organic matter decomposition, which gives them the characteristic black colour. With respect to water, agriculture in Cotacachi is rain dependent. The agricultural calendar is governed by rain cycles which correspond to the months of March and April and October and November. July and August are a period of intense sun and strong winds. Some communities have access to irrigation water from nearby streams flowing down from the higher zones, but increasingly longer and hotter summers limit water availability. As local people note, the weather has changed a lot and today it is not only hotter and drier but unpredictable and unmanageable. Climate change is also responsible for frequent frosts which, according to farmers, are responsible for much crop loss and have negative consequences for food security.

In higher zones, which start at about 2500 metres above sea level (masl) and comprise the *páramo*, communities cultivate Andean tubers (potatoes, *mashua*, *melloco*, *oca*, white carrot and sweet potato), legumes such as fava beans and *chochos*, quinua, some vegetables such as turnips and cabbage, and fruits such as blackberries. Potatoes grow bigger and better in this zone, which is why some families grow them commercially for the market. The higher zone also has the highest dairy production. In the *páramo*, many important medicinal plants and some wild fruits such as *mortiño* (Andean blueberry or *Vaccinium floribundum*), blackberries and goldenberries (*Physalis peruviana*), for example, are gathered for the home or market. Middle and lower communities, settled between 2000 and 2500 masl, are those with the highest agrobiodiversity given the more benign climatic conditions. However, like the lower zone, some soils in this area are drier and sandier, with less organic matter, lower water retention capacity, and more susceptible to erosion. Lower settlements have better climatic conditions for crop diversification but, due to their closeness to the urban centre, wage work has gained preeminence over agriculture. Today, farmers grow several varieties of maize, quinua, beans, squash, lentils, peas, sweet potatoes, white carrot, wheat, barley and various fruit trees such as avocado, capuli cherry, cherimoya, *chilguacan* (*Carica gouditiana*), *babaco* (*Carica pentagona*), *naranjilla* (*Solanum quitoense*), passionfruits, walnut and tree tomato, among others. The tendency, however, is towards simplification of intercropping patterns and the number of varieties cultivated.

The sustainability of agricultural practices of small farmers in mountains is related to sound management of soil, water and biodiversity interactions in order to propitiate ecological stability (Netting, 1993). Cultivation of various agricultural fields with mixed crops and different varieties, cover crops, crop rotation and fallow, and fertilization with manure from domestic animals are some of the management practices that have propitiated agricultural productivity. In addition, ritual practices are needed such as taking some soil and seeds to be blessed, making offerings to Mother Earth, and having faith in God to have a good harvest. Given the demanding effort and labour required in land preparation, planting, harvesting and maintenance of irrigation canals, is done in the form of *mingas* or cooperative work. Today, under the influence of diverse rural development programmes and in spite of high costs, commercial seeds, chemical fertilizers and pesticides are used to counteract low productivity levels or facilitate tasks given the unavailability of labour.

Intercropping is central to native Andean agriculture and traditionally included several varieties of maize mixed with various beans and quinua, fava, peas, squash, potatoes, lentils, sweet potatoes, white carrot, *chochos* and various Andean tubers. Introduced crops, such as wheat, barley and rye, have generally been planted alone. In mixed arrangements, maize is the principal crop in the diet and is the basis of many everyday and ceremonial dishes. According to Cesar Tapia from the National Institute of Agricultural Research (INIAP) (personal communication 2004), Cotacachi is characterized by the richness of native maize, bean and pepper varieties, many of which are still found in farmers' fields. Native varieties, called *chauchas*, are generally smaller and with shorter production cycles, but are more resistant. People with many children are said to be very productive like the *chaucha* seed. Diverse dishes are prepared with different native varieties; with respect to maize, an elder peasant noted that:

> The orange *chaucha*, red *chaucha*, black *chaucha*, *oritico chaucha*, *allpa mama chaucha*, are corn varieties from indigenous people. The thick yellow *cumba* corn is from the haciendas and takes longer to produce, but it is more useful for *mote* (hominy) and flour. Our *chaucha* corn is not so good for mote because it is very hard to peel and has a somewhat darkish color if made into flour, but it is tastier for making tostado.
>
> (Quitugo Workshop, 2003)

The *morocho chaucha* (white maize) is necessary for the traditional *colada de morocho*, a gruel that is served during funeral wakes. The white *canguil*, on the other hand, is used to make a fine flour consumed by toothless elders. Beans, or *porotos* as they are called in Quichua, include several varieties of the fast-growing *matambre* type (painkiller bean) such as *matambre pintado* (mottled), red *matambre*, white *matambre* and *poroto matambre*, as well as the larger and slower growing types such as the *poroto pupayón* or *popayán*, the red *poroto*, the *poroto bolón* and the *porotón*. In addition to propitiating ecological synergies and spreading out risk of crop failure, intercropping ensures not only a continuous supply of food, but also a variety of options for preparing different dishes that take advantage of the different growth stages of the crops. Today, mixed crops include one or two varieties of maize, a few varieties of beans, quinua, squash, *chochos* and fava beans, although younger farmers tend to limit intercropping to a single variety of maize and beans.

The growing season begins with the first rains, between September and October, although this varies depending on the altitudinal location of the communities. The tender grain harvest lasts from February to April, and these include the small *matambre* and *molón* beans, peas, *chochos*, squash, fava, *choclo* (sweet maize) and quinua. April coincides with the *Hari Pascua* or Men's Easter, a time associated with abundance of cultivated crops and wild plants that grow alongside the cultigens in the fields. Diverse dishes for daily use and ritual feasts such as the *fanesca* or 12-grain soup are prepared with these tender grains. The *fanesca* is partly the result of the strong influence of Catholicism in Ecuador; it is the national dish with regional and personal variations. Because of the Catholic prohibition against eating meat on Good Friday, the *fanesca* is made with fish but, due to economic limitations, indigenous people replace it with cheese or eggs. Noodles are now being added to this dish, which indicates the extent to which noodles and pasta have penetrated traditional ritual food.

Locros or non-flour-based soups, to which potatoes, tender grains and wild plants are added, are cooked at this time. *Zambo* and pumpkin soups (*locro de zambo* and *zapallo*) are favourite dishes in this season. Tender maize (*choclo*) is greatly liked for its sweet taste and tender texture; it is boiled, roasted or cooked in various preparations such as *chacha lusi* (mashed *choclo* soup) and maize breads wrapped with maize husks or *atzera* (*Canna indica*) leaves. *Choclo* breads are either cooked over a clay griddle like the *musiguita* or steamed like the *chucllu tanta* or *humitas*.

May is when *cao* or ripe grains are harvested and left to dry under the sun. *Chuchuca* soup, and *chucllu api* or *gallo api* are favourite soups made with *cao* maize. Once the maize is hard, the best seed is selected and stored for the rest of the year. To make it more digestible, dry maize is generally used in the form of flour for soups and breads. The basis of most thick soups is the *uchu jacu*, which is a mix of dry and toasted maize, fava, peas and beans ground into a nutritious flour. Dry maize is also soaked with lime to make *mote* (hominy) or until it germinates to make the ritual *chicha de jora* (germinated maize drink). June is the major harvest and feasting time when the *Inti Raymi* or sun solstice celebration is held. In addition to being an important moment for reasserting ethnic identity, community and sense of place through various religious rituals and social events, it is also a time to make offerings to the *Pachamama* (Mother Earth) for the current harvest and to ensure her blessings for the following agricultural season. Offerings are made with various varieties of maize, grains, fruits, bread figurines, alcohol and money strung into a structure made with reeds called a *castillo*.

As a harvest festival, *Inti Raymi* is also about gastronomic indulgence of food and drink as diverse dishes are prepared, displayed and shared. As the main and sacred crop, maize is prepared in various forms such as *mote*, *tostado* (toasted maize), boiled *choclo* and the very nutritious *chicha de jora*. The traditional food, called *cucayo*, consists of these maize preparations as well as potatoes, beans, various tubers, fava beans, rice and some type of animal protein whether cheese, roasted guinea pig or fried pork, and served with *zambo* or chilli pepper sauce.

During the following months of August, September and October, seed selection and storage, field preparation and planting take place. The Day of the Dead or *Huaccha Carai*, on 2 November, involves various ritual foods that are offered to the dead and shared with relatives and neighbours at the cemetery. Characteristic of this feast is the *pan de finados* or bread of the dead, which is made into various figurines: children, horses and doves. The *mazamorra* with *churus* or maize soup with earth snails, potatoes with *zambo* sauce, the *colada morada* or sweet maize gruel with wild blueberries and *champús* or maize drink are also traditional dishes consumed on this date. December and January coincide with the *Huarmi Pascua* or Woman's Easter, when only dry grains are available and this often is a period of famine and hunger. Traditionally wild potatoes or *arapapas* were consumed with various *yuyus* or wild plants sprinkled with *zambo* sauce or pepper sauce. According to the elders, a common dish for the New Year was *arapapa* cooked with matambre bean, and dressed with *zambo* seed sauce; it was said that this preparation would open people's stomachs because it was a new grain in the New Year (Quitugo Workshop, December 2003).

Homegardens

Frequently, smallholders' food security and well being depend on access to and control over natural resources in various public and private areas. In Cotacachi, indigenous and peasant nourishment depends on agricultural production in *chagras* (agricultural fields) primarily, but is complemented with products cultivated in homegardens as well as wild resources obtained in streams, lakes and forests. Homegardens contain native and introduced vegetables, fruits, aromatic herbs and spices for diverse culinary and medicinal uses. Wild species are tolerated and encouraged. Some homegardens are very diverse and multilayered, characterized by layering of trees, bushy and herbaceous species (see Piniero, Chapter 10, this volume). Others are limited to a few species for daily use in the kitchen and as remedies such as cabbage, cilantro, parsley, spearmint, wormseed and peppers.

Cows, goats, sheep, pigs, chickens and guinea pigs are kept in the areas surrounding the house, and are a primary source of fertilizer for homegardens and *chagras*. These animals provide protein for special occasions, and products derived from these minor species, such as milk, eggs, wool or manure, represent sporadic sources of cash. Domestic animals are also used as gifts during ceremonies and rituals.

The commons and wild foods

Reliance upon wild plants has been an important aspect of human nutrition and medicine (Etkin, 1994; Johns, 1996). In Cotacachi, wild plants gathered in forests, *páramos*, streams and agricultural fields alongside crops are strategic sources of micronutrients and vitamins, and key foods at times of scarcity and famine. Wild plants have been regularly consumed by indigenous populations in soups and with grains. When elders and women are asked to list traditional foods, they invariably name various wild plants such as *panra*, *sapi yuyu*, *alli yuyu*, *pima yuyu* (wild radishes and turnips), peppermint and wormseed, which combined with the basic maize, barley or quinua soups add to nutritional and taste diversity. Watercress and *bledo* (*Amaranthus*) are also nutrient-rich resources consumed during the winter months in salads, combined with potatoes and *zambo* sauce, or in omelettes. Wild fruits, on the other hand, are considered to be children's food, and are consumed by boys and girls while grazing the animals, fetching firewood or playing in the fields.

Hunting is no longer significant, but some people mention the occasional consumption of wild rabbits and birds from the *páramo*, or the eggs of these birds. Important to indigenous food traditions, snails and insects constitute sporadic sources of animal protein, calories and fat. Earth snails or *churus* are consumed traditionally during All Souls Day but are regularly found in the Otavalo market, and eaten with toasted maize and lemon. *Catzos* and *cusos*, or grubs of Coleoptera, are also eaten seasonally, fried or toasted, or employed as medicine for coughs. Farmers, however, say that these species are no longer as abundant as before, probably due to climate change.

Some communities possess community lots that are used to cultivate food which is then distributed, under the direction of community authorities, to those who participate in *mingas*. In Morochos, for instance, community members have used these lands collectively to plant the major agricultural crops such as maize, beans, quinua, fava beans, squash, *chochos* and tubers.

Exchanges

Intra-communal and extra-communal food exchanges are important to household food security and dietary diversification. In Cotacachi, traditionally, people from the upper and lower zones exchanged products between themselves. A typical example was the exchange of potatoes and tubers produced in the high zone for maize, beans and fruits grown in the middle and lower areas. Frequent among friends and relatives, these exchanges strengthened reciprocity bonds. As noted by Murra (1980), this is the essence of verticality and complementation in the Andes. According to local farmers, they no longer do this for fear of being looked down on as poor and needy, because communities are larger and people no longer have relatives or friends, or because they can go to the market to buy what they need. Nowadays, occasionally during harvest time, *mestizo* merchants go to the communities to barter primarily maize for store-bought goods.

Exchange of food is important during *mingas* or cooperative labour parties, religious and ritual ceremonies, and special events. For *mingas*, beverages and/or food must be provided for the workers, although workers can also contribute. *Mote*, potatoes, *tostado*, beans and rice are common items brought and spread over a long cloth on the floor around which people sit to eat, rest and talk. The major food event, however, is *Inti Raymi* or the sacred festival of the sun when food offerings are performed in every community, and harvested crops are cooked and exchanged during the ritual dances held in the communities and in the town of Cotacachi.

Rites of passage such as baptisms, weddings and wakes are associated with special meals, and are significant times for the consumption and exchange of animal protein. Various social agreements are sealed with food payments called *medianos*. There are two kinds of *medianos*: the salt one, which consists of a bowl containing various pounds of potatoes, several guinea pigs and a couple of hens. The sweet *mediano* is a bowl with bread figurines, a bunch of bananas and a gallon of alcohol. These food gifts are given by the groom's parents to the parents of the bride when they visit them to request the woman's hand. A smaller *mediano* is given to the bride who cannot eat from her parents' *mediano* because it would be as if she was eating her future husband's flesh. The various wedding godparents also receive a *mediano* in payment for their services. When the couple build their permanent house, a ceremony is held to baptise it. In this *huasi pichey* ritual, special food consisting of soup, guinea pigs, potatoes, chicken, bread and two different kinds of maize drink (*champús* and *chicha*) is given in payment to the house's godparents and the master builder.

During the *Huaccha Carai* or Day of the Dead, the bread, bananas and other foods that are brought to the cemetery to honour the dead are later shared with relatives and neighbours. There is a common local saying: 'During the *huaccha carai*, all the food must be given away at the cemetery, if it is brought back it's like donkey dung.' During regular funeral rites, food is exchanged in the same manner.

Markets

The indigenous food system is under the influence of larger socioeconomic processes such as urbanization, modernization and market integration, which not only introduce foreign products and new tastes, but also transform local cultures of production and consumption. People in the highlands have long been connected to the market and *mestizo* culture by means of wage labour and commerce due to Cotacachi's history as a textile and leather production centre, and its proximity to urban centres and regional markets such as Otavalo, Ibarra and Quito. In the past decades, however, the growth of industrial flower production and tourism in the area, international out-migration as well as the significant presence of diverse national and international institutions involved in development projects have propitiated new economic and social relationships and opportunities for local people. National education and other modernization programmes sponsored by the Ecuadorian state have also exposed indigenous people to *mestizo* practices, discourses and values.

The influence of these changes in local food habits can be observed in what people buy as well as their relative importance to the daily diet. During formal and informal conversations, people mentioned that salt, oil, sugar, *panela* (brown sugar), rice, oats, noodles, bread, potatoes, sodas and lard were among the most important food items purchased. Fruits and vegetables such as oranges, bananas and tomatoes are regularly bought. Treats include candies, sodas, juices, cookies, ice cream and bread. A favourite fast food among indigenous and *mestizo* youth is *salchipapa* (fried hot dogs and French fries).

The need for cash to purchase food, manufactured goods and medicines and to pay for utilities has become increasingly important for farmers. Household production is not a major source of income, but farmers sell milk, eggs, fruits, vegetables, tubers, honey, animals or meat in the local or regional markets when cash is needed. Sale of food in the communities as well as in Cotacachi and Otavalo is a cash-producing female activity. Bread, maize breads, soups, a popular snack of chochos, mote and onion, and a variety of fried foods such as potatoes, bananas, dough, pork meat and hot dogs are common foods sold by women.

Institutional food

Institutional food is part of the supplies provided by international aid agencies to national states as food relief for emergencies or for supplementary nutrition meal plans for children, mothers and the elderly. Commonly, food aid has been received at times of natural disasters such as earthquakes, floods, mudslides and crop loss, with consequent hunger and disease. In Cotacachi, food relief is memorable for indigenous people because it is associated with the introduction of new foods; people remember that during a specific period of crop loss, they received bulgur, ground maize, oil, 'white beans' (most probably soy) and powdered milk from 'that USA president that was killed,' referring to John F. Kennedy who was assassinated in 1963. The US flags in the tin cans were indicative of where the food came from. Others remember a time of famine when *mestizos* brought rice, noodles and plantain flour, which they had never seen and did not know how to eat or cook. Noodles were considered to be worms and rice was thought to be a kind of potato worm.

In Cotacachi, more steady food aid is distributed by public or private institutions in the form of nutritional supplements in nurseries, day care centres and primary schools. Institutional food follows western concepts of nutrition and attempts to correct nutritional deficiencies and imbalances. Food items distributed to the children are cereals, rice, lentils, oats, bread, cookies, fruits and juice. A special programme for the elderly, administered by the Municipality and supported with community labour, provides similar meals. The impact of these supplements has not been analysed from a nutritional and/or economic perspective in Cotacachi, but food aid critics argue that although they can bring some benefits to poor and hungry sectors of the population, they propitiate institutional dependency. An argument presented in Cotacachi by the nutritionist working for UNORCAC (personal communication 2004) was that the meals are distributed late, around 11 a.m., after children's most productive learning hours.

Local Food Classifications

Traditional food, special food, luxury food and famine food

Traditional food is associated with products and meals that identify *Cotacacheños* as indigenous people, with a particular peasant diet based on locally grown, filling and strength-giving products, prepared and consumed in a customary way. Indigenous food is a perfect example of slow food. Traditional food preparations involve time-consuming tasks such as soaking, drying, peeling, grinding (whether in the grinding stone or the manual mill), sieving and sifting. Cooking is done slowly in the stone hearth called the *tulpa* around which the family gathers to talk and tell stories. The smoke of the firewood is said to enhance the flavour of the food. Most foods are boiled or toasted, and meals are generally bland; hot or spicy meals are associated with *mestizo* food. Condiments include aromatic and medicinal cultivated and wild herbs such as annatto, wormseed, peppermint, onions, *yuyus* and squash seeds. Hot pepper sauces accompany many dishes. Food is served according to a defined hierarchy, starting with the father, followed by the oldest sibling, and ending with the mother. Meals are eaten slowly and quietly.

Special foods are related to those meals consumed during special occasions, that involve more time and labour, and that include animal protein. Ritual soups include *uchu caldo* (potato soup with meat), *uchu api* (maize soup with mote) and *champús* for weddings; white morocho *colada* (white maize gruel) for wakes; *chicha de jora* for *Inti Raymi*; and various maize breads for the Day of the Dead and the New Year.

Luxury food is food to 'show off', i.e. store bought or consumed in restaurants, mostly associated with *mestizo*s or rich people with purchasing power. Seasoned dishes, fast foods, meat or fish, dairy products, canned and manufactured goods, and sodas and juices stand out as luxuries. In contrast to special or luxury foods are the simple foods eaten at times of crop loss, scarcity and famine. These hunger foods are cultivated plants, such as onion, cilantro, wormseed, *achogcha* (*Cyclanthera pedata*), cabbage and wild *yuyus* whose aromatic, nutritional and medicinal properties enhance potato broths and other watery soups. Nowadays, plantain flour is considered a hunger food from which a sweet gruel is made. Commonly, plantain flour has been a low-cost nutritional supplement, given its high levels of carbohydrates and potassium.

Dry and wet

In addition to social and economic criteria for classifying food, sensory elements also play a role in classifications and preferences in indigenous cuisine. In the case of Cotacachi, texture is an important criterion. The texture of meals is related to the state of maturity of the crops: tender maize, and legumes such as peas, beans and fava are added to soups or are boiled and eaten with sauce or salt. When grains are dry, they are toasted and consumed with salt and lard or ground into flours to make thick soups. Soups are the most distinguishable indigenous dish; they are consumed in the morning, noon and evening, and two or three servings are the norm. Soups range from thick porridges (*mazamorra*, *colada*) made with cereal and legume flours, to soups (*locro*) with vegetables and potatoes, to broths (*caldos*) seasoned with a few herbs. Wet meals also include sweet meals (*mishki micuna*) in the form of porridges and gruels (*coladas*) cooked with aromatic herbs, cinnamon and/or milk and fruits, and sweetened with brown sugar (*panela*).

Dry meals (*chakishca micuna*) range from crunchy toasted maize and legumes to boiled maize and tubers, and to finely ground barley flour (*machica*). *Mote*, *tostado*, potatoes and rice are among the most representative *secos*. Dry meals also include the broad 'bread' (*tanta cuna* or *pan*) category that refers to various cornmeal preparations. Panes are consumed with a sweetened herb tea (*mishki jacu*) called *café*. Dishes containing wild plants or cultivated vegetables are denominated salads and classified as dry foods.

Food and snacks

To indigenous people of Cotacachi, food is synonymous with filling and strength-giving products represented by slow cooking and nourishing meals. Snacks, in contrast, are generally associated with fast food, children's food, *mestizo* food or junk food. Like treats, these are light foods that only provide energy for a short period. Some snacks consist of 'food to suck' such as juices, fruits, sugarcane, ice cream, candies and lollipops. Rice and noodles pose ambiguities because they meet the snack classification, and because they are non-filling but are also very significant foods for the daily diet. Rice is considered a snack if consumed in small amounts or is given to children as a treat, but constitutes a meal when served in large amounts and in combination with grains.

Warm and cool

Quechua concepts of health and disease are centred on balance and maintenance of strength (*fuerza*) and energy; well being is associated with balance in different levels (natural, social and spiritual) and is not limited to the individual but also to the collectivity (Ecuarunari, 1999). Thermal balance is important to health, and many illnesses are produced by excess or lack of heat or cold. According to the diagnostic, warm or cool treatments are made with plants with refreshing or energizing attributes. Cool plants heal internal infections that are manifested with fever and high temperature. Warm plants are used to cure illnesses produced by cold such as colds and pains. Similar to plants, food classification into warm and cool is related to the effects that they produce in the body rather than their actual temperature. Excess of warm or hot foods produces perspiration, gastric pain, constipation, diarrhoea, stomach inflammations and infectious processes. Spicy foods or foods that are hard to digest such as grains, greasy meats and alcohol are hot. Excess of cool foods can cool the organism down due to the loss of energy, and produce colic, apathy, weakness and pain. Rice and noodles, potatoes and fruits are cool and must be avoided during cool status such as after childbirth.

Nutritional value

Although a nutritional assessment of the indigenous diet was not part of the objectives of this study, on the ground observations and interviews indicate that there is a primacy of filling starchy cereals and carbohydrates, complemented by vegetable protein in the form of legumes and vegetables. The exceptional nutritional value of cereals such as quinua, legumes such as *chochos* and beans, tubers, vegetables and edible wild plants has been amply documented (National Research Council, 1989; Bermejo and León, 1994). However, according to nutritional experts, in the Ecuadorian highlands, the main causes of indigenous malnutrition are the lack of food, and the monotonous and imbalanced quality of the diet rich in carbohydrates from potatoes and flour with insufficient protein and micronutrient intake (Larrea and Freire, 2002). Iodine, iron and vitamin A deficiencies are a national and regional public health problem (Food and Agriculture Organization, 2001). Children are the most affected by malnutrition, and among its negative effects are the lower potential for learning due to a diminished cognitive development, sensory impairment and the inability to concentrate and perform complex tasks, all of which often go hand and hand with poor enrolment or irregular school attendance.

The Cotacachi canton does not have consolidated data on the nutritional status of the population, but preliminary information collected by UNORCAC indicates a high incidence of chronic and acute malnutrition (Pamela Baez, personal communication 2004). Despite the diversity of cereals, legumes, vegetables and cultivated and gathered fruits, deficiencies in the consumption of animal protein, fat, fruits and vegetables have been identified. This raises questions about the relationship between food security and the real use of cultivated and harvested biodiversity by local people, in the context of a tendency towards simplification of cropping patterns, loss of native agricultural varieties and less reliance on nutritious wild plants.

According to the cantonal Health Plan's recent health diagnostics in Cotacachi (Plan de Salud Cantonal, 2002), indigenous populations are not yet affected by obesity, diabetes and cardiovascular problems characteristic of poor and malnourished populations in the developing world, due to nutritional transition to diets composed of processed foods, sugars and saturated fats. What is less known, however, and varies from case to case, is the daily household food composition and intake, portion sizes, intra-household allocation, and the effects of seasonal variations on diets. However, according to the Ecuadorian nutritional profile elaborated by the Food and Agriculture Organization (2001), national consumption patterns follow the global trend towards Western diets, associated with an increase in obesity and diabetes among adolescents and women – which could eventually affect indigenous populations. In the past years, a decline has been seen in the availability of fruits, roots, tubers and legumes, followed by a significant increase in the availability of fat, cereals and derived products, meat and derived products, dairy products, eggs and sugars (Food and Agriculture Organization, 2001). Production and consumption of rice during the 1990s has also risen considerably. Changes in the Provision of Food Energy (Suministro de Energía Alimentaria, SEA) are related to larger levels of cereal, cooking oils and fats, sweeteners and dairy product imports, a trend that will probably increase with the new free trade agreements with the USA.

Environmental factors such as the seasonality of wild and cultivated resources, soil and genetic erosion, climate change and lack of irrigation water contribute to the decline of agricultural productivity. This, in turn, has a direct effect on household food availability and a diverse, adequate diet. However, other socioeconomic and cultural factors contribute to nutritional problems, such as unavailability of land and labour, lack of purchasing power, the surrounding health environment, the nutritional transition, feeding and health behaviours of key household members and increasing allocation of time to other activities different from food procurement and preparation.

To counter nutritional deficiencies in peasant and indigenous communities, diverse public and private initiatives have been implemented. Governmental and foreign donor food aid resources are channelled by the Municipality into nutritional supplementation in nurseries and elementary schools. Currently, a project on agroecological diversification with native varieties and economic alternatives based on local agrobiodiversity is being implemented between UNORCAC, UCODEP (an NGO) and the national gene bank – INIAP – with the support of the United States Department of Agriculture (USDA). Smaller initiatives include the *Finca de Futuros Ancestrales* (see Moates and Campbell, Chapter 3, this volume) and the homegarden diversification project implemented by community women with assistance from Jambi Mascaric and funds from SANREM.

Conclusion

This chapter presented results of an anthropological study of the food habits, diet and cuisine of indigenous people in Cotacachi. By means of the description of the food system, and the local food perceptions and classifications, I attempted to illustrate aspects of the complexity of a vital aspect of these communities' livelihoods, which are intimately related to their ethnic identity and the use of local biodiversity. Food is a sensory and material interface between nature and culture because food is at the crux of human–environment interactions: nature is not only socialized and contested (Descola and Palsson, 1996; Escobar, 1999; Brechin *et al.*, 2003) but is regularly ingested for sustenance and pleasure (Hladik *et al.*, 1993).

In the Andean highlands, food security and health have depended on access to different ecosystems or use spaces for the complementary utilization of wild species, cultivated plants and other resources. The broad exploitation of different ecological zones for food procurement and production indicates the importance of access not just to land but to the territory and the resources found therein. In this sense, territory can be defined as the concrete material space where a distinct culture, environmental relationships and indigenous livelihoods are created and recreated. The historical indigenous struggles for territorial rights and access and control over natural resources are still valid for those rural populations who greatly depend upon the land and forests.

Persistence of diversity in the food sources, productive systems and culinary preparations is related to the depth of their historical roots and their significance for the revitalization of Quichua identity. It is clear that food is a marker of identity and difference in Cotacachi. Social and ethnic differences between indigenous and *mestizos*, the poor and the wealthy, rural and urban populations are marked by food choices and preferences. Indigenous filling grains and slow food bland meals stand in contrast to *mestizo* fast-cooking light rice and noodles and luxury spicy dishes.

Generational tensions between older farmers and younger urban-oriented youth are also voiced in the language of food. The new preference of younger men and women for *mestizo* foods, acquired when they migrate to the city to work or due to increasing contact with *mestizo* values, concern their parents and the elders. According to local perceptions, these foods make young people too physically weak to work in the fields and too morally feeble to follow a traditional indigenous life style. For older generations, this lack of strength and willingness to work in agriculture is related to local productive decline, genetic erosion and food insecurity.

However, a broader look at the changes in productive processes, dietary repertoires and people's relationship to food shows that the transformations taking place in the socioeconomic and cultural landscapes of the rural Andes are also a consequence of national development policies (dollarization, economic adjustment, elimination of agricultural subsidies and liberalization of land markets) and globalized relationships of production and consumption (trade liberalization and food imports). The new Free Trade Agreements of the Americas (FTAA) will further impact national agricultural production, intellectual property rights, biodiversity, livelihoods and rural food security. In this light, a matter of concern is the potential for higher levels of malnutrition in the rural sierra due to the abandonment of nutritious locally grown foods for cheaper store-bought, often imported, products in accordance with the national trends towards more Western diets.

Because food habits are deeply ingrained aspects of a culture, local resistance to ongoing changes is expressed in the persistence of a diversified agricultural base that includes native crops, the maintenance of traditional culinary practices and the use of ritual dishes that strengthen social ties and reaffirm a sense of belonging. Concerns with changes in rural culture and livelihoods have situated food as a working priority for local indigenous organizations, the state and private institutions. Recently, initiatives have been implemented that address nutritional and environmental concerns through diversified agroecological and economic production, introduction of germplasm, nutritional education, food transformation and processing projects, and nutritional supplements with foreign food aid.

As Bebbington (1996) notes, one of the strengths of the Ecuadorian indigenous movement has been the crafting of political demands in the name of culture and difference. In this sense, food as a substantial aspect of indigenous identity and well being has begun to figure more prominently in the indigenous 'development with identity' agenda as well as in the social movements' claims for cultural, political and productive autonomy. The active participation of local communities and leaders from Cotacachi in these movements and mobilizations for food security and food sovereignty point to the powerful links that exist between food, culture, identity, biodiversity and social justice. If we follow the saying that 'we are what we eat' (or what we do not eat), to enquire about the material, political and symbolic dimensions of food becomes more relevant in the context of new trade agreements whose negative effects on the environment and agricultural production are debated in the national public arena. Although a total rejection of foreign foods or food aid is not the case, it is necessary to explore more the implications of these new measures on the food situation of the most vulnerable populations such as the indigenous people of the rural sierra.

References

Alchon, S. (1997) The great killers in precolumbian America. A hemispheric perspective. *Latin American Population History Bulletin* No. 27. Department of History, University of Minnesota.

Appadurai, A. (1981) Gastro-politics in Hindu South Asia. *American Ethnologist* 8, 494–511.

Archetti, E. (1997) *Guinea-pigs: Food, Symbol, and Conflict of Knowledge in Ecuador.* Berg, Oxford, UK.

Bebbington, A. (1993) Modernization from below: an alternative indigenous development? *Economic Geography* 69, 274–292.

Bebbington, A. (1996) Movements, modernizations, and markets. Indigenous organizations and agrarian strategies in Ecuador. In: Peet, R. and Watts, M. (eds) *Liberation Ecologies. Environment, Development, Social Movements.* Routledge, London, pp. 86–109.

Bermejo, J., Hernándo, E. and León, J. (eds) (1994) *Neglected Crops: 1492 from a Different Perspective.* Plant Production and Protection Series No. 26. FAO, Rome.

Bourque, N. (2001) Eating your words: communicating with food in the Ecuadorian Andes. In: Hendry, J. and Watson, C.W. (eds) *An Anthropology of Indirect Communication.* Routledge, London, pp. 85–100.

Brechin, R.S., Wilshusen, P.R., Fortwangler, C. and West, P.C. (2003) *Contested Nature. Promoting International Biodiversity with Social Justice in the Twenty-first Century.* State University of New York Press, Albany, New York.

Brush, S. (1980) Potato taxonomies in Andean agriculture. In: Brokensha, D.W., Warren, D.M. and Werner, O. (eds) *Indigenous Knowledge Systems and Development.* University Press of America, Landham, Maryland, pp. 37–47.

Casagrande, J. (1981) Strategies for survival: the Indians of highland Ecuador. In: Whitten, N.E. (ed.) *Cultural Transformations and Ethnicity in Modern Ecuador.* University of Illinois Press, Urbana, Illinois, pp. 260–277.

Coe, S. (1994) *America's First Cuisines.* University of Texas Press, Austin, Texas.

Corr, R. (2002) Reciprocity, communion, and sacrifice: food in Andean ritual and social life. *Food and Foodways* 10, 1–25.
Counihan, C. and van Esterik, P. (1997) *Food and Culture. A Reader.* Routledge, New York.
Descola, P. and Palsson, G. (1996) *Nature and Society. Anthropological Perspectives.* Routledge, London.
Ecuarunari (1999) *Manual de la Medicina de los Pueblos Kichuas del Ecuador.* Proyecto de investigacion y sistematizacion, Ecuarunari-Codenpe. Crear Grafica, Quito, Ecuador.
Escobar, A. (1999) After nature: steps to an antiessentialist political ecology. *Current Anthropology* 40, 1–30.
Etkin, N. (1994) *Eating on the Wild Side. The Pharmacologic, Ecologic, and Social Implications of Using Noncultigens.* University of Arizona Press, Tucson, Arizona.
Fischler, C. (1995) *El (H)omnivoro. El gusto, la cocina y el cuerpo.* Editorial Anagrama, Barcelona, Spain.
Food and Agriculture Organization (2001) *Perfil Nutricional del Ecuador.* FAO, Rome.
Goodman, A., Dufour, D. and Pelto, G. (2000) *Nutritional Anthropology: Biocultural Perspectives on Food and Nutrition.* Mayfield Publishing, San Francisco, California.
GRAIN (2001) GMOs found in food aid to Latin America. *Seedling* 18, 2.
Harris, M. and Ross, E.B. (1987) *Food and Evolution. Toward a Theory of Human Food Habits.* Temple University Press, Philadelphia, Pennsylvania.
Harrison, R. (1989) *Signs, Songs, and Memory in the Andes. Translating Quechua Language and Culture.* University of Texas Press, Austin, Texas.
Hastorf, C. and Johannessen, S. (1993) Pre-Hispanic political change and the role of maize in the Central Andes of Peru. *American Anthropologist* 95,113–137.
Hladik, C.M., Hladik, A.M., Linares, O.F., Pagezy, H., Koppert, G.J.A. and Froment, A. (1993) *Tropical Forests, People and Food. Biocultural Interactions and Applications to Development.* Man and the Biosphere Series. UNESCO. Parthenon Publishing Group, UK.
Johns, T. (1996) *The Origins of Human Diet and Medicine.* University of Arizona Press, Tucson, Arizona.
Larrea, C. and Freire, W. (2002) Social inequality and child malnutrition in four Andean countries. *Revista Panamerican de Salud Publica* 11, 356–364.
Messer, E. (1989) Methods for studying determinants of food intake. In: Pelto, G.H., Pelto, P. and Messer, E. (eds) *Research Methods in Nutritional Anthropology.* United Nations University, Tokyo, pp. 1–32.
Mintz, S. (1985) *Sweetness and Power. The Place of Sugar in Modern History.* Penguin Books, New York.
Murra, J. (1980) *The Economic Organization of the Inka State.* JAI Press, Greenwich, Connecticut.
National Research Council (1989) *Lost Crops of the Incas: Little-known Plants of the Andes with Promise for Worldwide Cultivation.* Ad Hoc Panel of the Advisory Committee on Technology Innovation, Board on Science and Technology for International Development, National Research Council. National Academy Press, Washington, DC.
Netting, R.McC. (1993) *Smallholders, Householders: Farm Families and the Ecology of Intensive, Sustainable Agriculture.* Stanford University Press, Stanford, California.
Oficina de Planificación de la Presidencia de la República–Food and Agriculture Organization (2001) *Perfil Nutricional del Ecuador.* Lineamientos de política sobre seguridad alimentaria y nutrición. Oficina de Planificación de la Presidencia de la República, Quito, Ecuador.
Ohnuki-Tierney, E. (1993) *Rice as Self. Japanese Identities Through Time.* Princeton University Press, Princeton, New Jersey.
Orlove, B.S. (1987) Stability and change in highland dietary patterns. In: Harris, M. and Ross, E.B. (eds) *Food and Evolution. Toward a Theory of Human Food Habits.* Temple University Press, Philadelphia, Pennsylvania, pp. 481–517.
Patiño, V.M. (1984) *Historia de la Cultura Material en la America Equinoccial. La Alimentacion en Colombia y en los Paises Vecinos.* Tercer Año de la Segunda Expedicion Botanica. Biblioteca Cientifica de la Presidencia de la Republica, Bogota.
Piperno, D.R. and Pearsal, D.M. (1998) *The Origins of Agriculture in the Lowland Neotropics.* Academic Press, San Diego, California.
Quitugo Workshop (2003) Workshop on traditional crops held in the Quitugo community, on December 16, moderated by Rosita Ramos and Kristine Skarbø.
Ramirez, M. and Williams, D.E. (2003) *Guia Agro-Culinaria de Cotacachi, Ecuador y Alrededores.* IPGRI-Americas, Cali, Colombia.

Salomon, F. (1986) *Native Lords of Quito in the Age of the Incas: the Political Economy of North Andean Chiefdoms*. Cambridge Press, London.

Salman, T. and Zoomers, A. (ed.) (2003) *Imaging the Andes. Shifting Margins of a Marginal World.* CEDLA. Latin American Series. Asant, The Netherlands.

Seremetakis, N. (1994) *The Senses Still.* University of Chicago Press, Chicago, Illinois.

Stanish, C. (2001) The origin of state societies in South America. *Annual Review of Anthropology* 30, 41–64.

Sistema Integrado de Indicadores Sociales del Ecuador-SIISE (2003) Electronic document.

Weismantel, M.J. (1988) *Food, Gender, and Poverty in the Ecuadorian Andes*. University of Pennsylvania Press, Philadelphia, Pennsylvania.

World Bank (2004) *Ecuador.* Poverty Assessment. Poverty Reduction and Economic Management Sector Unit. Report No. 27061-EC.

Part III

Soils, Water and Sustainability

A Cotacachi farmer prepares the soil of her garden for planting (photo: Sanrem Files).

The chapters in this section explore the vital natural resources of soils and water, the complex interaction between the two and how local decision making affects their sustainability. In Chapter 12, Franz Zehetner and William Miller characterize the soil fertility status and nutrient limitations in different zones of 41 Andean communities around the Cotacachi volcano. The volcanic ash soils have the potential to support agriculture – whether traditional or modern – but the soil fertility needs rejuvenation. Zehetner and Miller present results of controlled experiments seeking possible solutions to restoring and maintaining soil fertility as a key first step towards sustainable agriculture. Organic nutrient sources are preferable over inorganic sources not only for enhancing the quality of the soil but also because they can be produced locally at a lower cost and are more compatible with local culture.

Zehetner, Miller and Zapata continue the soil–water analysis in Chapter 13 on the potential benefits to crop production by expanding existing irrigation systems to a broader area of the watershed. Irrigation is available to some lower-zone communities but is rare in the higher zones where the poorest inhabitants live. They approach this problem through simulation modelling to quantify possible improvements in 'wet-season and dry-season maize production', assuming irrigation is available. Scientific models are heuristic devices which are like stories: they do not present a reality but rather are useful representations for guiding further study and challenging existing ideas. A good model compresses time so that decision makers can understand future results of a given action, in this case the building of an irrigation system. The modelling exercise shows that water stress is rarely a limiting factor in crop growth in the rainy season, and expansion of the irrigation system is unnecessary for this cropping period. However, water stress is significant in the dry season and would benefit enormously from irrigation. They recommend that more efficient water use would be feasible through installation of a low-cost sprinkler and irrigation system.

Water in Cotacachi has multiple uses and plays many cultural roles: a vital resource without which humans and their animals and crops could not live; a ceremonial and religious object called forth in prayers or used to purify the soul through ritual baths. Water is also a politically charged resource that is increasingly scarce, contaminated and challenging for management among increasingly diverse users. As a result, water is at the centre of resource conflict between stakeholders. With climate change and growing demand, water will continue to be a major contested issue involving civil, judicial, political and planning bodies.

Water resource scientists Jenny Aragundy and Xavier Zapata Ríos present in Chapter 14 results of their research aimed at determining water contamination sources as well as problem-solving alternatives. Forty-five per cent of the dwellings in Cotacachi do not have access to a sanitary sewage system. Water from the rivers is not suitable for human consumption and is linked to illnesses caused by water-borne diseases, parasites, insects and other infections. Among the factors causing this contamination are grazing near watering sites, lack of sanitation facilities, poor hygiene and sanitary habits, laundry practices and disposal of solid waste into the water supply. Basic and simple solutions exist for solving these problems, but a political will must be in place to create a safe and reliable water supply.

Economists Fabián Rodríguez and Douglas Southgate in Chapter 15 address the question: are rural households willing to pay for improvement in the quality and quantity of water supply (potable and irrigation)? Using a technique called contingent valuation (CV), participants in their study respond to simulated transactions in a hypothetical market setting. These transactions reveal what people are willing to pay for non-market goods and services provided by the natural environment. Results of the study show that rural households are hypothetically willing to pay about 50% more than what they are currently paying for drinking water that is of higher quality and reliable. The policy implication is that a decentralized approach of encouraging local people to buy into improved local water systems is not only desirable but feasible.

In Chapter 16, water resource scientists Sergio Ruiz-Córdova, Bryan Duncan, William Deutsch and Nicolás Gómez present research on community-based water monitoring (CBWM) groups in Cotacachi. CBWM focuses on involving local citizens in water monitoring activities, thereby empowering them to play an active role in protecting their water and other natural resources. The method of CBWM involves first identifying sampling sites based on local identification, followed by physical–chemical monitoring by citizens themselves. Bacteriological monitoring of potable water occurs on a quarterly basis. Results showed that drinking water is not safe in 75% of the communities studied, and 80% of the sampling sites tested positive for coliform bacteria. These results, collected by Cotacacheños themselves, have helped raise public awareness of water conditions and are being incorporated into a canton-wide natural resource plan.

12 Toward Sustainable Crop Production in Cotacachi: an Assessment of the Soils' Nutrient Status

Franz Zehetner and William P. Miller
University of Georgia, Department of Crop and Soil Sciences, 3107 Plant Science, Athens, GA 30602-7272, USA

Introduction

Archaeological finds near Otavalo have revealed that the Andean region of northern Ecuador has been inhabited for thousands of years (Athens, 1999). Pollen analysis in Lago San Pablo sediments has confirmed that the volcanic ash soils of the area had been cultivated by early maize agriculturalists as long as 4200 years ago (Athens, 1999). Volcanic ash soils have the reputation of being fertile and highly productive, allowing for high human-carrying capacity. They were conducive to the development of early civilizations in Central and South America (Lauer, 1993) as well as in Asia (Shoji *et al.*, 1993).

The high natural fertility of volcanic ash soils has been attributed to the abundance of nutrients in the soil parent material, the development of thick humus horizons containing large amounts of organic nitrogen, free drainage, high plant-available water-holding capacity and a deep, unrestricted rooting zone (Shoji *et al.*, 1993). However, not all soils derived from volcanic deposits can supply the large amounts of nutrients and water necessary for vigorous plant growth. The inherent fertility of volcanic ash soils is strongly dependent on texture and composition of their parent material, on the nature, intensity and duration of its alteration by the processes of weathering, and on the magnitude of organic matter accumulation in the course of soil development.

According to their chemical composition, volcanic ashes are classified into the rock types rhyolite, dacite, andesite, basaltic andesite and basalt (Shoji *et al.*, 1975). The contents of silicon decrease in this order, whereas the concentrations of calcium, magnesium, iron and other micronutrients increase. However, total phosphorus and potassium contents are higher in rhyolitic compared with basaltic materials (Shoji *et al.*, 1993). The nutrients contained in the soil parent material are largely unavailable for plant roots until liberated in the course of mineral weathering and stored in a more available form on mineral and organic surfaces. Parent material texture is an important factor influencing the rate of weathering and nutrient release. The finer the texture, the greater the surface area exposed to the ambient solution and the higher the rate of chemical weathering (Shoji *et al.*, 1993). With increasing intensity and duration of weathering and soil development, the clay contents of volcanic ash soils increase,

 Development with Identity: Community, Culture and Sustainability in the Andes (ed. R.E. Rhoades)

which in turn enhances their ability to retain plant nutrients against leaching and to store plant-available water. For volcanic deposits in New Zealand, Lowe (1986) showed that tephra younger than 3000 years had < 5% clay, deposits of 3000–10,000 years contained 5–10% clay and 10,000–50,000-year-old tephra had clay contents of 15–30%.

The clay mineralogy of volcanic ash soils is largely determined by the amount of rainfall and leaching. Active amorphous constituents, such as allophane and aluminium–humus complexes, dominate the clay fraction where the climate is moist and soil solution silicon is removed by leaching, whereas halloysite is found as the dominant clay mineral in drier, silicon-rich environments (e.g. Parfitt *et al.*, 1983; Parfitt and Wilson, 1985). These differences in clay mineralogy have important bearings on nutrient cycling. Active amorphous constituents are responsible for the strong sorption of phosphate often observed in volcanic ash soils (e.g. Parfitt, 1989; Wada, 1989; Shoji *et al.*, 1993) that makes phosphorus sparingly available for plants and microorganisms. Active amorphous constituents further react with soil organic matter, leading to its stabilization and enhanced protection against microbial decomposition (Parfitt *et al.*, 1997, 2001; Gijsman and Sanz, 1998). This results in humus accumulation and the presence of appreciable quantities of organic nitrogen in these soils. However, until mineralized (converted from organic to mineral form in the course of organic matter decomposition), this nitrogen is not available for plant uptake, and, due to the above-mentioned protective effects, nitrogen mineralization is comparatively slow in soils with active amorphous constituents (Shoji *et al.*, 1993; Parfitt *et al.*, 2001).

In order for volcanic ash soils to remain sustainably productive, proper management practices are a prerequisite. Continuous cropping without adequate inputs leads to nutrient mining and declining productivity. In the Andean ecoregion, fallow rotation systems have traditionally been practised to restore soil fertility and avoid outbreaks of pests and diseases (Sarmiento *et al.*, 1993; Schad, 1998; Pestalozzi, 2000; Phiri *et al.*, 2001). However, increasing population pressure, competing land use demands and the incorporation of elements from market-oriented agriculture have been changing these traditional systems (Sarmiento *et al.*, 1993; Phiri *et al.*, 2001). Using a modelling approach, de Koning *et al.* (1997) assessed the sustainability of Ecuadorian agroecosystems based on their soil fertility status. They calculated nutrient balances for different land use types and found net nitrogen and potassium losses, which were higher for cropland than for grassland. Nutrient depletion was more severe in the Andean than in the coastal region.

Crop yield data collected by the Ecuadorian Centro Andino de Acción Popular (CAAP) indicate comparatively low maize yields in the Cotacachi area (Field, 1991), and members of the Cotacachi communities have identified decreasing soil fertility as a threat to their subsistence (UNORCAC, 1999). In this study, we analyse the fertility status of the soils in the Cotacachi area and identify nutrient limitations in different zones of the area. We use crop growth modelling to examine the long-term effects of nitrogen fertilization and residue management on maize yields, and discuss possible avenues for restoring and maintaining soil fertility as the basis of sustainable agricultural production.

Methodology

Soil sampling

A total of 145 cultivated fields in 41 Andean communities around Cotacachi volcano were randomly selected for soil analysis and geo-referenced. The locations of the sampled fields are marked by solid dots in Fig. 12.1. Depending on the size of the fields, between ten and 20 soil samples were taken from the plough layer (0–15 cm depth) and mixed to obtain one composite sample of each field. The samples were air-dried and passed through a 2 mm sieve prior to laboratory analyses.

Soil analysis

Particle size distribution was determined with the hydrometer method according to Bouyoucos (1962). Soil organic matter was measured by wet oxidation with the Walkley–Black method (Soil Survey Staff, 1996), and pH was measured in H_2O and 1 M NaF (Soil Survey Staff, 1996). Cation exchange capacity (CEC) and exchangeable Ca^{2+}, Mg^{2+}, K^+ and Na^+ were determined with 1 M NH_4OAc buffered at pH 7 (Soil Survey Staff, 1996). Phosphorus was extracted with 0.5 M $NaHCO_3$ buffered at pH 8.5 (Olsen *et al.*, 1954). The fertilizer P availability index was determined according to Sharpley *et al.* (1984, 1989) for three different soils (sites labelled A, B and C in Fig. 12.1) after incubation with various amounts of P for 30 and 180 days, respectively.

Crop growth modelling

The Decision Support System for Agrotechnology Transfer (DSSAT, version 3.5) was used to simulate maize (*Zea mays* L.) growth in the area under study. DSSAT is an integrated modelling platform that comprises several crop simulation models and databases, as described in Jones *et al.* (1998). It operates on a field scale with a daily time step and is capable of long-term simulations. The CERES model used within DSSAT to simulate maize growth models phenological and morphological development, biomass accumulation and partitioning, soil water balance and soil nitrogen transformations (Jones *et al.*, 1998). Calibrated genetic coefficients for the locally grown maize variety *Chaucho Mejorado* (INIAP-122) were obtained from Bowen (2000; personal communication). Using these

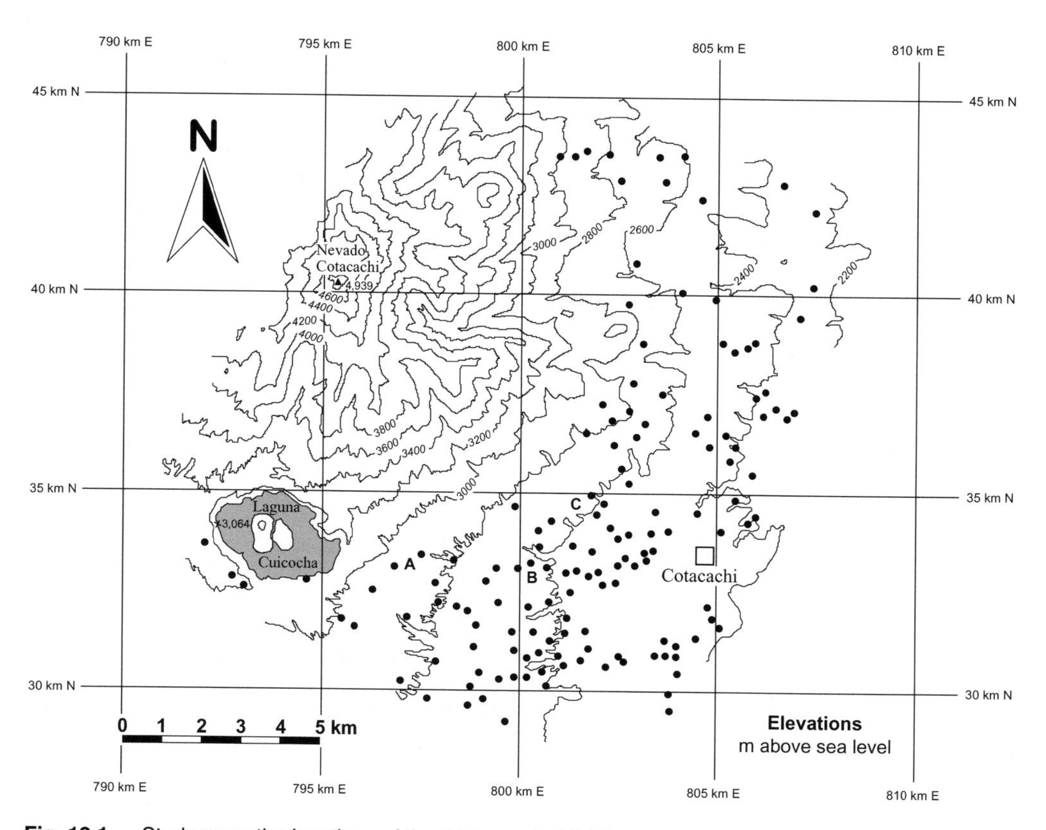

Fig. 12.1. Study area: the locations of the 145 sampled fields are marked with solid dots; at sites A, B and C, the fertilizer phosphorus availability index was determined.

coefficients, simulated grain yields corresponded to measured values for maize grown in the study area during the 2000–2001 rainy season. In order to analyse long-term effects of land management, 30 years of weather data were randomly generated with DSSAT's weather generator WGEN (Richardson and Wright, 1984) based on an existing 30-year weather data set from the nearby town of Otavalo and known altitudinal variations of climatic parameters.

Parent Materials and Soil Texture

The earliest classifications of soils were based on an assessment of their ease of cultivation and management (Russell, 1988), which is largely determined by their particle size distribution or textural class. To this day, soil texture is among the most widely used farmers' criteria of local soil classification systems (Talawar and Rhoades, 1998). The distributions of topsoil sand and clay contents in the area under study are shown in Figs 12.2 and 12.3, respectively.

The studied volcanic ash soils generally exhibit high sand and low clay contents and are therefore easy to cultivate and not prone to compaction; however, there are marked textural variations within the study area. In the southern part, sand contents are considerably higher and clay contents considerably lower than in the northeastern part. This is probably due to different duration of soil development in the two areas. The soils in the southern part have formed on the 3000-year-old Cuicocha deposits, whereas the soils in the northeastern part have formed on deposits older than 40,000 years and have thus been exposed to the actions of physical and chemical weathering for a longer time. In the course of weathering, sand- and silt-sized primary minerals are fractured

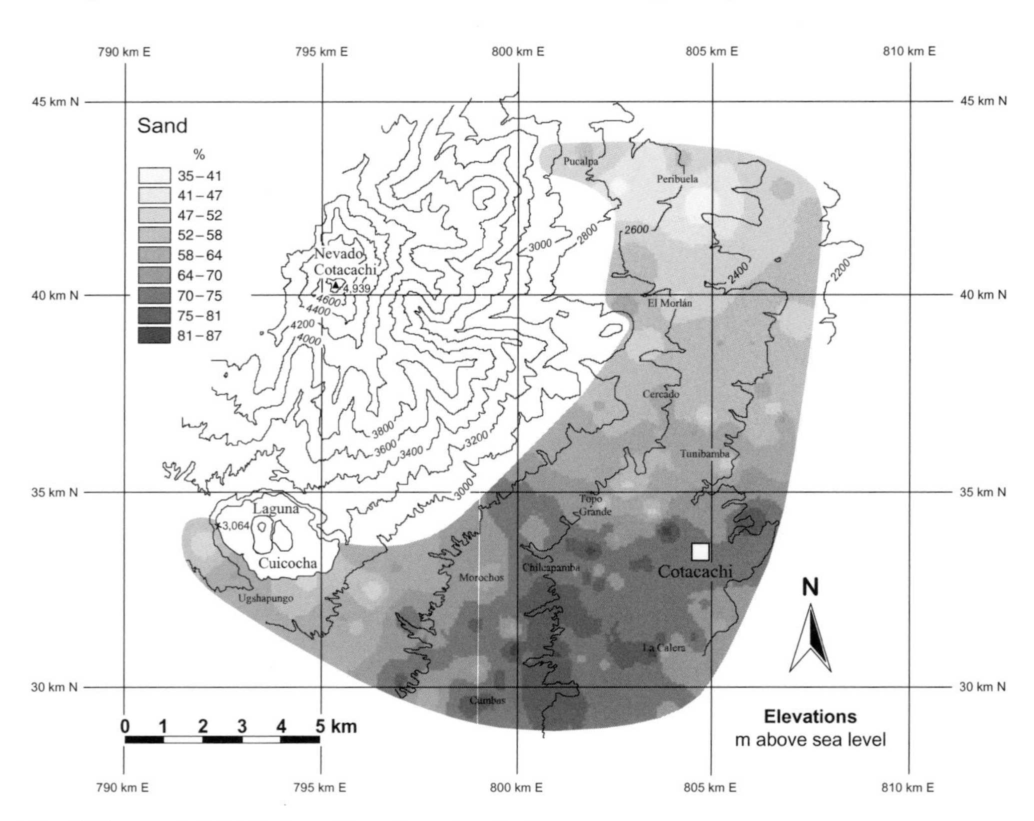

Fig. 12.2. Spatial distribution of topsoil sand contents.

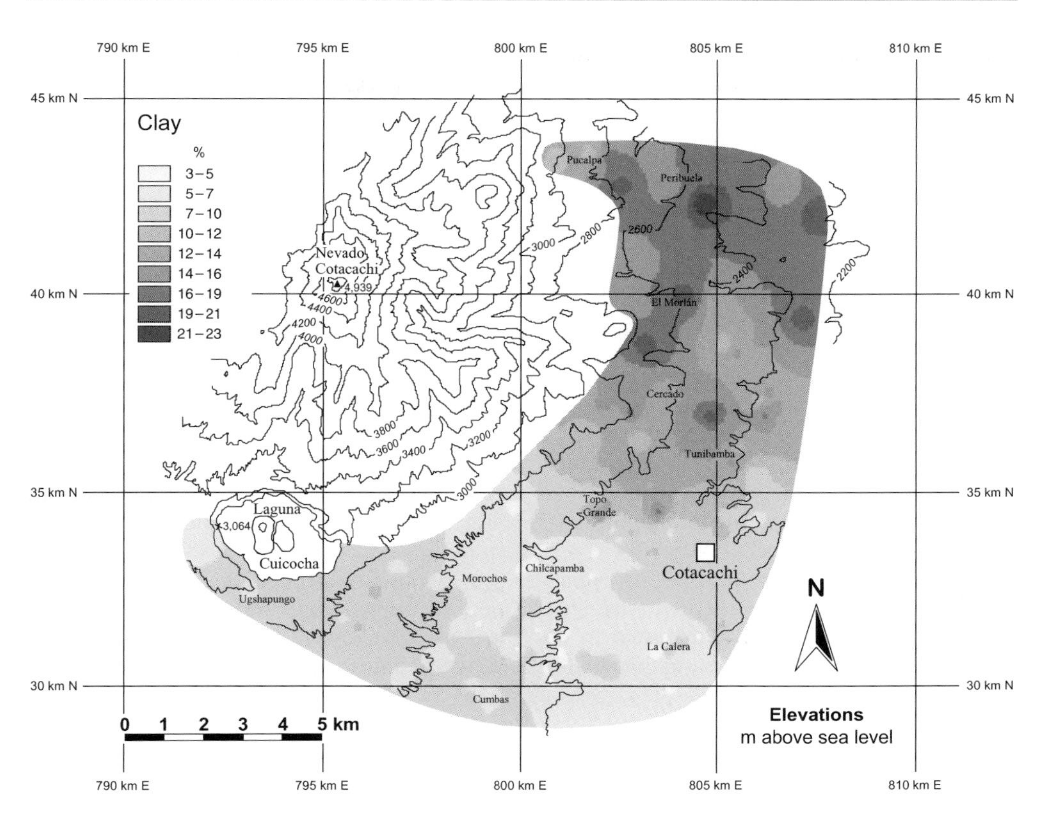

Fig. 12.3. Spatial distribution of topsoil clay contents.

and dissolved, and new clay-sized secondary minerals are formed. As a result, soil texture becomes finer with progressive soil development. The soil age–clay content relationship in the area under study is similar to that observed for volcanic deposits in New Zealand (Lowe, 1986).

The secondary or clay minerals formed during weathering are characterized by small particle sizes, charged surfaces and high reactivity, and thus dramatically change the soils' physical and chemical behaviour. The higher clay contents on the older parent materials in the northeastern part lend the soils higher water-holding capacity and greater ability to retain nutrients against leaching as water percolates through the soil. On the other hand, the sandy soils encountered in the southern part, especially at lower elevations, cannot store much plant-available water and are prone to leaching losses of mobile nutrient ions, such as nitrate and potassium.

Soil Organic Matter and Nitrogen

Soil organic matter content is considered the most critical indicator of soil quality as it affects a variety of biological, chemical and physical characteristics influencing a soil's productive capacity (Havlin *et al.*, 1999). The importance of soil organic matter in the farmers' view is reflected by soil colour being a central criterion in local soil classifications (Talawar and Rhoades, 1998).

Some of the functions of soil organic matter are as follows:

- Provides plant nutrients (N, P, S and most micronutrients) that become available as it decomposes.
- Increases a soil's exchange capacity and ability to retain nutrients against leaching.
- Enhances the plant availability of micronutrients by forming soluble complexes.

- Improves a soil's water-holding capacity.
- Stabilizes soil structure, thus promoting infiltration and making the soil more resistant to erosion.

Organic matter in the Cotacachi soils

The distribution of topsoil organic matter contents in the area under study is shown in Fig. 12.4. In both the younger soils of the southern part and the older soils of the northeastern part, organic matter contents show dramatic increases with elevation on the volcano. This altitudinal accumulation is probably the result of slowed microbial decomposition due to lower temperatures and lower pH values at higher elevations. The presence of active amorphous constituents above 2700 metres above sea level (masl) may have further contributed to the accumulation by protecting organic matter against microbial decomposition. The stabilizing effects of active amorphous constituents on soil organic matter have been demonstrated by Parfitt *et al.* (1997), who reported organic matter decreases upon conversion from pasture to cropland that were considerably higher in an Inceptisol than in an Andisol, the latter containing active amorphous constituents.

The low organic matter contents of the sandy low-elevation soils in the southern part of the study area further contribute to their low water storage and nutrient retention capacity. On the other hand, the ability to store plant-available water and to retain nutrients against leaching is greatly enhanced by organic matter accumulation in high-elevation soils.

Organic matter is an important source of plant nutrients, with the vast majority of soil nitrogen stored in organic compounds. The organic-rich high-elevation soils of the

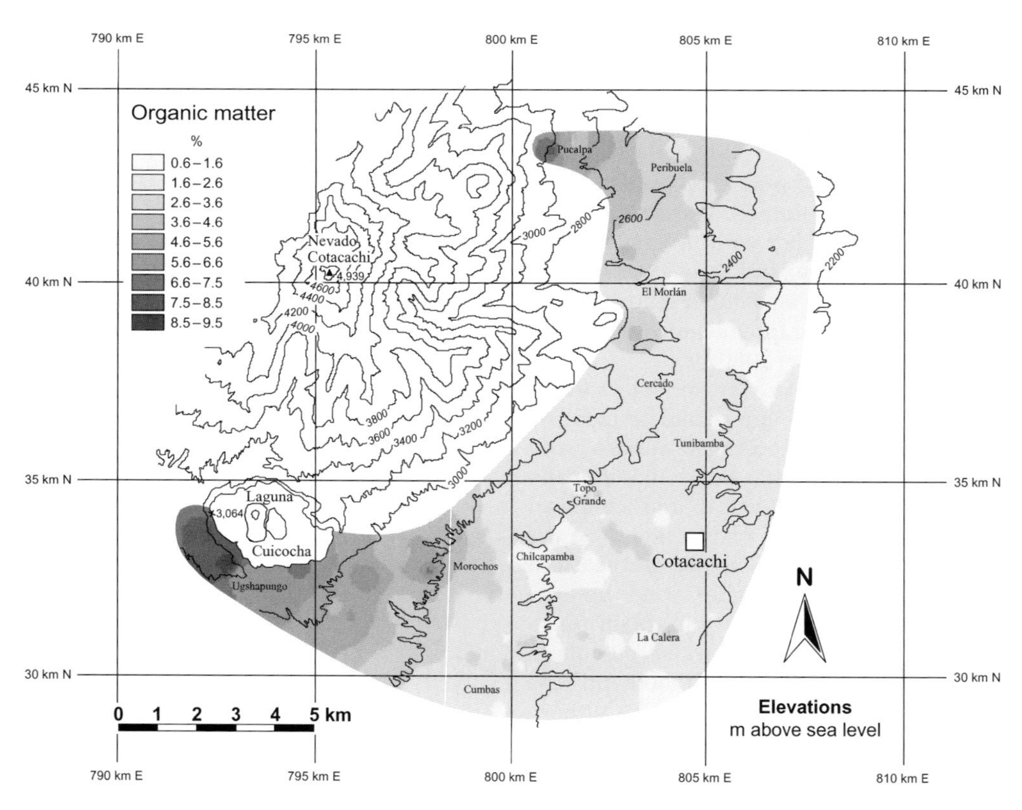

Fig. 12.4. Spatial distribution of topsoil organic matter contents.

study area therefore have a considerable pool of nitrogen. However, in order for it to become available for plant uptake, organic nitrogen needs to be converted to mineral forms (mineralized) in the course of organic matter decomposition. In high-elevation soils, microbial decomposition of organic matter and the rate of nitrogen mineralization are slowed.

Nitrogen cycling and crop growth modelling

We used the DSSAT model to simulate nitrogen cycling and analyse the long-term effects of residue management and nitrogen fertilization on maize yields in two agroecological zones. The N transformations simulated by the model are presented in Fig. 12.5. The two sites chosen for the modelling have markedly different soil properties. The Topo Grande site is located in the traditional maize zone at 2550 masl. The soils of the area are Entisols with low organic matter contents and devoid of active amorphous constituents. The Morochos site lies 200 m higher towards the upper limit of the maize zone. The soils are Andisols with high organic matter contents and active amorphous constituents.

An annual maize–fallow rotation was simulated for a duration of 30 years at both sites. The simulation conditions were as follows: the maize crop was planted at the onset of the rainy season in October and harvested in the summer dry season, as traditionally practised by the local farmers. Between harvest and the next planting, the land was fallow for several months. Four treatments were simulated: (i) the soil was never fertilized and all crop residues were removed after harvest; (ii) the soil was never fertilized but all crop residues were returned to the soil; (iii) the soil was fertilized with 25 kg/ha of N and all crop residues were returned to the soil; and (iv) the soil was fertilized with 50 kg/ha of N and all crop residues were returned to the soil.

Nitrogen fertilization was conducted with urea and chicken manure, which both resulted in the same grain yields and were therefore not presented separately. The default soil organic matter mineralization rate was used for the Entisol, and was multiplied by 0.2 for the Andisol, as suggested by Godwin and Singh (1998), to account for the slowed decomposition in this soil type.

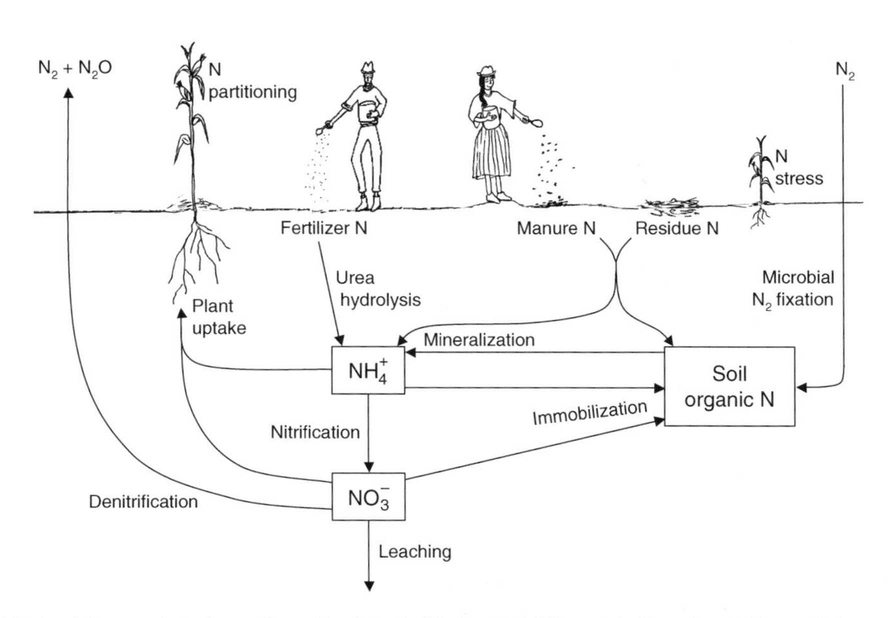

Fig. 12.5. Nitrogen transformations simulated with the DSSAT model. (Drawing: I-Sheng Zehetner.)

Nitrogen cycling was modelled while all other factors were assumed not limiting.

Model outputs

Predicted grain yields over the 30-year simulation period are shown in Figs 12.6 and 12.7 for the Topo Grande and Morochos sites, respectively. When the fields were never fertilized and all crop residues were removed after harvest, maize yields declined at both sites from about 2000 kg/ha to little over 1000 kg/ha after 30 years of cultivation. By returning crop residues to the soil, maize yields were generally maintained above 2000 kg/ha over the 30-year simulation period, and yield declines were less pronounced. Maize yields were initially higher in Topo Grande but showed a steeper decline over time relative to Morochos. At both sites, maximum yields of around 3000 kg/ha were obtained by fertilizing with 25 kg/ha of N in inorganic or organic forms and returning crop residues to the soil.

The Andisol of the Morochos site has a larger nitrogen pool, which was maintained at a high level over the 30-year simulation period. Its lower mineralization rate provided a slow release of mineral nitrogen, which resulted in less leaching losses and more efficient plant uptake and nutrient cycling. The Entisol in Topo Grande has a higher mineralization rate resulting in more rapid release of mineral nitrogen. However, most of the nitrogen mineralized in this sandy soil was lost from the nutrient cycle through leaching, which led to a steady depletion of the soil's N reserves over the 30-year simulation period. Residue management and manuring did not efficiently counteract this depletion.

Implications for management

Our data suggest that crop yields can be significantly improved by returning the residues to the soil after harvest. The peasant farmers in the area under study commonly remove crop residues and use them as stock feed, fuel or roofing material. Some of the exported nutrients may be returned in the form of animal manure, but most appear to be lost from the production systems. Solutions to this problem would greatly

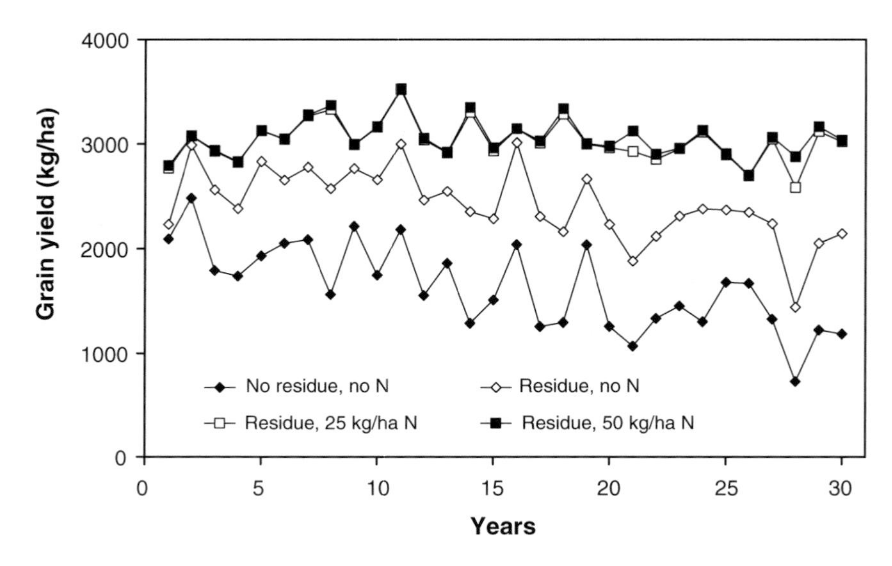

Fig. 12.6. Simulated grain yields in an annual maize–fallow rotation with different nitrogen inputs from residue incorporation and fertilization for Topo Grande (soil type: Vitrandic Udorthents; elevation: 2550 m); nitrogen cycling was simulated, other factors were assumed not limiting.

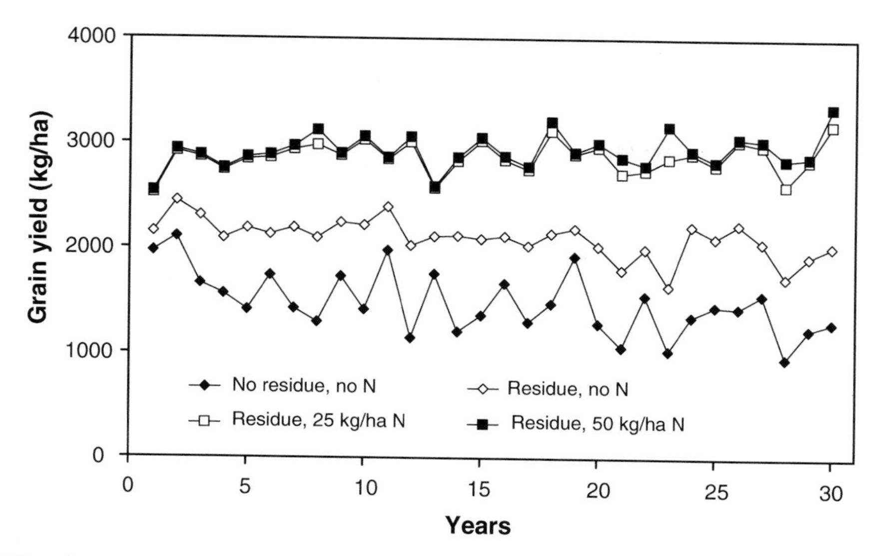

Fig. 12.7. Simulated grain yields in an annual maize–fallow rotation with different nitrogen inputs from residue incorporation and fertilization for Morochos (soil type: Humic Udivitrands; elevation: 2750 masl); nitrogen cycling was simulated, other factors were assumed not limiting.

contribute towards agricultural sustainability in the area.

However, recycling crop residues alone cannot sustain high productivity of the soils in the study area. Additional N inputs are needed to compensate for slowed N mineralization in high-elevation soils and leaching losses in low-elevation soils, to supply the N necessary for maximum crop yields and to sustain the soil N reserves. There are several avenues by which N may be added to the soils. The most obvious, in the eyes of a Western agronomist anyway, is the use of inorganic fertilizers. However, in Ecuador, these are mostly imported and therefore relatively expensive and beyond the economic means of resource-poor peasant farmers. Besides, the use of inorganic fertilizers may not be most obvious to them. Traditionally, animal-based farming and shifting cultivation involving long fallow cycles have been practised in the Andean ecoregion to restore soil fertility (Sarmiento *et al.*, 1993; Pestalozzi, 2000; Phiri *et al.*, 2001). However, in the Cotacachi communities, theft of livestock has led many peasant farmers to give up animal-based farming, and limited amounts of available land preclude long fallow cycles.

Leguminous plants, which have the ability to bind atmospheric N, have been widely used all over the world to restore soil nitrogen fertility. The traditional intercropping of maize and beans in Andean agriculture (which we could not simulate due to model limitations) may improve the soils' N status to some extent, and crop rotations including legumes, sometimes practised by the farmers in the study area, may increase the N supply for the following crop (Bossio and Cassman, 1991; Nieto-Cabrera *et al.*, 1997). A particularly promising practice for adding N and organic matter to the soil is the management of short-term fallow systems with planted herbaceous or woody legumes, so-called improved fallows (Phiri *et al.*, 2001).

Another source of N is organic fertilizers, such as chicken manure or worm compost. These have the advantage of adding other plant nutrients and organic matter along with N, and to buffer acidity that arises from plant uptake and nitrification of ammonium. The addition of organic matter would improve the soils' nutrient retention and water-holding capacity, which is critically important for the sandy low-elevation soils in the southern part of the study area. Because of its slower decomposition, composted material is more suitable to serve this purpose.

Nitrogen loss by leaching is a common problem in well-drained soils of humid climates and a major contributor to soil fertility decline in many soils. Våje *et al.* (2000) found considerable leaching losses of N in a Tanzanian volcanic ash soil, and de Koning *et al.* (1997) estimated that leaching losses significantly contribute to N output from Andean cropland areas. If inorganic fertilizers are used, band placement near the plant roots and partitioning into several smaller applications may prove successful in enhancing their use efficiency and prevent excessive leaching losses. If organic amendments are used, composted material is preferable over fresh manure since it decomposes much more slowly, thus releasing nitrogen at a slow and steady rate, which makes it less susceptible to leaching.

Amorphous Materials and Phosphorus

Active amorphous constituents, such as allophane and aluminium–humus complexes, are typical weathering products in volcanic ash soils; however, they require humid conditions for their formation (e.g. Parfitt *et al.*, 1983; Parfitt and Wilson, 1985). The dominant presence of these constituents in the clay fraction of volcanic ash soils is indicated by pH (NaF) values > 9.4 (Wada, 1980). The distribution of topsoil pH (NaF) values in the area under study is shown in Fig. 12.8. Irrespective of the soil parent material, the presence of active amorphous constituents shows a clear altitudinal pattern. Their formation is noticeably favoured by the high rainfall–low evapotranspiration environment at higher elevations, and they are virtually absent below 2700 masl.

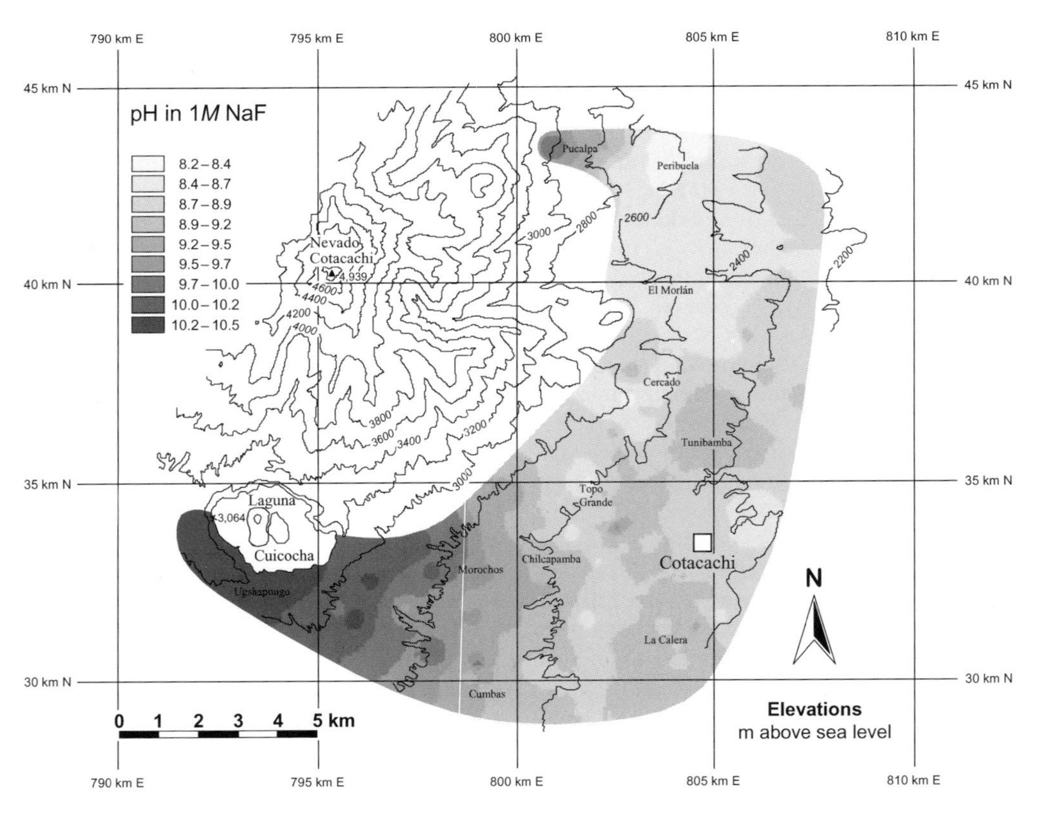

Fig. 12.8. Spatial distribution of topsoil pH (NaF); values > 9.4 indicate the dominant presence of active amorphous constituents in the clay fraction (Wada, 1980).

Phosphate sorption

Active amorphous constituents are known for their ability to bind phosphate ions strongly (e.g. Parfitt, 1989; Wada, 1989; Shoji *et al.*, 1993), making phosphorus sparingly available for plants and microorganisms. In soils with active amorphous constituents (Andisols), P is often the growth-limiting nutrient for agricultural crops (Shoji *et al.*, 1993). Apart from being directly yield limiting, low P levels may further limit biological N fixation (Vitousek, 1999), thus impairing the soil N status.

To characterize phosphate sorption in the soils of the study area, we analysed how much fertilizer P was still plant available 30 and 180 days after application (Figs 12.9 and 12.10, respectively) in three different soils (labelled A, B and C in Fig. 12.1) along a transect from high to low pH (NaF). The slopes of the straight lines in Figs 12.9 and 12.10 denote the fractions of fertilizer P that are plant available after the respective amount of time. Soil A, an Andisol located at 2900 masl, shows the highest phosphate sorption with 31% of fertilizer P available after 30 days, decreasing to 19% after 180 days. Soil B is an Inceptisol located near the pH (NaF) boundary of 9.4 that separates soils with and without dominant presence of active amorphous constituents. It shows intermediate phosphate sorption with 47% of fertilizer P available after 30 days, which only slightly decreases to 42% after 180 days. Soil C, an Entisol in the zone devoid of active amorphous constituents, shows the least phosphate sorption. Around three-quarters of fertilizer P are plant available 30 and 180 days after application.

Phosphate sorption in the area under study is largely controlled by active amorphous constituents, which predominate at higher elevations. The low soil pH (H_2O) values at these elevations further increase the positive charge of colloid surfaces, thus enhancing phosphate sorption in high-elevation soils. This may lead to phosphorus deficiencies in the high zones of the study area that have traditionally been cultivated with potato and other Andean tubers.

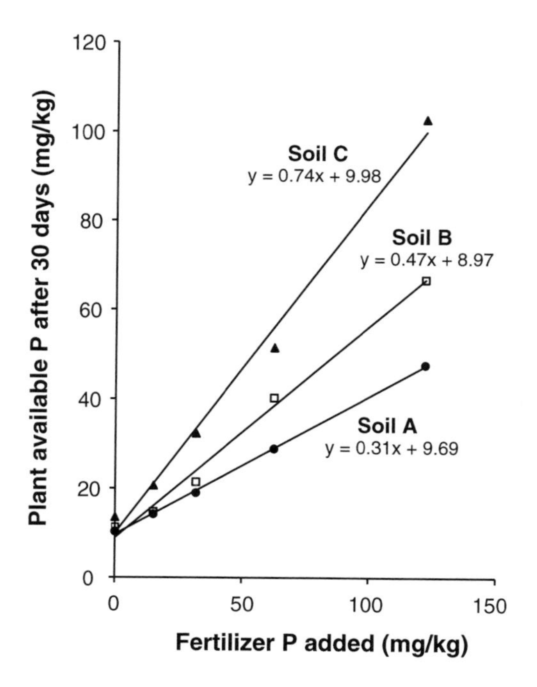

Fig. 12.9. Fertilizer phosphorus availability 30 days after application in three different soils.

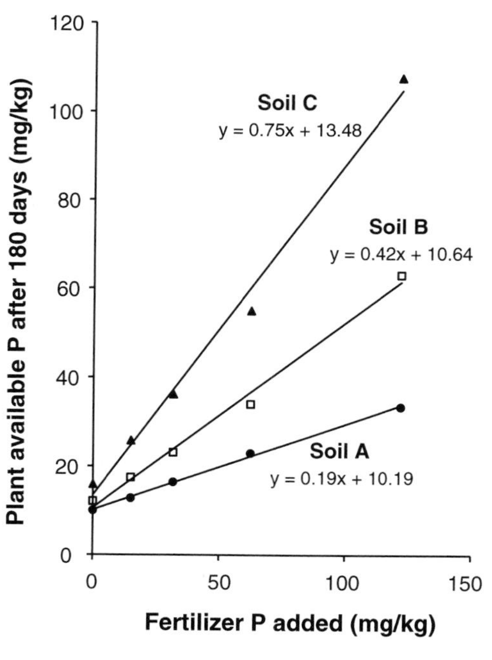

Fig. 12.10. Fertilizer phosphorus availability 180 days after application in three different soils.

Status of phosphorus in the Cotacachi soils

The distribution of bicarbonate-extractable phosphorus (Olsen-P; Olsen *et al.*, 1954) in the area under study is shown in Fig. 12.11. A soil's Olsen-P content is a measure of its phosphorus-supplying capacity for plant growth. According to the Ecuadorian Instituto Nacional Autónomo de Investigaciones Agropecuarias (INIAP), soils in the Andean region that have Olsen-P values < 10 mg/kg are considered low in phosphorus and require high amounts of added P for maximum crop yields. Between 10 and 20 mg/kg of Olsen-P is regarded as intermediate, requiring less additions of fertilizer P, and soils with Olsen-P > 20 mg/kg supply sufficient amounts of phosphorus for optimum crop growth.

As expected, high-elevation soils generally show the lowest values of bicarbonate-extractable phosphorus, which is probably associated with the strong sorption of phosphate by active amorphous constituents, as discussed above. However, Olsen-P appears to be higher in the community of Ugshapungo above 3000 masl than in the zone between 2800 and 3000 masl (Fig. 12.11). This may be a residual effect of high fertilizer applications in past decades, when potato was grown with high inputs in this area. Continuous heavy application of phosphorus fertilizers leads to the occupation of P sorption sites, thus enhancing recovery of fertilizer P (Havlin *et al.*, 1999) and increasing soil phosphorus to levels that may exceed plant requirements (Shoji *et al.*, 1993).

Apart from altitudinal variations, if we only look at low-elevation areas, bicarbonate-extractable phosphorus appears to be higher in the younger soils of the southern part compared with the older soils in the

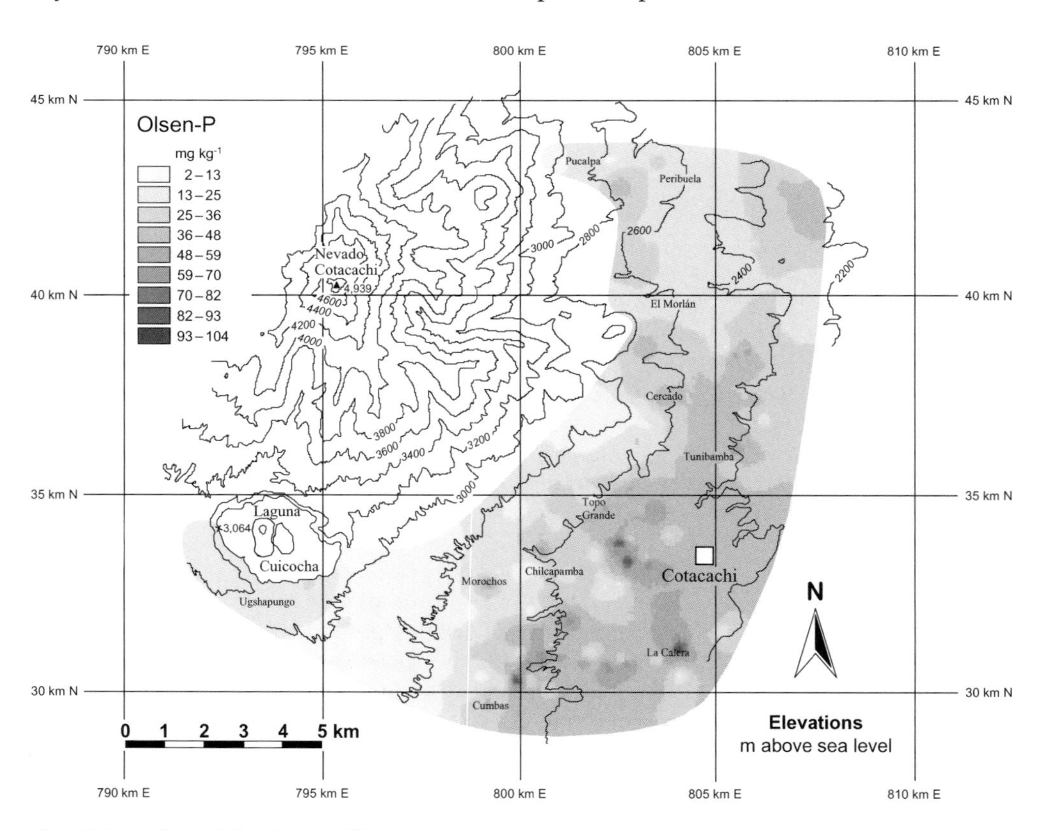

Fig. 12.11. Spatial distribution of bicarbonate-extractable phosphorus (Olsen-P; Olsen *et al.*, 1954) in topsoils.

northeastern part of the study area. This could be caused by differences in parent material P contents and time of soil development. Shoji *et al.* (1993) suggested that plant availability of phosphorus decreases rapidly as weathering proceeds in volcanic ash soils due to the relatively rapid dissolution of apatite, the primary phosphorus-bearing mineral in volcanic deposits, and conversion of P into more insoluble forms.

Implications for management

Phosphorus does not appear as yield limiting in the area under study as commonly assumed and often reported for volcanic ash soils. The young low-elevation soils in the southern part of the study area generally supply sufficient amounts of P for optimum crop growth. Phosphorus additions seem necessary to ensure maximum yields at higher elevations, especially when – as traditionally – potato is grown in the high zone, and may prove beneficial in some low-elevation zones of the northeastern study area.

Phosphorus is fairly immobile and is not easily lost from the root zone by leaching; however, as shown in Figs 12.9 and 12.10, added fertilizer P may be rendered unavailable through strong sorption by active amorphous constituents. For this reason, recovery of fertilizer P by agricultural crops is commonly $< 20\%$ in Andisols (Shoji *et al.*, 1993). Band applications near the plant roots and the use of P sources with low solubility may delay the reaction of phosphate with active amorphous constituents, thus resulting in enhanced availability for plant uptake.

The use of organic P sources, such as chicken manure or organic wastes, may prove particularly beneficial in improving the soils' P status. Mazzarino *et al.* (1997) reported higher P utilization efficiency and plant uptake, higher residual Olsen-P and lower P sorption in the soil if organic amendments were used instead of inorganic fertilizer. Apart from supplying phosphorus that is mineralized upon decomposition, organic amendments can increase P availability by a second mechanism. Organic anions formed during the decomposition of organic inputs can compete with phosphate ions for sorption sites on active amorphous constituents and thereby increase plant availability of phosphorus (Iyamuremye and Dick, 1996; Iyamuremye *et al.*, 1996).

Cation Exchange Capacity and Potassium

An important soil fertility characteristic is the storage capacity of exchangeable nutrient cations (essentially Ca^{2+}, Mg^{2+} and K^{+}) on the charged surfaces of soil particles. As plant roots take nutrients from the soil solution, they are resupplied from these surfaces by exchange reactions. In the soils of the study area, most of the inherent charge arises from organic matter and active amorphous constituents. In both cases, the charge is variable, which means it is highly dependent on the ambient conditions, primarily soil pH and electrolyte concentration. As soil pH increases, the variable charge surfaces become increasingly negative, thus favouring the storage of nutrient cations.

Cation exchange capacity in the Cotacachi soils

The distribution of topsoil CEC (measured at pH 7) in the area under study is shown in Fig. 12.12. However, the conventional method of determination at pH 7 can drastically overestimate the CEC of variable charge soils at natural pH and electrolyte levels (Parfitt, 1980). The values in Fig. 12.12 may therefore be viewed as estimates of a potential CEC that illustrate the spatial distribution of charge-bearing soil constituents in the area. These may or may not be effective in retaining nutrient or pollutant cations under field conditions depending on the natural soil pH values. At pH 7, the highest amounts of negative charge are found in high-elevation soils and probably originate from organic matter

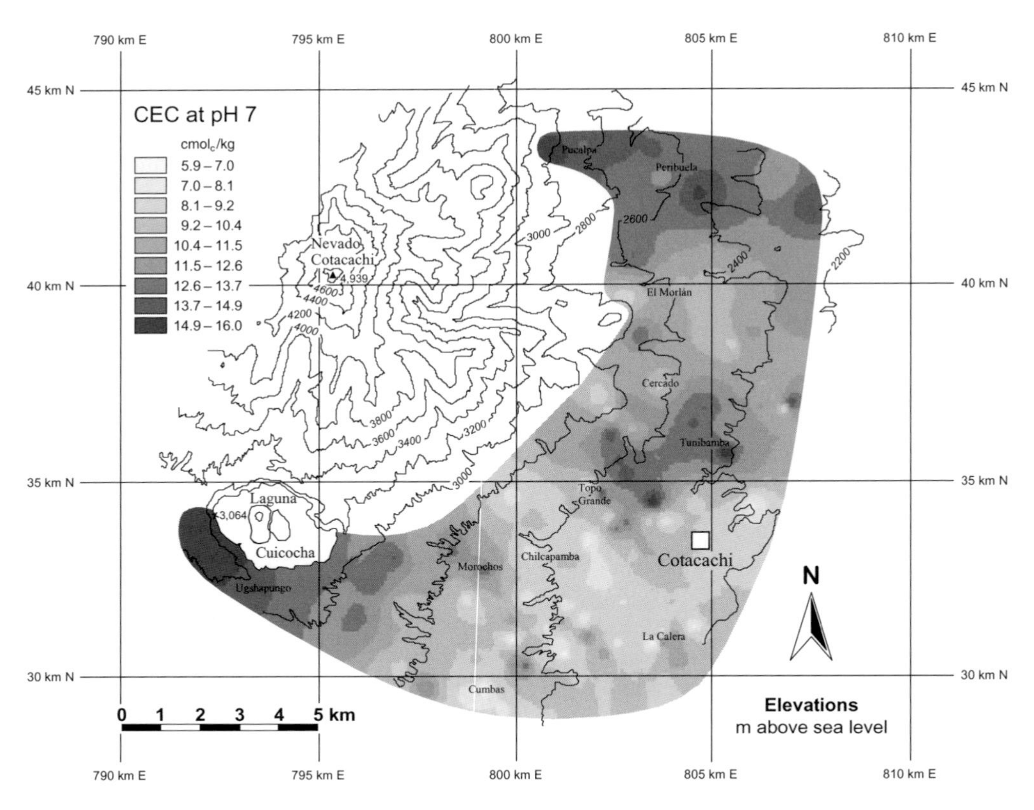

Fig. 12.12. Spatial distribution of topsoil cation exchange capacity (CEC) at pH 7.

and active amorphous constituents. The older soils in the northeastern part of the study area generally exhibit higher negative charge than the younger soils in the southern part, which may arise from higher clay contents in the older, more developed soils.

Since exchangeable aluminium is very low in the soils of the study area, probably because of strong bonding with organic matter and the formation of aluminium–humus complexes, the sum of exchangeable bases (Ca^{2+}, Mg^{2+}, K^+ and Na^+) may serve as a good estimate of effective CEC. The spatial distribution of the sum of exchangeable bases, presented in Fig. 12.13, shows a very different pattern, with absolute values considerably lower than CEC at pH 7. The majority of the negative charge observed in the high-elevation soils of the southern study area at pH 7 is absent at the low natural soil pH values, resulting in low amounts of exchangeable bases (Fig. 12.13). Effective CEC slightly increases with pH in the lower parts of the southern study area; however, due to the sandy textures, absolute values are still very low. The finer textured soils in the northeastern part of the study area show considerably higher effective CEC, which results in better ability to retain nutrients against leaching, to filter pollutants and to buffer acidity.

Exchangeable potassium

The amount of exchangeable potassium held on the charged surfaces of soil particles is a reliable indicator of K availability to plants (Shoji *et al.*, 1993). The distribution of exchangeable K in topsoils around the study area is shown in Fig. 12.14. The young, sandy soils in the southern study area have considerably lower exchangeable K than the older, finer-textured soils of the northeastern part. Exchangeable K is particularly high in the zone just north of the

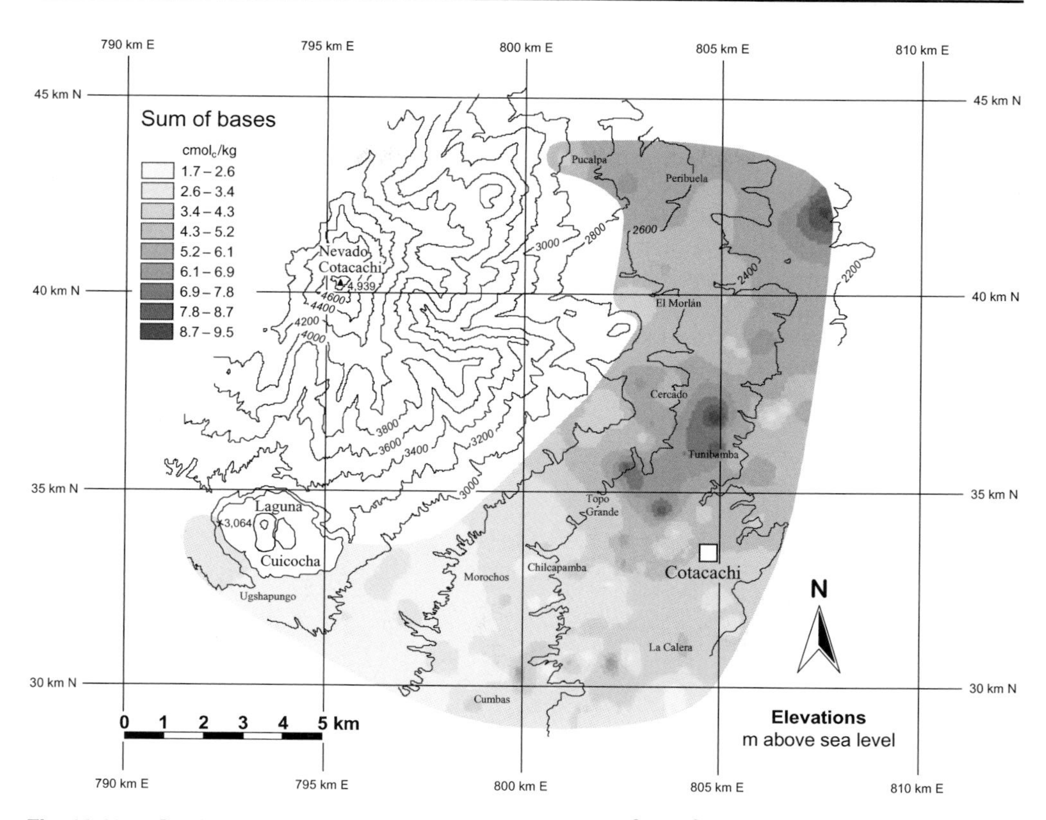

Fig. 12.13. Spatial distribution of the sum of exchangeable Ca^{2+}, Mg^{2+}, K^+ and Na^+ in topsoils.

town of Cotacachi (Fig. 12.14). Differences in age and composition of parent materials are likely to be responsible for these variations within the area under study. With concentrations of about 1.3% as K_2O, the Cuicocha deposits of the southern study area are on the lower end of K contents reported for volcanic parent materials (Shoji *et al.*, 1993). Moreover, there has not yet been enough time for weathering in these 3000-year-old deposits to form significant amounts of clay and exchange surfaces for the retention and storage of potassium ions. As with nitrogen, leaching losses of potassium may be significant in these sandy soils. On the other hand, de Koning *et al.* (1997) attributed net K losses from Andean cropland areas mainly to removal with harvested products.

Exchangeable potassium contents below 0.3 $cmol_c$/kg may be considered as low, resulting in potassium deficiencies in common upland crops (Shoji *et al.*, 1993). However, in field trials conducted during the 2000–2001 rainy season in the southern part of the study area, maize yields showed marked responses to additions of fertilizer K on soils that had 0.6 $cmol_c$/kg of exchangeable potassium.

Exchangeable calcium and magnesium

The spatial distributions (maps not shown) of exchangeable Ca and Mg exhibit similar patterns to those observed for exchangeable K, with higher values in the northeastern compared with the southern part of the study area. The soils generally have adequate ratios of Ca^{2+}, Mg^{2+} and K^+ on the exchange surfaces and low exchangeable acidity, which prevents nutrient imbalances and ensures sufficient supply of Ca and Mg to agricultural crops.

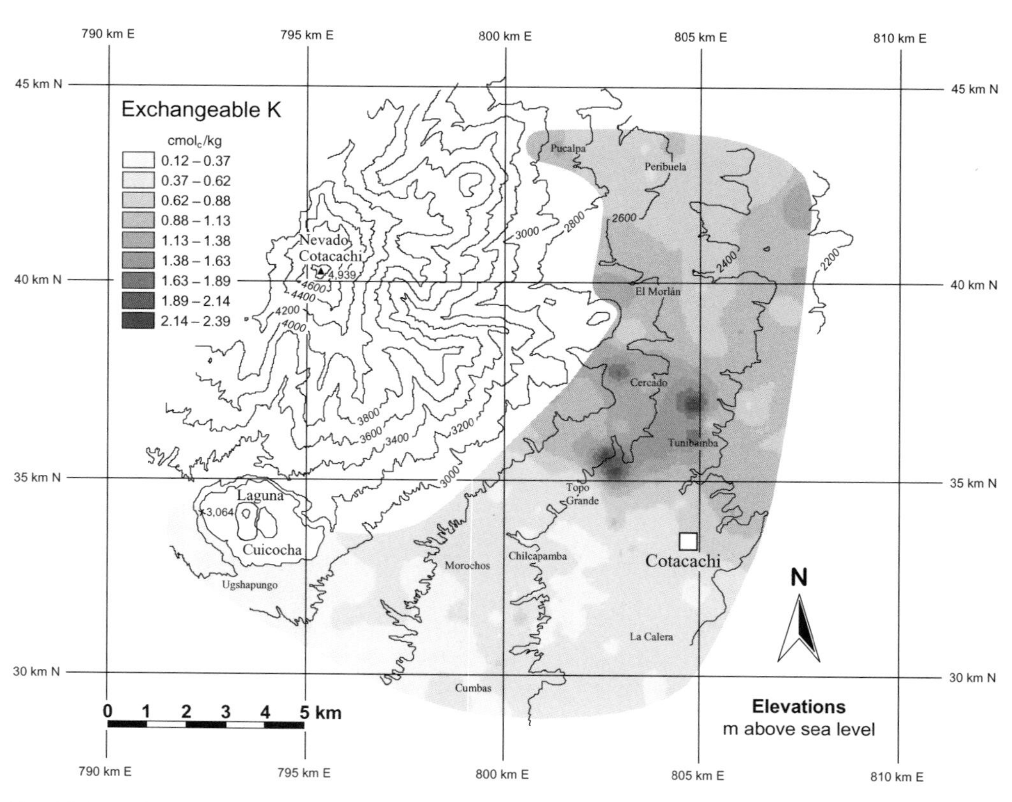

Fig. 12.14. Spatial distribution of exchangeable potassium in topsoils.

Implications for management

The distribution of exchangeable potassium suggests that K deficiencies may limit crop growth in the southern part of the study area. Such a conclusion is supported by our field experiments that showed clear yield increases of maize upon K fertilization in that area. Additions of potassium seem necessary therefore for adequate crop growth in the young soils of the southern study area. If potassium is added through inorganic fertilizers, band placement near the plant roots and partitioning into several smaller applications is recommended to enhance recovery and avert excessive leaching losses. Organic fertilizers may prove more beneficial in the long term for reasons discussed in the context of N and P fertilization. It should be mentioned here that the low CEC values of the low-elevation soils in the southern study area could be increased by management practices that foster the build-up of soil organic matter, such as the continued application of organic amendments or the incorporation of improved fallows in crop rotations. The use of deep-rooting woody legumes in such fallows could mobilize potassium and other nutrients from the subsoil and recycle them back into the topsoil, available for the following crop.

Soil Acidity and pH

Soil acidity and associated aluminium toxicity may limit root development and thus impair normal growth of agricultural crops. The distribution of active acidity (pH in H_2O) in topsoils around the study area is shown in Fig. 12.15. In both the younger soils of the southern part and the older soils of the northeastern part, pH (H_2O) values

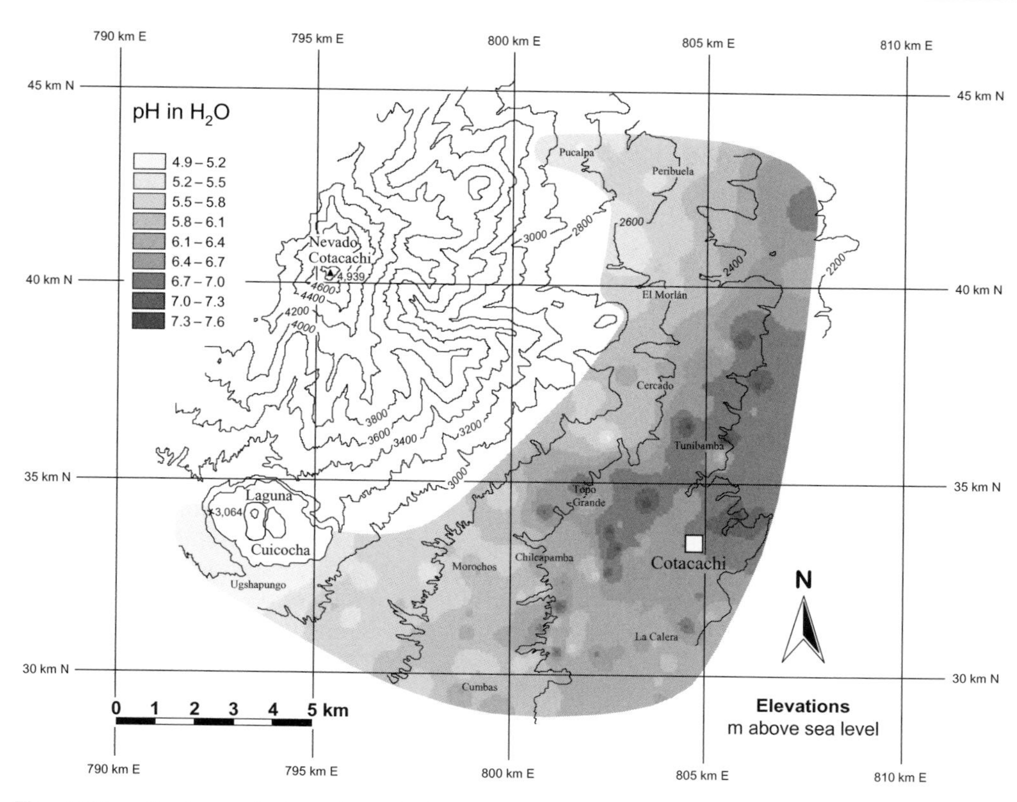

Fig. 12.15. Spatial distribution of topsoil pH (H_2O).

decrease substantially with increasing elevation on the volcano. This altitudinal pattern is likely to be the result of greater acidity inputs with increased precipitation, greater leaching of basic cations and greater activity of organic acids as a result of slowed organic matter decomposition at higher elevations.

Most agricultural crops grow normally above pH 5.5 (Havlin *et al.*, 1999) and become increasingly sensitive to elevated levels of aluminium as the pH drops below this value. However, as mentioned in the context of CEC, exchangeable Al is very low in the soils of the study area, even at pH 5 – presumably a result of strong bonding with organic matter and formation of Al–humus complexes. It is therefore unlikely that soil solution Al reaches levels toxic to plant roots even in high-elevation soils.

The variable charge characteristics of the soils under study make pH an important determinant of the soils' nutrient retention properties. Management practices that affect soil pH are therefore likely to have a substantial effect on the fate of plant nutrients in the soil (Sollins *et al.*, 1988). The application of liming materials at higher elevations would increase CEC and thus enhance the soils' storage capacity for potassium. It could further reduce phosphate sorption and make P more available for plant uptake. Moreover, most microbially mediated reactions are restricted under acidic conditions. Liming has been shown to enhance mineralization of organic N and to increase N fixation by leguminous plants (Havlin *et al.*, 1999).

Thus, the application of liming materials would have several beneficial effects in the high-elevation soils of the study area. However, for the cultivation of potato, the soil pH values between 5 and 5.5 in the high zone are desirable since *Streptomyces scabies*, the disease organism that causes scab in potatoes, is much better adapted to a neutral than to an acid soil (Russell, 1988).

Conclusion

The volcanic ash soils in the Andean communities around the Cotacachi volcano have the potential to support traditional subsistence agriculture as well as thriving market-oriented production of niche crops, vegetables and fruits. In either case, the key to agricultural sustainability is the restoration and maintenance of soil fertility. Nutrient losses need to be minimized, and nutrients that leave the production systems through harvested products need to be replaced. The soils show marked variations within the study area and, depending on the specific location, different factors may become limiting to crop growth and therefore deserve special attention.

The young soils on the Cuicocha deposits of the southern study area have inherently low potassium contents and would greatly benefit from additions of this nutrient. In high-elevation soils, strong sorption of phosphorus on active amorphous constituents may become yield limiting and necessitate increased P inputs for optimum crop growth. Nitrogen may be mineralized too slowly to satisfy crop requirements in high-elevation soils, whereas leaching losses may impair adequate N supply in low-elevation soils. The nitrogen status of all soils would greatly benefit from returning residues to the soil after harvest, including leguminous plants in crop rotations, and managing short, improved fallows.

Generally, organic nutrient sources seem preferable over inorganic for a number of reasons. First, they simultaneously add several nutrients including micronutrients to the soils and thus prevent potential nutrient imbalances. Second, they enhance the plant availability of phosphorus and micronutrients by the formation of organic anions that compete with phosphate for sorption sites and form soluble complexes with micronutrient cations. Third, they increase the soils' organic matter contents and so have beneficial effects on nutrient storage, water retention, infiltration capacity, etc. Fourth, they can be produced in and by the communities themselves, are therefore more readily and cheaply available, and create less dependency on uncontrollable external factors. Fifth, they may be more compatible with local traditions and hence be more widely accepted.

References

Athens, J.S. (1999) Volcanism and archaeology in the northern highlands of Ecuador. In: Mothes, P. (ed.) *Actividad Volcánica y Pueblos Precolombinos en el Ecuador.* Ediciones Abya-Yala, Quito, Ecuador, pp. 157–189.

Bossio, D.A. and Cassman, K.G. (1991) Traditional rainfed barley production in the Andean highlands of Ecuador: soil nutrient limitations and other constraints. *Mountain Research and Development* 11, 115–126.

Bouyoucos, G.J. (1962) Hydrometer method improved for making particle size analyses of soils. *Agronomy Journal* 54, 464–465.

de Koning, G.H.J., van de Kop, P.J. and Fresco, L.O. (1997) Estimates of sub-national nutrient balances as sustainability indicators for agro-ecosystems in Ecuador. *Agriculture, Ecosystems and Environment* 65, 127–139.

Field, L. (1991) *Sistemas Agrícolas Campesinos en la Sierra Norte.* Centro Andino de Acción Popular, Albazul Offset, Quito, Ecuador.

Gijsman, A.J. and Sanz, J.I. (1998) Soil organic matter pools in a volcanic-ash soil under fallow or cultivation with applied chicken manure. *European Journal of Soil Science* 49, 427–436.

Godwin, D.C. and Singh, U. (1998) Nitrogen balance and crop response to nitrogen in upland and lowland cropping systems. In: Tsuji, G.Y., Hoogenboom, G. and Thornton, P.K. (eds) *Understanding Options for Agricultural Production.* Kluwer Academic Publishers, Dordrecht, The Netherlands, pp. 55–77.

Havlin, J.L., Beaton, J.D., Tisdale, S.L. and Nelson, W.L. (1999) *Soil Fertility and Fertilizers. An Introduction to Nutrient Management*, 6th edn. Prentice Hall, Upper Saddle River, New Jersey.
Iyamuremye, F. and Dick, R.P. (1996) Organic amendments and phosphorus sorption by soils. *Advances in Agronomy* 56, 139–185.
Iyamuremye, F., Dick, R.P. and Baham, J. (1996) Organic amendments and phosphorus dynamics, III: phosphorus speciation. *Soil Science* 161, 444–451.
Jones, J.W., Tsuji, G.Y., Hoogenboom, G., Hunt, L.A., Thornton, P.K., Wilkens, P.W., Imamura, D.T., Bowen, W.T. and Singh, U. (1998) Decision support system for agrotechnology transfer: DSSAT v3. In: Tsuji, G.Y., Hoogenboom, G. and Thornton, P.K. (eds) *Understanding Options for Agricultural Production*. Kluwer Academic Publishers, Dordrecht, The Netherlands, pp. 157–177.
Lauer, W. (1993) Human development and environment in the Andes: a geoecological overview. *Mountain Research and Development* 13, 157–166.
Lowe, D.J. (1986) Controls on the rates of weathering and clay mineral genesis in airfall tephras: a review and New Zealand case study. In: Colman, S.M. and Dethier, D.P. (eds) *Rates of Chemical Weathering of Rocks and Minerals*. Academic Press, Orlando, Florida, pp. 265–330.
Mazzarino, M.J., Walter, I., Costa, G., Laos, F., Roselli, L. and Satti, P. (1997) Plant response to fish farming wastes in volcanic soils. *Journal of Environmental Quality* 26, 522–528.
Nieto-Cabrera, C., Francis, C., Caicedo, C., Gutiérrez, P.F. and Rivera, M. (1997) Response of four Andean crops to rotation and fertilization. *Mountain Research and Development* 17, 273–282.
Olsen, S.R., Cole, C.V., Watanabe, F.S. and Dean, L.A. (1954) *Estimation of Available Phosphorus in Soils by Extraction with Sodium Bicarbonate*. USDA Circular No. 939. USDA, Washington, DC.
Parfitt, R.L. (1980) Chemical properties of variable charge soils. In: Theng, B.K.G. (ed.) *Soils with Variable Charge*. New Zealand Society of Soil Science, Offset Publications, Palmerston North, New Zealand, pp. 167–194.
Parfitt, R.L. (1989) Phosphate reactions with natural allophane, ferrihydrite and goethite. *Journal of Soil Science* 40, 359–369.
Parfitt, R.L. and Wilson, A.D. (1985) Estimation of allophane and halloysite in three sequences of volcanic soils, New Zealand. In: Fernandez Caldas, E. and Yaalon, D.H. (eds) *Volcanic Soils: Weathering and Landscape Relationships of Soils on Tephra and Basalt*. Catena Supplement 7. Catena Verlag, Cremlingen, Germany, pp. 1–8.
Parfitt, R.L., Russell, M. and Orbell, G.E. (1983) Weathering sequence of soils from volcanic ash involving allophane and halloysite, New Zealand. *Geoderma* 29, 41–57.
Parfitt, R.L., Theng, B.K.G., Whitton, J.S. and Shepherd, T.G. (1997) Effects of clay minerals and land use on organic matter pools. *Geoderma* 75, 1–12.
Parfitt, R.L., Salt, G.J. and Saggar, S. (2001) Effect of leaching and clay content on carbon and nitrogen mineralisation in maize and pasture soils. *Australian Journal of Soil Research* 39, 535–542.
Pestalozzi, H. (2000) Sectoral fallow systems and the management of soil fertility: the rationality of indigenous knowledge in the High Andes of Bolivia. *Mountain Research and Development* 20, 64–71.
Phiri, S., Barrios, E., Rao, I.M. and Singh, B.R. (2001) Changes in soil organic matter and phosphorus fractions under planted fallows and a crop rotation system on a Colombian volcanic-ash soil. *Plant and Soil* 231, 211–223.
Richardson, C.W. and Wright, D.A. (1984) *WGEN: a Model for Generating Daily Weather Variables*. USDA-ARS, ARS-8, Washington, DC.
Russell, E.W. (1988) *Russell's Soil Conditions and Plant Growth*, 11th edn. Wild, A. (ed.). Longman Scientific & Technical, Essex, UK.
Sarmiento, L., Monasterio, M. and Montilla, M. (1993) Ecological bases, sustainability, and current trends in traditional agriculture in the Venezuelan High Andes. *Mountain Research and Development* 13, 167–176.
Schad, P. (1998) Humusvorräte und nutzbare Wasserspeicherkapazitäten der Böden als differenzierende Faktoren in der traditionellen hochandinen Landwirtschaft der Charazani-Region (Bolivien). *Forstwissenschaftliches Centralblatt* 117, 176–188.
Sharpley, A.N., Jones, C.A., Gray, C. and Cole, C.V. (1984) A simplified soil and plant phosphorus model: II. Prediction of labile, organic, and sorbed phosphorus. *Soil Science Society of America Journal* 48, 805–809.
Sharpley, A.N., Singh, U., Uehara, G. and Kimble, J. (1989) Modeling soil and plant phosphorus dynamics in calcareous and highly weathered soils. *Soil Science Society of America Journal* 53, 153–158.

Shoji, S., Kobayashi, S., Yamada, I. and Masui, J. (1975) Chemical and mineralogical studies on volcanic ashes. I. Chemical composition of volcanic ashes and their classification. *Soil Science and Plant Nutrition* 21, 311–318.

Shoji, S., Nanzyo, M. and Dahlgren, R.A. (1993) *Volcanic Ash Soils: Genesis, Properties and Utilization.* Developments in Soil Science No. 21. Elsevier, Amsterdam, The Netherlands.

Soil Survey Staff (1996) *Soil Survey Laboratory Methods Manual.* Soil Survey Investigations Report No. 42. USDA-NRCS, Washington, DC.

Sollins, P., Robertson, G.P. and Uehara, G. (1988) Nutrient mobility in variable- and permanent-charge soils. *Biogeochemistry* 6, 181–199.

Talawar, S. and Rhoades, R.E. (1998) Scientific and local classification and management of soils. *Agriculture and Human Values* 15, 3–14.

UNORCAC (1999) *El Autodiagnóstico de la Unión de Organizaciones Campesinas e Indígenas de Cotacachi (UNORCAC).* Cotacachi, Ecuador.

Våje, P.I., Singh, B.R. and Lal, R. (2000) Leaching and plant uptake of nitrogen from a volcanic ash soil in Kilimanjaro region, Tanzania. *Journal of Sustainable Agriculture* 16, 95–112.

Vitousek, P.M. (1999) Nutrient limitation to nitrogen fixation in young volcanic sites. *Ecosystems* 2, 505–510.

Wada, K. (1980) Mineralogical characteristics of Andisols. In: Theng, B.K.G. (ed.) *Soils with Variable Charge.* New Zealand Society of Soil Science, Offset Publications, Palmerston North, New Zealand, pp. 87–107.

Wada, K. (1989) Allophane and imogolite. In: Dixon, J.B. and Weed, S.B. (eds) *Minerals in Soil Environments*, 2nd edn. SSSA Book Series No. 1. SSSA, Madison, Wisconsin, pp. 1051–1087.

13 Plant–Water Relationships in an Andean Landscape: Modelling the Effect of Irrigation on Upland Crop Production

Franz Zehetner,[1] William P. Miller[1] and Xavier Zapata Ríos[2]
[1]*University of Georgia, Department of Crop and Soil Sciences, 3107 Plant Science, Athens, GA 30602-7272, USA*; [2]*SANREM–Andes Project, PO Box 17-12-85, Quito, Ecuador*

Introduction

Plants use large amounts of water during growth, and daily water use of an actively growing crop may be several times its own mass (Wild, 1988). Inadequate water availability especially during the development and fertilization of the reproductive organs can drastically limit crop production. In the inter-Andean valleys of northern Ecuador, irrigation systems have long been used to minimize drought risk and secure the production of food crops during dry periods. However, not all Andean communities have access to irrigation water. Historically, increasing population pressure has forced many mountain farmers to move higher up the volcanic slopes and cultivate marginal land under rainfed conditions.

In the Cotacachi area, only some of the communities at lower elevations have access to irrigation water, but local community members and officials of the local water authorities (*juntas de agua*) have been seeking to expand existing irrigation systems to a wider area. In this chapter, we analyse the potential benefits to crop production that such expansion would bring about in different zones of the area that currently do not have access to irrigation water. We use simulation modelling to quantify the improvement of wet-season and dry-season maize production under the assumption that irrigation water is available.

Methodology

The simulation model

The Decision Support System for Agrotechnology Transfer (DSSAT, version 3.5) was used to simulate rainfed and irrigated maize (*Zea mays* L.) production in the area under study. DSSAT is an integrated modelling platform that comprises several crop simulation models and databases (Jones *et al.*, 1998). It operates on a field scale, with a daily time step, and is capable of long-term simulations. The maize simulation model within DSSAT models phenological and morphological development, biomass accumulation and partitioning, soil water balance, and soil nitrogen transformations (Jones *et al.*, 1998). Input information about fields, soils, cultivars, weather and

 Development with Identity: Community, Culture and Sustainability in the Andes (ed. R.E. Rhoades)

management is required for the maize simulation model.

Water balance and crop growth modelling

Two sites in different agroecological zones of the study area were chosen for the simulation modelling, both presently without access to irrigation water. The Topo Grande site is located at 2550 metres above sea level (masl), has a mean annual precipitation of about 900 mm and soils with low water-holding capacity. The Morochos site lies at 2750 masl, receives about 1100 mm of mean annual rainfall, and has soils with higher water-holding capacity.

The fields and soils around both sites were described, and soil samples were analysed in the laboratory using standard methods. Calibrated genetic coefficients for the locally grown maize variety *Chaucho Mejorado* (INIAP-122) were obtained from Bowen (2000, personal communication), and validated in field trials during the 2000–2001 rainy season. In order to analyse long-term trends, 30 years of weather data were randomly generated with DSSAT's weather generator WGEN (Richardson and Wright, 1984) based on an existing 30-year weather data set from the nearby town of Otavalo and known altitudinal variations of climatic parameters.

For both the Topo Grande and Morochos sites, an annual maize–fallow rotation was simulated with and without irrigation for a duration of 30 years. In a first model run, the maize crop was grown during the rainy season as traditionally practised by the local farmers. Maize was planted at the onset of the rainy season in October and harvested in the summer dry season. In a second model run, maize was grown over the summer dry season, as sometimes practised when irrigation is available. Maize was planted towards the end of the rainy season in April and harvested in the short winter dry period. Between harvest and the next planting, the land was fallow for several months. The water balance (Fig. 13.1) was modelled while other factors were assumed not limiting.

Model outputs

Predicted grain yields for maize grown during the rainy season are shown in Figs 13.2 and 13.3 for the Morochos and Topo Grande sites, respectively. At neither site did rainy

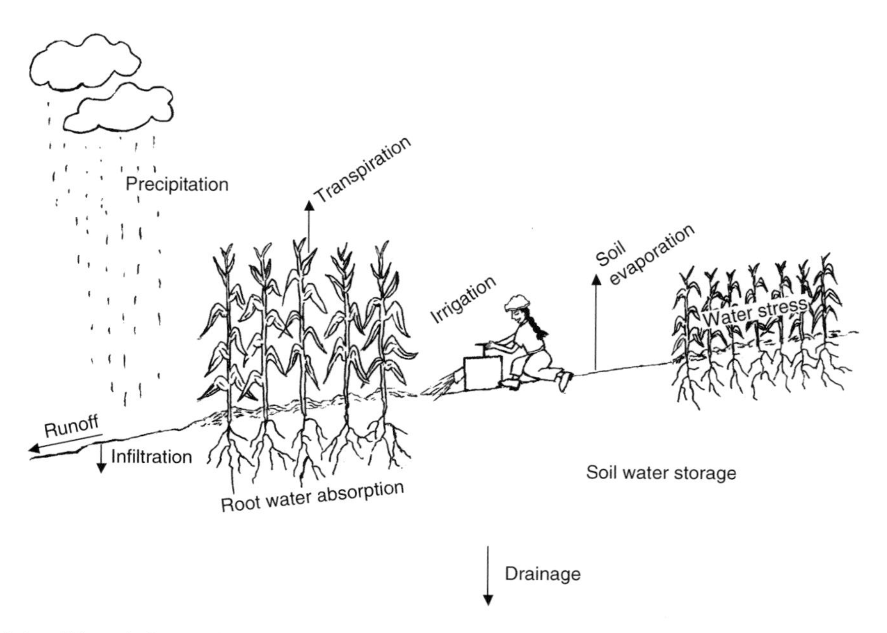

Fig. 13.1. Water balance simulated with the DSSAT model. (Drawing: I-Sheng Zehetner.)

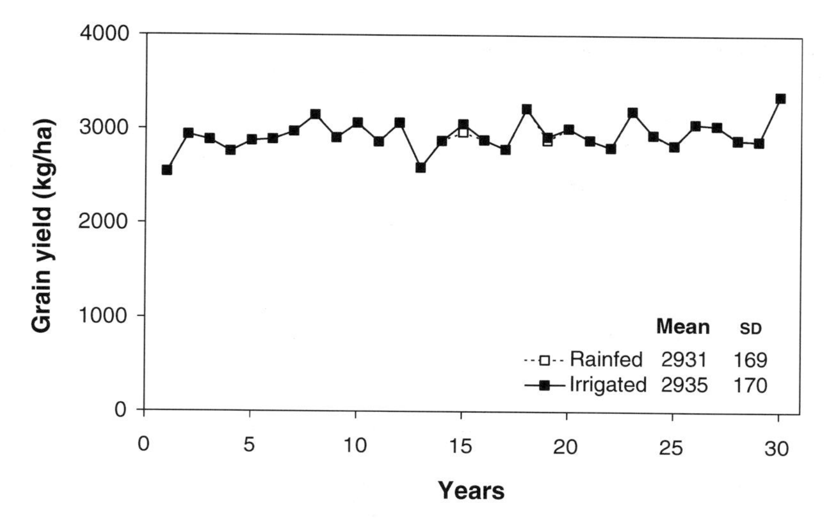

Fig. 13.2. Simulated grain yields in an annual maize–fallow rotation (maize grown during the rainy season) with and without irrigation for Morochos (soil type: Humic Udivitrands; elevation: 2750 masl); the water balance was simulated, other factors were assumed not limiting; SD = standard deviation.

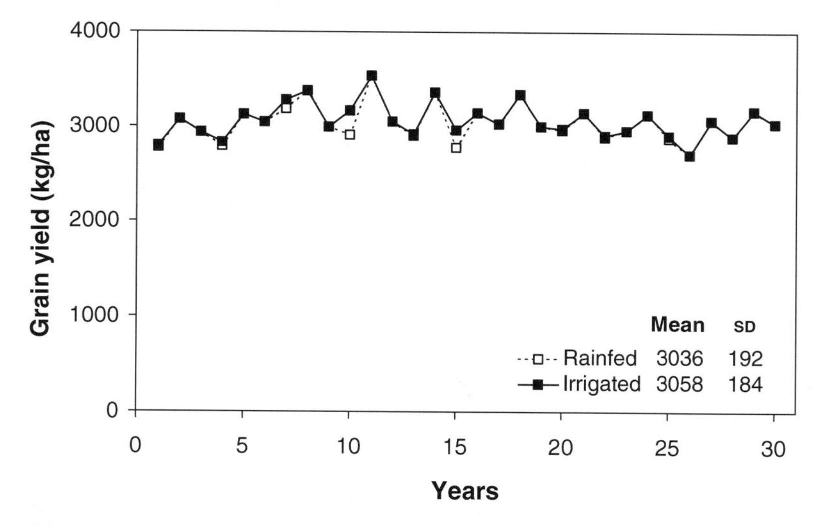

Fig. 13.3. Simulated grain yields in an annual maize–fallow rotation (maize grown during the rainy season) with and without irrigation for Topo Grande (soil type: Vitrandic Udorthents; elevation: 2550 masl); the water balance was simulated, other factors were assumed not limiting; SD = standard deviation.

season water availability limit crop growth during the 30-year simulation period. Consequently, irrigation had little or no effect on maize yields, which fluctuated around 3000 kg/ha at both sites.

The situation was different when maize was grown over the summer dry season, as shown in Figs 13.4 and 13.5 for the Morochos and Topo Grande sites, respectively. Maximum maize yields, provided there were sufficient amounts of plant-available water, were higher due to increased growing energy during the sunny summer months. In Morochos, rainfed cultivation of

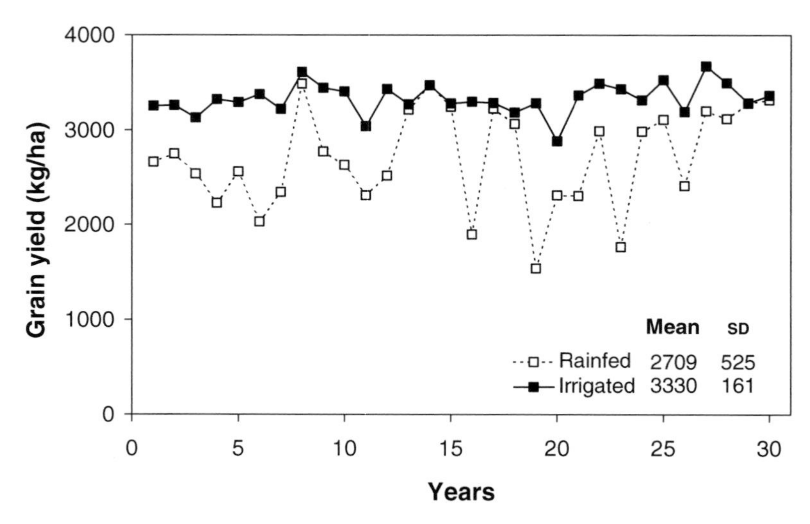

Fig. 13.4. Simulated grain yields in an annual maize–allow rotation (maize grown over the dry season) with and without irrigation for Morochos (soil type: Humic Udivitrands; elevation: 2750 masl); the water balance was simulated, other factors were assumed not limiting; SD = standard deviation.

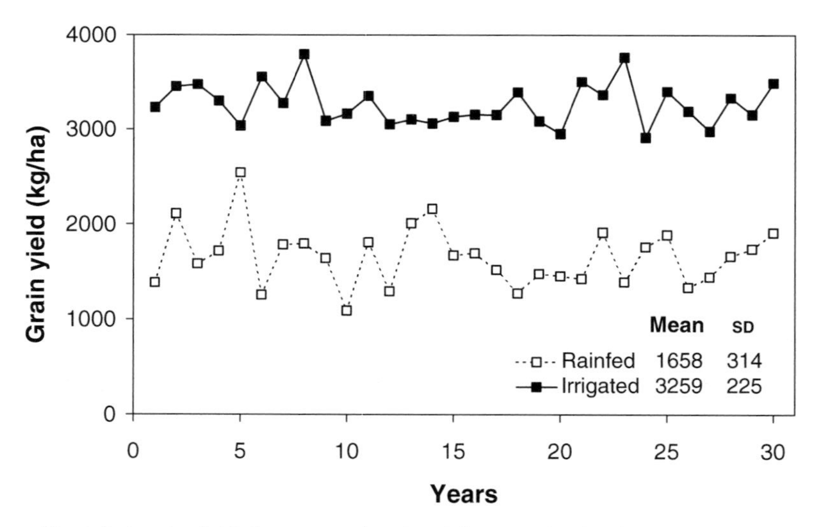

Fig. 13.5. Simulated grain yields in an annual maize–fallow rotation (maize grown over the dry season) with and without irrigation for Topo Grande (soil type: Vitrandic Udorthents; elevation: 2550 masl); the water balance was simulated, other factors were assumed not limiting; SD = standard deviation.

maize resulted in near maximum yields only in wet years, and drought stress significantly lowered crop production in dry years. Irrigation resulted in an average yield increase of > 600 kg/ha (Fig. 13.4). In Topo Grande, crop growth was drastically limited by water stress in each of the 30 simulated years, and maize yields could on average be doubled with irrigation (Fig. 13.5).

Cumulative probability functions of potential yield increase upon irrigation of maize grown over the summer dry season are presented in Fig. 13.6. Irrigation would increase maize yields by about 600 kg/ha in Morochos and by > 1500 kg/ha in Topo

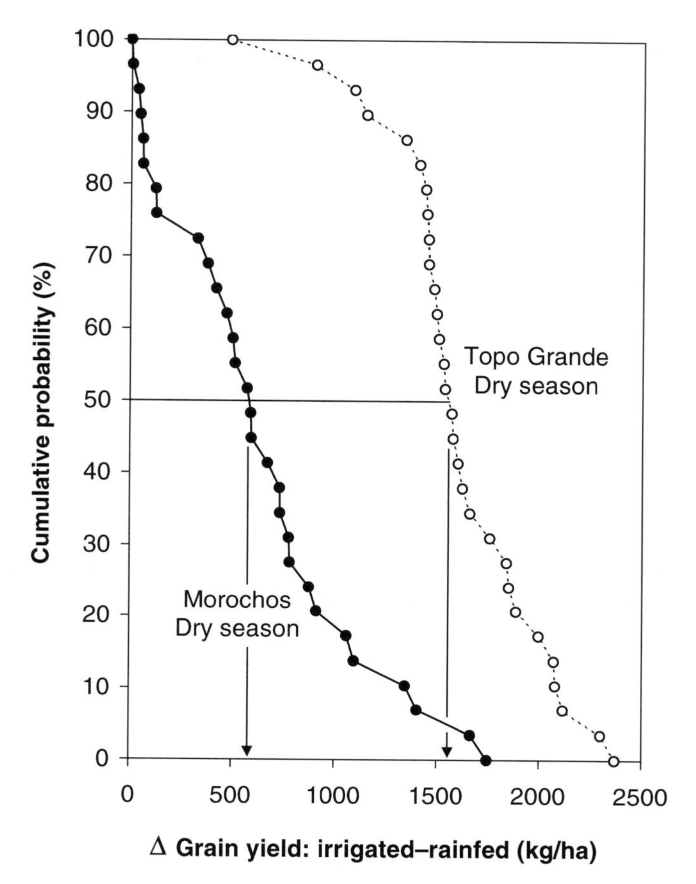

Fig. 13.6. Cumulative probability functions showing expected grain yield increases upon irrigation for maize grown over the dry season in Morochos and Topo Grande.

Grande with a 50% probability, or statistically every other year. In Topo Grande, the chances to increase maize yields by > 1400 kg/ha are > 80% (Fig. 13.6).

Management Implications

Our data suggest that water stress will seldom (only in exceptionally dry years) be limiting to crop growth during the rainy season, even in the drier low-elevation zones of the study area. An expansion of the irrigation systems seems therefore unnecessary if crops are grown during the rainy season and the land is fallow during the dry season. On the other hand, water stress can drastically limit crop growth during the dry season, especially in the drier low-elevation zones of the study area, where the soils are sandy and low in organic matter. Availability of irrigation water would therefore greatly benefit dry-season crop production, particularly in these low-elevation zones.

However, the feasibility of expanding present irrigation systems may be constrained by the following:

- Conflicting demands by the growing urban population, industries and emerging floriculture enterprises in the valley.
- Decisions not to irrigate despite sufficient water supply, as reported by Gilot *et al.* (1997) for nearby Urcuquí. There, many farmers pursue extra-agricultural occupations outside of their villages and seemingly take the risk not to irrigate, thus saving time and/or money

for labour. The occupational situation is similar in Cotacachi, where many peasant farmers leave their communities in the pursuit of jobs in the urban centres.

- Physical limits, such as total streamflow and location (elevation) of fields relative to water sources. In the dry season, water availability is very limited in the Cotacachi area. At higher elevations the streambeds are empty, and at lower elevations most of the available water is already used for irrigation. Local water officials have sketched ambitious proposals involving the construction of long canals and large reservoirs to bring irrigation water from high-elevation zones outside of the study area. While the feasibility of such undertakings is questionable, there are conceivable measures that could improve the situation in the short term. Currently, furrow irrigation is the prevalent irrigation practice, whereby large amounts of water are often used wastefully. More efficient water use and change to sprinkler and drip irrigation could allow an expansion of irrigated agriculture in the Cotacachi area.

Conclusion

Various factors may be limiting to crop production in the Cotacachi area. One such factor, which is quite apparent if one visits the area in the dry season, is water availability. The volcanic landscape in the Cotacachi area is characterized by its verticality, with climate and soils showing pronounced altitudinal variations. The zones most likely to be affected by drought are the low-elevation zones of the area. There, potential water stress caused by lower rainfall and higher evapotranspiration is aggravated by soils with low organic matter contents and sandy textures, which therefore have little water storage capacity. Simulation modelling suggests that, while water availability during the rainy season is generally sufficient for rainfed agriculture, an expansion of existing irrigation systems would considerably improve dry-season crop production, particularly in the low-elevation zones.

Simulation modelling proved a powerful tool for evaluating the benefits to upland crop production that availability of irrigation water would bring about in different zones of the study area. It could serve as a valuable decision support tool, helping local authorities devise irrigation systems that make the best use of the limited water resources. It is important to note that in the present study, the simulated climate was based on past weather patterns and did not take into account potential future climate change. However, the crop growth model used in this study could be linked to climate change models, and simulate the effects of irrigation under various climate change scenarios.

References

Gilot, L., Calvez, R., Le Goulven, P. and Ruf, T. (1997) Evaluating water delivery in tertiary units. Part 2: a case study, Urcuqui, a farmer-managed irrigation system in the Andes. *Agricultural Water Management* 32, 163–179.

Jones, J.W., Tsuji, G.Y., Hoogenboom, G., Hunt, L.A., Thornton, P.K., Wilkens, P.W., Imamura, D.T., Bowen, W.T. and Singh, U. (1998) Decision support system for agrotechnology transfer: DSSAT v3. In: Tsuji, G.Y., Hoogenboom, G. and Thornton, P.K. (eds) *Understanding Options for Agricultural Production.* Kluwer Academic Publishers, Dordrecht, The Netherlands, pp. 157–177.

Richardson, C.W. and Wright, D.A. (1984) *WGEN: a Model for Generating Daily Weather Variables.* USDA-ARS, ARS- 8, Washington, DC.

Wild, A. (ed.) (1988) *Russell's Soil Conditions and Plant Growth*, 11th edn. Longman Scientific & Technical, Harlow, UK.

14 Water Quality and Human Needs in Cotacachi: the Pichavi Watershed

Jenny Aragundy[1] and Xavier Zapata Ríos[2]

[1]*SANREM–Andes Project, Ciudadela Jardines del Pichincha, Pasaje B, N63-204, Quito, Ecuador,* [2]*SANREM–Andes Project, PO Box 17-12-85, Quito, Ecuador*

Introduction

In the Andean zone of Cotacachi, water is a basic resource for agricultural, industrial and domestic activities. The region's inhabitants are in constant interaction with water resources, and their behaviour impacts distribution and quality. Population growth, the loss of glaciers on Cotacachi volcano and inefficient water systems are among the principal reasons for the decline in the available per capita water supply. Data generated during the 5-year water monitoring project carried out by SANREM–Andes also indicate that the water is contaminated by bacteria, thus creating an additional obstacle to a safe water supply. In light of this situation, residents are interested in determining contamination sources and alternatives that will help mitigate water problems. In this chapter, we examine these problems and seek solutions in the context of an analysis of the Pichaví river watershed.

The Study Area

The Andean zone of Cotacachi County covers 21,902 ha and includes the towns of Cotacachi, Quiroga and Imantag, along with agroindustries and 42 farming communities. The area receives an average of 1259 mm of precipitation per year, providing an average available volume of 11,300 m^3 per inhabitant per year. The primary sources of water for the population in question are the micro-watersheds of the Pichaví, Pichambiche and Yanayacu rivers (Fig. 14.1). Water generated by these three watersheds does not satisfy the demands of the rural and urban populations and is contaminated by bacteria.

The watershed of the Pichaví river, located in the southern part of the Andean zone of Cotacachi, measures 3730 ha and provides water to a number of communities as it drains into the Ambi river. The basin includes the territories of the communities of Morochos, Arrayanes, Morales Chupa, Quiroga, El Ejido, San Miguel, La Banda, Ashambuela, Yanaburo, Piava San Pedro, Tunibamba, Perafán and Piava Chupa. The watershed also provides water to other communities served by the Chumaví and San Nicolás water systems. The watershed includes a range of altitudes, from 2300 metres above sea level (masl) at the mouth of the Ambi river to 4939 masl on the peak of Cotacachi volcano.

Methodology

In order to determine the causes of contamination in the Pichaví river and to propose

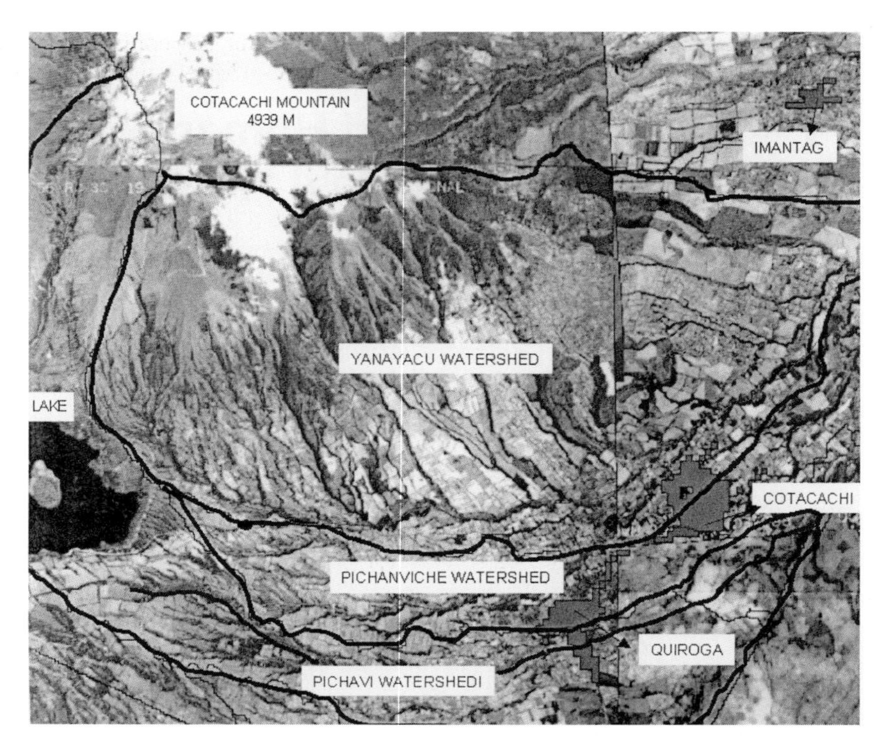

Fig. 14.1. Major watersheds in the Andean zone of Cotacachi County. Aerial photos by the Instituto Geográfico Militar (IGM).

mitigation measures, human and animal behaviours were studied through a series of activities.

First, the water supply and services available in dwellings in Cotacachi county were analysed using the latest census of the Instituto Nacional de Estadísticas y Censos (2001).

Second, the quality of water in the Pichaví river and the results of the 5-year monitoring project carried out by the SANREM team were analysed. The water quality analysis included physical, chemical and bacteriological parameters (see Ruiz-Córdova *et al.*, Chapter 16, this volume). The analysis of physical and chemical parameters included study of the following variables: water and air temperature, pH, alkalinity, hardness and dissolved oxygen. The bacteriological parameters included: total coliforms and *Escherichia coli*.

The third activity involved fieldwork throughout the watershed in order to identify hygiene practices, sanitary systems and means of solid waste disposal, supply points for water for domestic use and irrigation, and other key variables. Observations and location were noted with the aid of the global positioning system (GPS) for future analysis. Information gleaned from observation was supported by interviews with residents of three representative communities in the basin: San Nicolás in the high area; San Antonio del Punge in the middle range; and La Calera in the lower zone. The size of the sample in each of the three communities was determined by the following formula:

$$n = \frac{Z^2\pi(1-\pi)}{(p-\pi)^2} \qquad \text{(Webster, 2000)}$$

in which: n (level of confidence) = 95%, Z = 1.96, probability of error ($p - \pi$) = 3% and an assumed expected proportion (π) =5%.

The sample size of each community was determined based on population data (number of inhabitants) for the three communities (San Nicolás, 60; San Antonio del

Punge, 400; La Calera, 1200; UNORCAC, 2004, personal communication), by the following formula:

$$n = \frac{N * Z^2\pi(1-\pi)}{(p-\pi)^2 * (N-1) + Z^2\pi(1-\pi)}$$

(Fernández, 1996)

where N is the population of each community.

The values of n for San Nicolás, San Antonio del Punge and La Calera are 45, 135 and 174, respectively. Sanitary and hygienic habits found in individual dwellings were assumed to be similar, and thus we divided the number of inhabitants by 5 persons per dwelling. Consequently, the number of interviews for each community is as follows: San Nicolás, nine; San Antonio del Punge, 27; and La Calera, 35.

The basic questionnaire contained 30 items divided into four sections. The first of these was devoted to general data on families, including the name of the community, ethnic identity and number of individuals who lived in the dwelling. The second section focused on the availability and characteristics of water used for domestic purposes. The third section dealt with matters related to grazing, including types of animals, where they graze and for how long. The final section was devoted to the sanitary and hygiene practices of the individual interviewed and those of his or her family; questions included the type of sanitary unit available (latrine, septic tank, sewer or none); how and how often family members wash their hands; what they use for ablution; the most frequent kinds of sickness experienced; and the handling of solid wastes.

Finally, the data obtained were analysed and interpreted with the aid of the geographic information systems (GIS) in order to visualize the different variables in geographical terms.

Results

According to the National Institute of Meteorology and Hydrology, 70% of surface waters in Ecuador are contaminated (Instituto Nacional de Meteorología e Hidrología INAMHI, 1963–2000). This is due, among other reasons, to the fact that only 57% of the nation's population has access to sanitary solutions (Asociación Interamericana de Ingeniería Sanitaria y Ambiental y Asociación Ecuatoriana de Ingeniería Sanitaria y Ambiental AIDIS/AEISA, 2000). This does not mean that all dwellings having access to sanitary solutions treat their domestic and/or industrial wastewaters prior to discharge into a watercourse. By way of example, in Cotacachi county, 45% of dwellings do not have access to any sanitary solution, 16% have septic tanks and 38% are connected to a sewage system (Table 14.1) (Instituto Nacional de Estadísticas y Censos, 2001).

The various sources of untreated sewage affect the quality of surface and groundwaters in the county. Table 14.2 provides data on the sources of water used in dwellings in the county.

Water quality in the Pichavi watershed

The results presented below contain data for samples taken from four control points: point 1, the spring in Quiroga; point 2, the bridge from Cuicocha Centro and San José del Punge; point 3, in Quiroga; and point 4, the stadium in Cotacachi.

The analysis of samples from the Pichaví river indicate that the physical–chemical condition of the water is good (Fig. 14.2). Average pH is 7, with little variation related to the site from which samples were taken or the time of sampling. The average water temperature of 17°C also varies little. The value of dissolved oxygen rises as the water runs its course in the river and diminishes due to the respiration of aquatic animal species, decomposition of organic matter and chemical reactions. Thus, at control point 1 (Quiroga spring), values of 4 mg/l were found, and at control point 4 (Cotacachi stadium) the figure was 6 mg/l. Alkalinity measures, on average 220 mg/l in the basin, represent the ability of water to neutralize acids and buffer any change in the water's pH.

Table 14.1. Sewage elimination systems and sanitary services available in dwellings in canton Cotacachi.

	Type of dwelling					
Services available in dwellings	Urban	%	Rural	%	Total	%
Total dwellings	1904		6359		8263	
Sewage removal system						
Connected to public sewage disposal grid	1743	92	1371	22	3114	38
Open pit	64	3	830	13	894	11
Septic tank	54	3	432	7	486	6
Other	43	2	3726	59	3769	46
Toilet						
Private bathroom	1603	84	1523	24	3126	38
Common bathroom	193	10	607	10	800	10
Latrine	10	1	1302	21	1312	16
None of the above	98	5	2927	46	3025	37
Shower						
Private use	1482	78	1846	29	3328	40
Common shower	256	13	265	4	521	6
None	166	9	4248	67	4414	53

Source: Instituto Nacional de Estadisticas y Censos (2001).
Percentages have been rounded, therefore the total sum may not add up 100%.

Table 14.2. Water supply to dwellings in canton Cotacachi.

	Type of dwelling					
Services available in dwellings	Urban	%	Rural	%	Total	%
Total dwellings	1904		6359		8263	
Water supply						
Water piped into dwelling	1625	85	1457	23	3082	37
Water piped or hose outside dwelling, inside the building, lot or parcel	236	12	3220	51	3456	42
Pipes or hose outside the building, lot or parcel	16	1	651	10	667	8
No water via pipe or hose	27	1	1031	16	1058	13
Water comes from:						
Public water system	1865	98.0	3535	55.6	5400	65.4
Well	12	1	277	4	289	4
River, spring	19	1	2390	38	2409	29
Tanker	0	0	42	1	42	1
Other	8	0	115	2	123	2

Source: Instituto Nacional de Estadisticas y Censos (2001).
Percentages have been rounded, therefore the total sum may not add up 100%.

Finally, water hardness is 200 mg/l, on average, a high figure due, primarily, to the content of calcium and magnesium salts in the volcanic soils of the watershed. In general, when hardness exceeds 150 mg/l, water-softening treatment may be necessary (Spellman, 2000). High values for water hardness mean that there may be blockage in pipes and soap will not readily produce foam. All of these characteristics were verified through interviews with community residents.

The Pichaví river, however, is characterized by a high degree of contamination due to faecal coliforms, with the MPN values for *E. coli* (most probable number) registering > 10,000 colonies/100 ml (Fig. 14.3). At lower elevations along the river, where

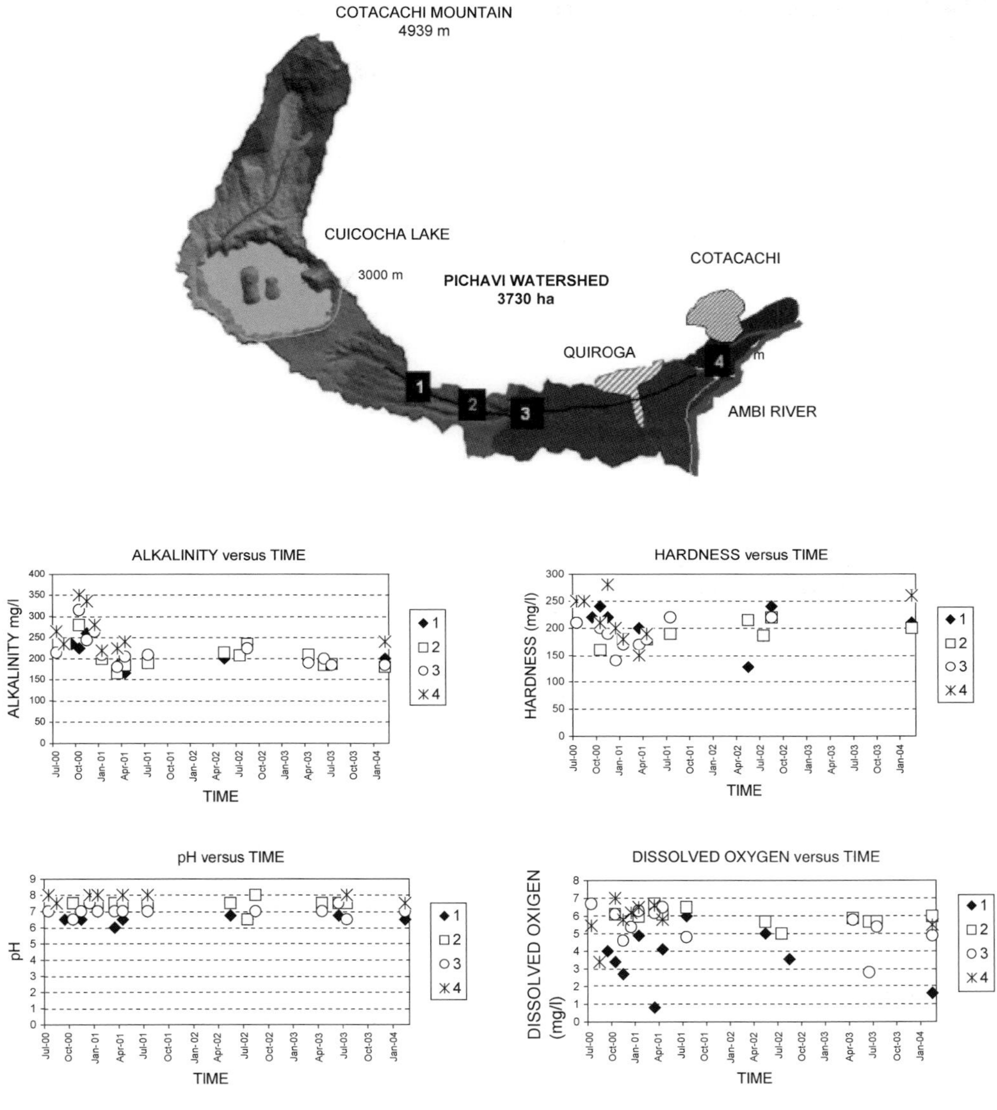

Fig. 14.2. Results of water quality analyses. Physical–chemical parameters. Source: Zapata Ríos (2003); and Sergio Ruiz-Górdova *et al.*, Chapter 16, this volume).

water has been exposed to human activity, the bacteriological content is higher. This means that water from the Pichaví is not suitable for human consumption or irrigation as there is the risk of illnesses such as diarrhoea (due to cholera, dysentery and non-specific diarrhoea), parasitic infection (e.g. ascariasis and trichuriasis), skin and eye infections (e.g. scabies, conjunctivitis and trachoma), and infections transmitted by lice, mosquitoes and flies (e.g. typhoid transmitted by lice, malaria, dengue fever and yellow fever) (Yzunsa, 2003). A study of the incidence of anaemia and parasites undertaken in Imbabura indicates a strong correlation between the presence of intestinal parasites and anaemia in children < 5 years of age (Table 14.3).

In addition, at the provincial level in Imbabura, a study by Instituto Nacional de Estadísticas y Censos (2001) indicates that 16% of all deaths in the province are due to

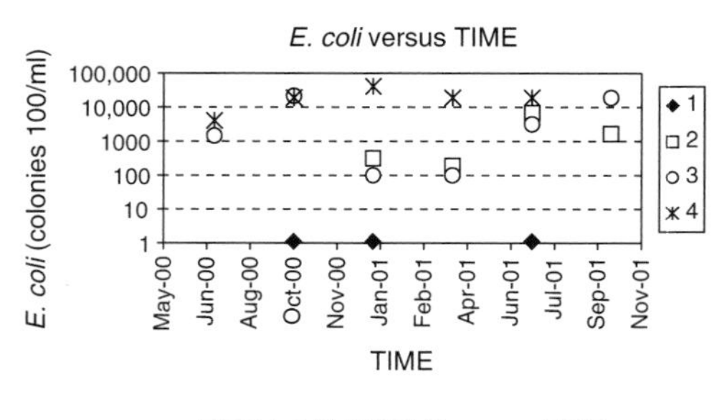

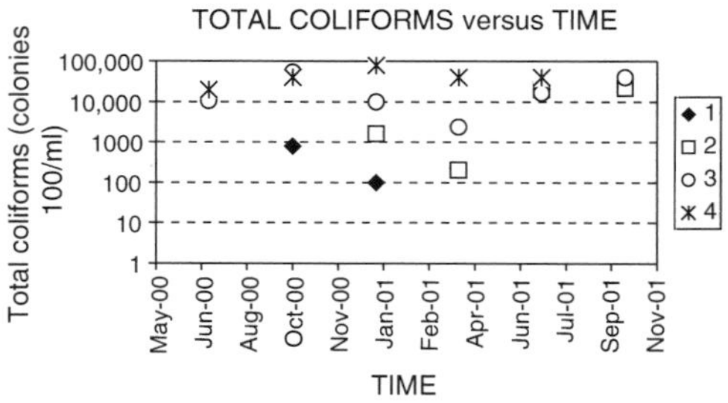

Fig. 14.3. Results of water quality analyses. Bacteriological parameters. Source: Zapata Ríos (2003); and Sergio Ruiz-Górdova *et al.*, Chapter 16, this volume).

waterborne diseases including those caused by parasites and hepatitis A.

Fieldwork and interviews

The third stage of this study involved fieldwork and interviews. Thirteen sites from which water for human consumption and irrigation was utilized were identified and major sources of contamination were determined. As Fig. 14.4 indicates, only control point 1 (Quiroga water source) provides water suitable for recreational use and human consumption, with the presence of *E. coli* measured at < 10 MPN/100 ml. Water sampled at the remaining control points is not suitable for human consumption or recreation since, according to the US Environmental Protection Agency (EPA) (Deutsch *et al.*, 2003), water with > 600 MPN/100 ml should not be used for recreational purposes. In addition, EPA requires that water with *E. coli* values > 2000 MPN/100 ml be treated prior to use. Thus, all communities using water below the Quiroga water source, women who do family laundry in the river and children who play in the water are exposed to high concentrations of faecal coliforms and the risks represented by said exposure.

Factors and human behaviours leading to the contamination of the Pichaví riverbed identified in the course of this study are as follows.

Grazing

Given the lack of designated grazing areas and livestock watering sites, the banks of the Pichaví river are used for these purposes. Livestock faeces are present throughout the basin. Specifically, the presence of faeces was confirmed in the Chumaví ravine, at the headwaters of the river that flows down to the Cuicocha lagoon, at points from which water is taken for domestic use and irrigation, and at the confluence of the Pichaví and Ambi rivers (Fig. 14.5). Grazing practices were confirmed in the course of interviews with residents from the communities of La Calera, San Antonio del Punge and San Nicolás, in which 26, 70 and 55%, respectively, of interviewees said that they grazed their animals (cattle, sheep and pigs) on the banks of the Pichaví river and in the ravines through which it flows (Fig. 14.6). This fact is the primary cause of contamination by faecal coliforms and

Table 14.3. Incidence of anaemia and parasites in children under 5 years of age in the province of Imbabura.

	Incidence of anaemia	Incidence of parasites	Lacking water system	Open air disposal of sewage water	Open air garbage disposal
Afro-Ecuadorians	72	84	66	67	62
Indigenous	63	84	58	28	58
Mestizos	54	80	49	16	56

Source: Benson Institute (2001).

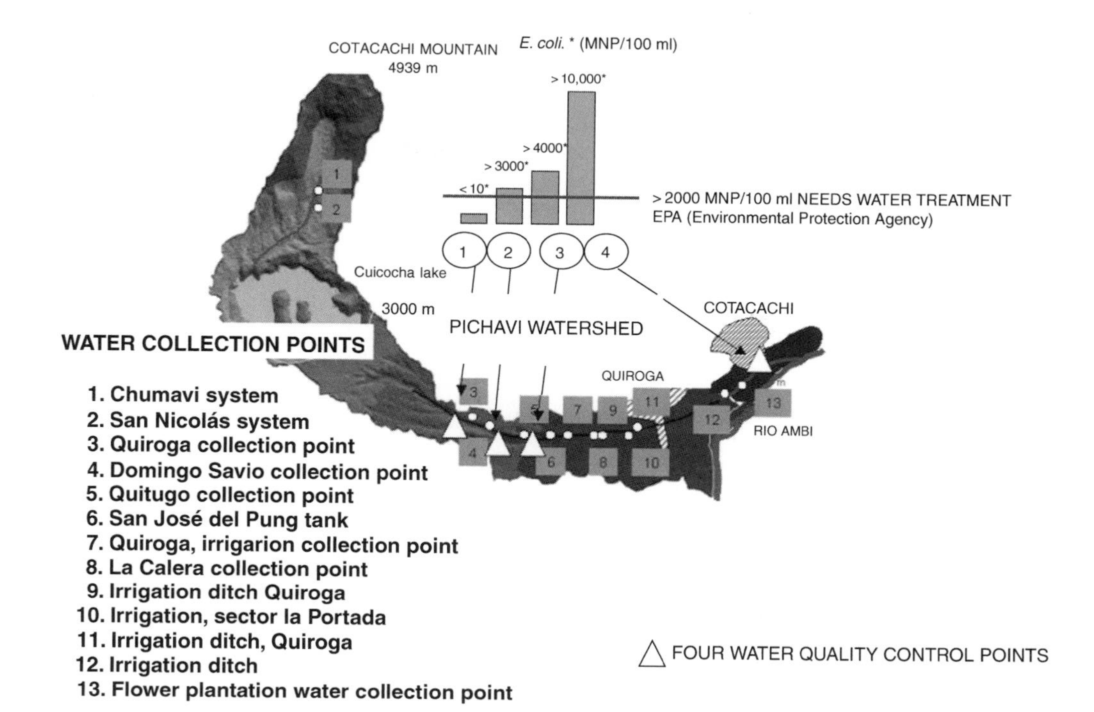

Fig. 14.4. Water collection points for human use and irrigation and water quality control points.

Fig. 14.5. Grazing, La Calera (Photo: Jenny Aragundy).

explains their presence in the water analyses undertaken by SANREM. Furthermore, it contributes to fertilization of the water through the addition of nutrients.

Sanitation

Existing sanitation technologies in the region are: connection to a sewage system, septic tanks and latrines. However, for a significant number of residents in the Pichavi river watershed, these options are not available. Interviews reveal that among the sampled communities, only Calera (60% of those interviewed) had access to the sewer grid. In the community of San Antonio del Punge, 50% of those interviewed had latrines. In San Nicolás, 100% of interviewees had no access to any of the options mentioned (Fig. 14.7), in part because residents used the portable latrines they were given to enlarge their dwellings or as store rooms for maize.

Each of the sanitation options available in the study area affects the water of the watershed in a different way. The disposal of untreated or insufficiently treated wastewater through the sewage system contributes pathogens, nutrients, organic matter and other contamination to the river. An excess of nutrients (principally phosphorus and nitrogen) contributes to the eutrophication of the water flow, creating an environment ideal for the proliferation of algae and other kinds of aquatic plants. In addition to absorbing the nutrients, these plants

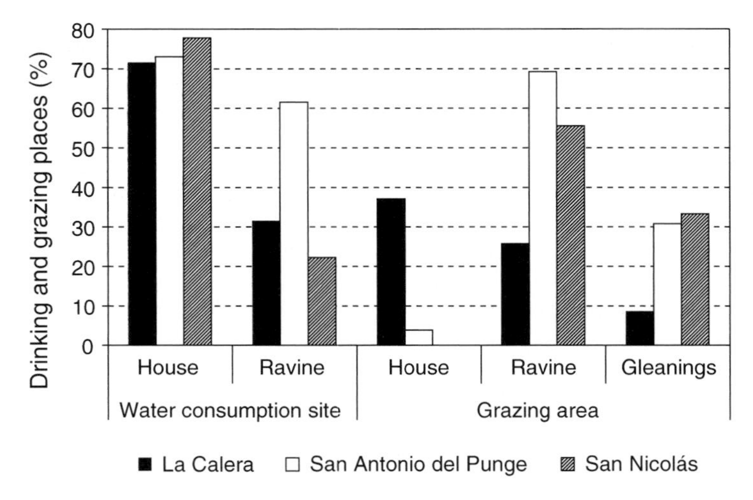

Fig. 14.6. Drinking and grazing places for three communities in the Pichavi watershed.

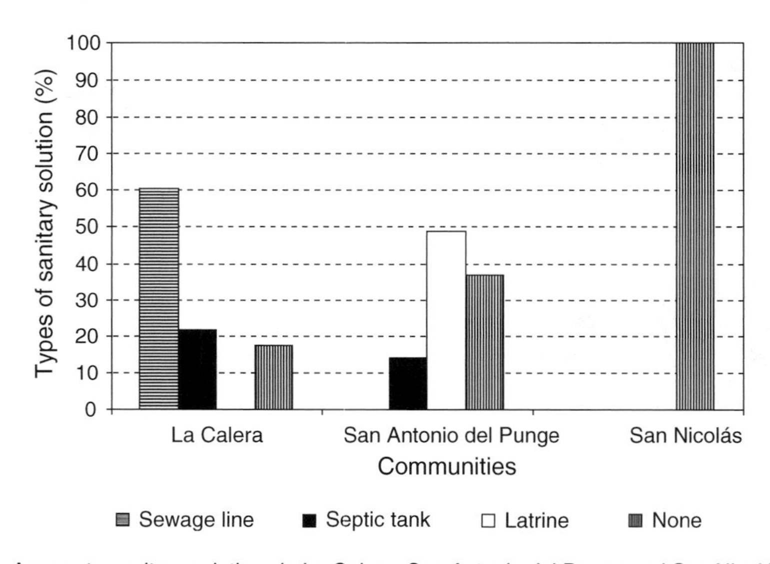

Fig. 14.7. Access to sanitary solutions in La Calera, San Antonio del Punge and San Nicolás.

consume oxygen available in the water, leaving no excess oxygen to assist in the recovery of the body of water. This is the case in the area of La Victoria (Fig. 14.8).

The septic tanks and latrines in the area are drop and storage sanitation systems. If a tank or well is not waterproof, deposited wastes can filter into and contaminate the groundwater. Residents who have no sanitary facilities use the open countryside as a toilet. This practice facilitates vectors which disseminate disease, contributing to the proliferation of illnesses caused by waterborne agents, such as diarrhoea, cholera, hepatitis, parasitic infections and others.

Hygiene and sanitary habits

Hygiene and sanitary habits contribute directly to the spread of diseases. Thus, while good quality water may be available in abundance, the risk of diseases will exist if residents do not wash their hands after defecation or urination, do not use soap, wipe their hands with leaves or dirt after defecation, or do not wash their hands before eating or cooking or after changing a baby's diaper.

Another habit that affects disease transmission is the way people clean themselves after defecation. As indicated in Fig. 14.9, interviewees mentioned using leaves, rags, used paper and toilet paper for this purpose. The material used varies according to the sanitary facilities available in the different communities. In La Calera, for example, 63% of those interviewed say they use toilet paper for ablution, while in San Antonio del Punge only 33% do so. In San Nicolás, where no sanitary facilities exist and where residents relieve themselves outside, no one uses toilet paper. Ninety per cent of interviewees in San Nicolas use leaves and 10% use rags for ablution.

Boiling water can be a way to break the disease transmission chain. In La Calera, San Antonio del Punge and San Nicolás, 44, 91 and 100%, respectively, of those interviewed do not boil drinking water.

Laundry practices

While a large number of residents in the communities of Cotacachi county have running water, many people do not wash clothes at home; instead they use nearby pools of water (Fig. 14.10). This is due, in part, to the fact that only some homes have running water while those without fill tubs at their neighbours' taps (Fig. 14.11). In addition,

Fig. 14.8. Wastewater discharge, Sector La Victoria (Photo: Jenny Aragundy).

Fig. 14.10. Washing clothes in the community of San Jorge (Photo: Jenny Aragundy).

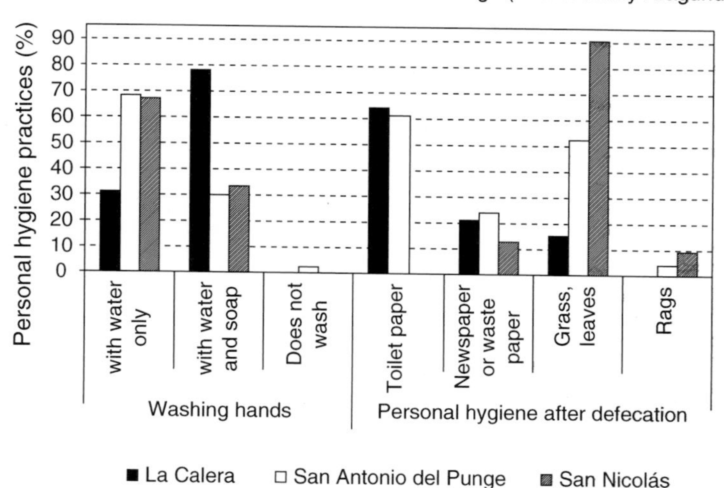

Fig. 14.9. Personal hygiene practices in the communities of La Calera, San Antonio del Punge and San Nicolás.

although some have running water in their dwellings, through force of habit they continue to wash clothes in streams. In La Calera, for example, approximately 26% of those interviewed wash clothes at home or in the river (Fig. 14.12). People also take certain food items to the river for washing, such as chochos, which have to be washed for days in order to remove the legume's bitter coating. Detergents and soaps used in laundering clothes add phosphorus to surface waters, causing an excess of this nutrient and increasing eutrophication.

Disposal of solid wastes

A variety of solid wastes have been deposited along the length of the Pichaví river. These include plastic bottles, bits of clothing, rubble, organic residues and residues of animal flesh from a tannery's wastewater discharged near the stadium in Cotacachi (Fig. 14.13). In addition, although the community of La Calera has a solid waste collection service, 16% of those interviewed stated that they threw their inorganic trash into ravines or irrigation ditches and 43% said that they burned it (Fig. 14.14). In the communities of San Antonio del Punge and San Nicolás, where there is no waste collection service, 25 and 5%, respectively, of those interviewed throw solid wastes into ravines or irrigation ditches; the remaining wastes for these communities are burned.

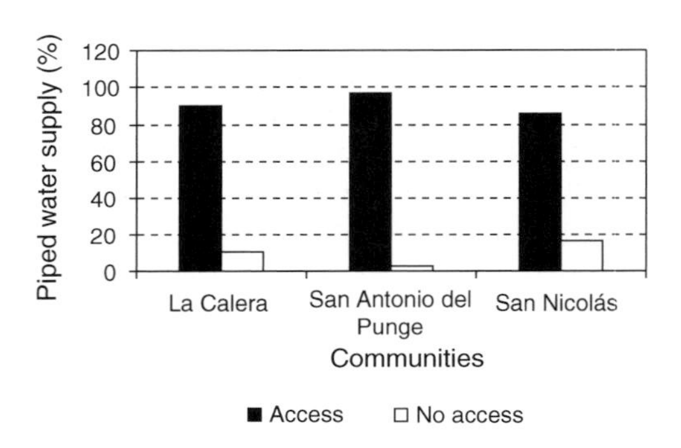

Fig. 14.11. Piped water supply in the communities La Calera, San Antonio del Punge and San Nicolás.

Fig. 14.13. Solid wastes, Pichaví river, Quiroga (Photo: Jenny Aragundy).

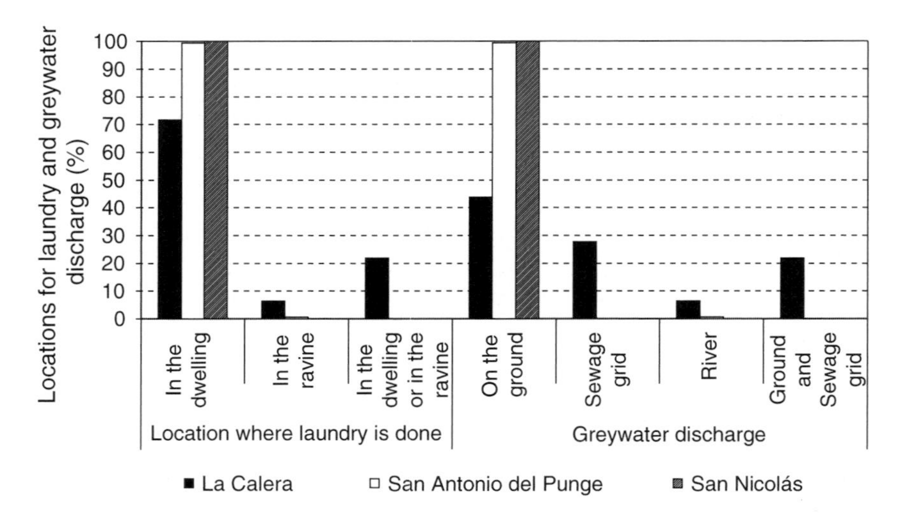

Fig. 14.12. Locations for laundry and greywater discharge.

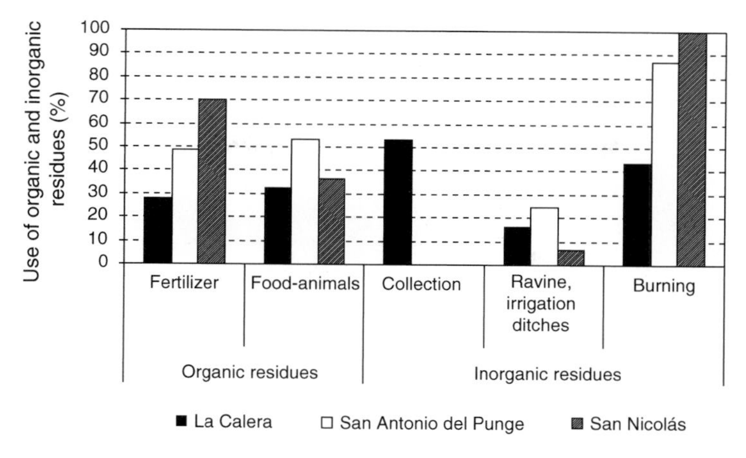

Fig. 14.14. Use of organic and inorganic residues in La Calera, San Antonio del Punge and San Nicolás.

Iron bacteria

Theodore Wolf in his 1892 *Geography and Geology of Ecuador* stated that iron had been detected in the water in Cotacachi. During the fieldwork along the Pichaví river, iron bacteria were found at the source of the river, in the collection systems in the communities of Domingo Sabio and Quiroga, in the tank at La Calera and along the length of the river. This type of bacterium is not harmful to human beings, but it produces an unpleasant odour, adds a yellowish tinge to water and corrodes pipes (Edstrom Industries, 2003), phenomena described by community residents as the major problems observed in piped water.

Remediation Measures and Watershed Protection

Actions for the remediation and protection of the watershed along with proper water use and management have been included in a municipal ordinance declaring Cotacachi as an 'ecological canton' (Consejo Municipal de Cotacachi, 2000). However, these legal instruments have yet to be applied and enforced. According to article 41 of the ordinance, areas for the protection of watersheds are to be established, and these will include 50 m along both banks, and along ravines, lagoons and springs. As a consequence, grazing, burning or the harvesting of grass are not allowed in these areas which are to be forested with tree species native to the area (Fig. 14.15).

Article 42 of the same ordinance prohibits contamination of water sources in a watershed by declaring: 'The contamination of rivers and watersheds with substances noxious to human health and the normal development of aquatic flora and fauna is forbidden' (Consejo Municipal de Cotacachi, 2000). Tanks for the collection of water for human use and the banks of rivers may not be used for washing clothes or grazing animals, both activities that cause eutrophication.

Establishment of grazing areas in the Pichaví river basin

According to article 147 of the General Regulations for the implementation of Ecuador's national law governing the use and management of water, in reference to the establishment of watering areas for animals, 'Any person, community or settlement with no access to water necessary for watering animals has the right to create watering areas for this purpose after taking measures necessary to avoid contamination of the body of water. . .' (Codified Legislation, 2003). This also includes the right to drive livestock along existing paths to

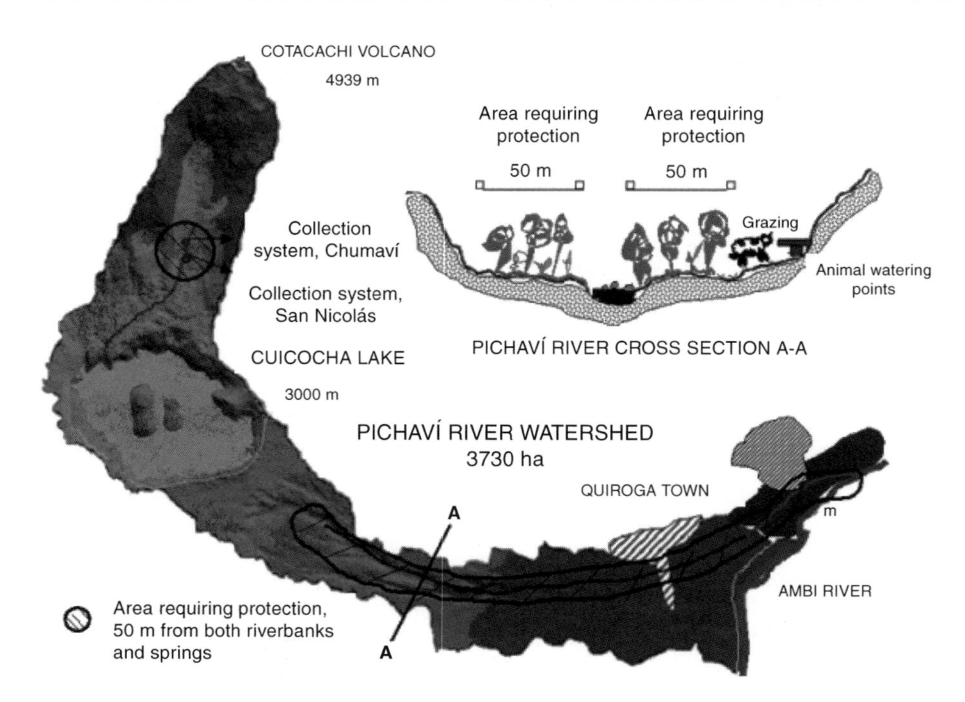

Fig. 14.15. Boundaries of areas requiring protection in the Pichaví river watershed.

drink at the site in question during fixed periods. The most frequently used grazing areas in the Pichaví watershed are the riverbanks and grasslands near water collection points. Thus, it is necessary to define grazing areas provided with watering points. Based on field reconnaissance, there are sufficient areas in the watershed that could be used for grazing in the communities of Domingo Sabio, San José del Punge and San Borja. Approximately 237 ha along the river could be used for this purpose, providing water troughs for the animals and fences.

Sanitation improvements in the Pichaví river basin

The following measures are required in order to improve sanitary conditions in the watershed:

- Periodic testing of the quality of wastewater treated in the existing sewage treatment tanks (e.g. Cuicocha and Quiroga) to ensure that water returned to the river is not a source of contamination.
- Implementation of sewage treatment systems in communities with sewage lines but without plants for treatment of wastewater. Technologies selected should be easy to operate and maintain in order to ensure long-term operation. Appropriate technologies include: constructed wetlands, i.e. subsurface flow systems, free surface flow systems; trickling filters combined with sedimentation ponds; and anaerobic or facultative precipitation ponds.
- Providing solutions to sanitation problems in communities where these are lacking. All residents interviewed in San Antonio del Punge and San Nicolás, and 60% of interviewees in La Calera, state that they throw greywater into the ground. Approximately 90% of those interviewed in San Antonio del Punge and San Nicolás, and 70% in La Calera use guano (animal manure) to fertilize crops (Fig. 14.16). The most

common sicknesses in these communities include diarrhoea, skin problems, parasitic infections and stomach aches, all related to deficient sanitary conditions (Fig. 14.17).

Our research indicates that the most appropriate sanitation solution would be ecological sanitation. Ecological sanitation involves the application of *in situ* sanitation measures and the separate treatment of wastewater (greywater, yellow water (urine) and brown water (faeces)), thus achieving an integral sanitation solution. For example, greywater is treated and reused for irrigation. Clean water should not be used to transport excrement. Excrement could be treated to produce fertilizers that contribute to improved soil quality and crop yields and thus to improvement in the residents' economic situation.

The recommended ecological sanitation technology includes urine-diverting dry toilets (Fig. 14.18). Treatment of faeces involves the application of a drying material after each use. The drying material facilitates dehydration of faeces and increases the pH. This is an aerobic

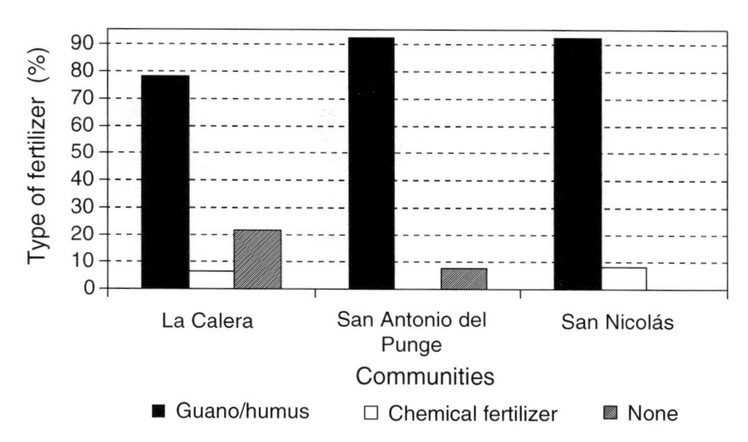

Fig. 14.16. Type of fertilizer used in the communities La Calera, San Antonio del Punge and San Nicolás.

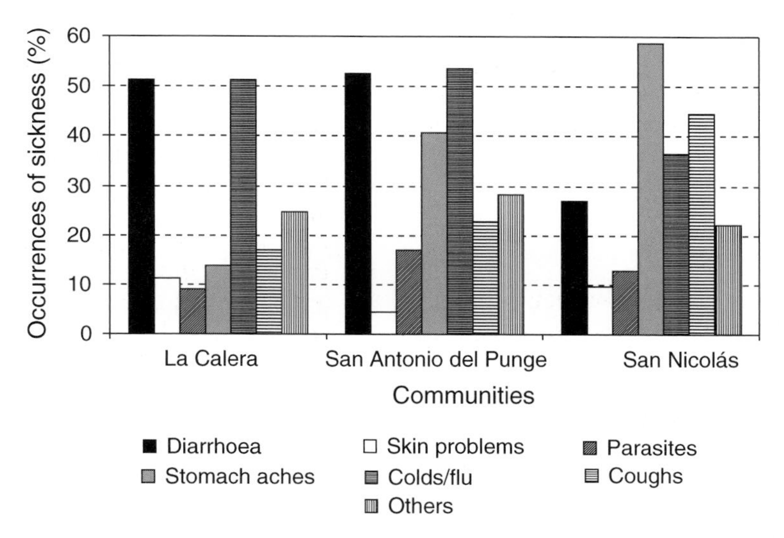

Fig. 14.17. Most frequent illnesses in the communities La Calera, San Antonio del Punge and San Nicolás.

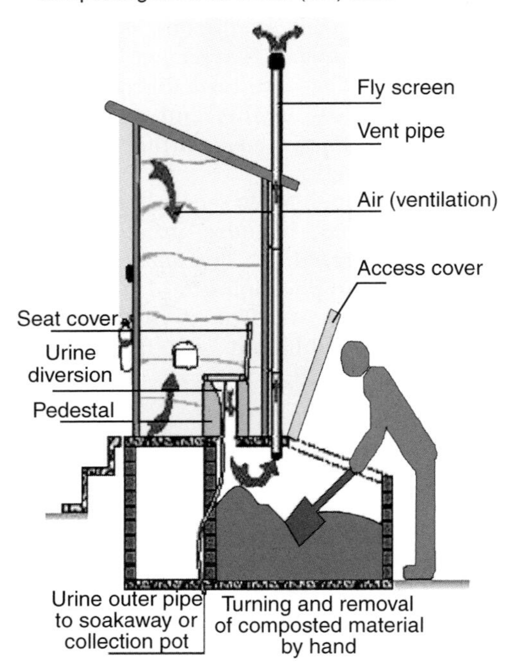

Fig. 14.18. Urine diversion dry toilet. Source: Mang (2003).

technology and does not produce noxious odours. The toilet can be located inside or outside the dwelling, and is designed on the basis of family size. A family of six generates five sacks and 7000 l/year of fertilizer rich in nitrogen, phosphorus and potassium.

Solid waste management

Solid wastes should be managed in a way that avoids dumping on riverbanks and sites near water sources. In an improved system, organic and inorganic wastes are already separated prior to dumping, with the former used as fertilizer or animal feed. The same separation process should be encouraged for inorganic wastes which are both recyclable and non-recyclable in order to decrease the volume of waste dumped, buried or burned by residents who do not have a waste collection service.

Education

Awareness campaigns should be promoted in the following areas: (i) care and maintenance of watersheds in order to improve the quality of water for human consumption and irrigation, and to protect the health of residents; and (ii) sanitation (treatment of excreta and residues) and hygienic and sanitary habits, such as washing hands with water and soap and the proper disposition of materials used for ablution in order to break the transmission cycle of waterborne diseases and those which propagate in unsanitary environments (Fig. 14.19).

Final Remarks

- An awareness campaign is needed regarding the importance of caring for and protecting the watershed.
- Legislation is required for the protection of water and enforcement of compliance.
- The importance of having a closed cycle sanitation system, a cycle in which wastewater is treated prior to disposal or ecological sanitation technologies, should be divulged.
- The success of sanitation measures depends on the level of acceptance by the community. If dry toilets are to be built in communities where residents lack sanitation facilities, advocacy workshops should first be organized to inform residents about how they work, and their advantages and disadvantages. An example of a failed system can be appreciated in the community of San Nicolás. Residents who received portable latrines used them for construction material instead of the purpose for which they were intended.
- The needs of communities, and how to satisfy them, must be identified with residents in order to ensure project success.
- The grazing of farm animals in areas near watersheds should be avoided in order to prevent contamination by faecal coliforms.

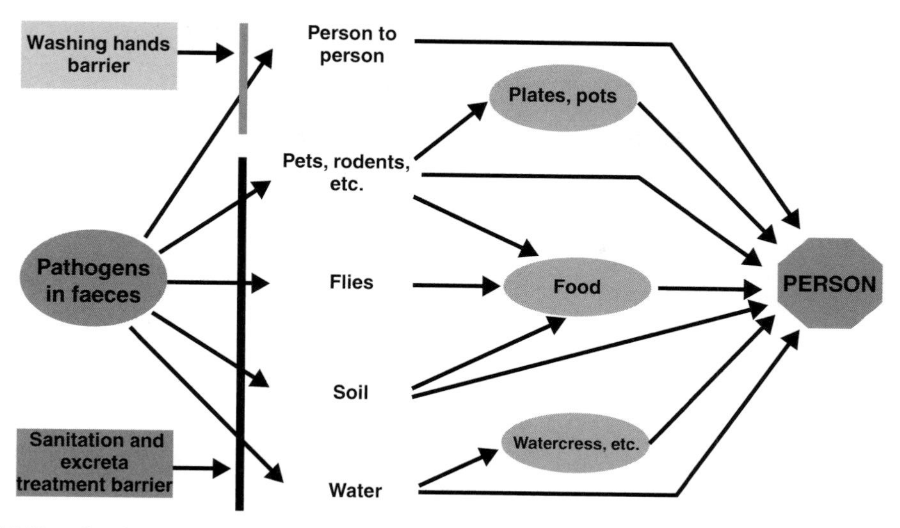

Fig. 14.19. Barriers to the transmission of waterborne diseases and those found in an unsanitary environment.

- The laundering of clothes in rivers, irrigation ditches, collection tanks and other water sources used for domestic purposes and irrigation should be avoided. If necessary, laundry areas equipped with water treatment systems should be constructed.
- Solid wastes should be disposed of according to the canton's Municipal Ordinance.
- Education campaigns should be undertaken in order to encourage changes in hygiene and sanitary habits of the population.
- Water quality monitoring data collected by SANREM indicates that the situation in the Pichaví river watershed is similar to that of other microwatersheds in the region, including those of the Yanayacu and Pichambiche rivers. Thus, the recommendations included in this study for the Pichaví can be applied to other watersheds in the region.

References

Asociación Interamericana de Ingeniería Sanitaria y Ambiental y Asociación Ecuatoriana de Ingeniería Sanitaria y Ambiental AIDIS/AEISA (2000) *Evaluación Nacional de los Servicios de Agua Potable, Alcantarillado y Desechos Sólidos.* Ingeniería sanitaria y ambiental, No. 7. Guayaquil, Ecuador 2.

Benson Institute (2001) Making Recommendations in Ecuador. Available at: http:/benson.byu.edu/ Members/kma/Rev23Pagauay/view

Consejo Municipal de Cotacachi (2000) Ordenanza que declara a Cotacachi como 'Cantón Ecológico'. Cotacachi, Ecuador. Artículos 41 y 42.

Deutsch, W., Duncan, B. and Ruiz, S. (2003) *Manual de Certificación Básica. Monitoreo Bacteriológico del Agua. Guía para Ejecutar Proyectos Participativos de Monitoreo del Agua en Comunidades de la Región Andina. Manual.* University of Auburn, Auburn, Alabama.

Edstrom Industries (2003) Iron and Iron Bacteria in Water. Available at: http://www.edstrom.com

Fernández, P. (1996) Determinación del tamaño muestral. Available at: http://www.fisterra.com/mbe/investiga/9muestras/9muestras.htm

Mang, H. (2003) Ecosan: Cerrando el Ciclo en el Manejo de Aguas Residuales y Saneamiento. GTZ-Ecosan presentation.

Instituto Nacional de Estadísticas y Censos (INEC) (2001) Datos del Censo de noviembre 2001.

Instituto Nacional de Hidrología y Meteorología (INAMHI) (1963–2000) Datos de precipitación para Estación Cotacachi.
Legislación Codificada (2003) Reglamento general para la aplicación de la ley de Aguas, Normas pertinentes del texto unificado de la legislación secundaria del Ministerio de Agricultura y Ganadería en Ley de Aguas, Reglamento y Legislación conexa, 4th edn. Corporación de Estudios y Publicaciones. Quito, Ecuador. art. 147.
Spellman, F. and Drinan, J. (2000) *The Drinking Water Handbook.* Technomic Publishing Co. Inc., Lancaster, Pennsylvania.
Webster, A. (2000) *Estadística Aplicada a los Negocios y la Economía*, 3rd edn. Irwin; McGraw–Hill, Santa Fé de Bogotá, Colombia, pp. 183–185.
Wolf, T. (1892) *Geografía y Geología del Ecuador.* Brockhaus, Leipzig, Germany.
Ysunza, A. (2003) Saneamiento ambiental y salud, centro de capacitación integral para promotores comunitarios, Oaxaca, México. Paper presented at Ecosan Workshop, El Tránsito, El Salvador.
Zapata, X. (2003) Calidad del agua en las cuencas del Pichaví, Pichambiche y Yanayacu. Paper presented at the annual Asamblea de Unidad Cantonal. Cotacachi, Ecuador.

15 Local Resolution of Watershed Management Trade-offs: the Case of Cotacachi

Fabián Rodríguez[1] and Douglas Southgate[2]

[1]*PO Box 17-10-7193, Quito, Ecuador,* [2]*Department of Agricultural, Environmental and Development Economics, The Ohio State University, 2120 Fyffe Road, Columbus, OH 43210, USA*

Introduction

Various environmental trade-offs arise in the Cotacachi watershed. Of particular concern to the local population is the declining quality and diminished reliability of water supplies for household and agricultural uses resulting from deforestation, poor pasture management and related problems in the upper reaches of the drainage basin. Water systems are often plagued by poor planning, which is based on erroneous assumptions about the needs and desires of rural populations. In many areas, there are no markets for water resources, and therefore no ways to evaluate the costs and benefits of system improvement. Elsewhere, as in Ecuador, there are markets, but prices are distorted by subsidies and other policies.

Economists have developed various techniques for evaluating natural resources in the absence of price signals, including what is known as contingent valuation (CV). This methodology calls for respondents to participate in simulated transactions in a hypothetical market setting, designed to reveal what people are willing to pay for non-market goods and services provided by the natural environment. The purpose of this study is to determine what, if anything, local people are willing to pay for the sort of watershed management that would make water supplies cleaner and less subject to interruption. Estimates of willingness to pay (WTP) are obtained using CV.

Literature Review

Throughout the developing world, water supplies are inadequate, in terms of quantity as well as quality. Among others, Whittington *et al.* (1990a) has examined the problem of wrongful planning of water system supplies and limited availability of service for potential users. They argue that poor planning is largely the result of inadequate information about potential users, which leads to erroneous assumptions about the needs and desires of rural populations. Additionally, there is no market or limited market for this commodity, which means that no price exists that can be used to evaluate the project's benefits. Another possibility, which arises regularly in Ecuador, is that there are markets, but prices are distorted by subsidies and other public policies.

Economists have developed various techniques for evaluating natural resources in the absence of efficient price signals. One of these is CV, in which individual participants in a survey make transactions in a hypothetical, though realistic, market setting. These simulated transactions take various forms, although all of them are designed to reveal what people are willing to pay for non-market goods and services provided by the natural environment (Haab and McConnell, 2002).

CV has been carried out frequently to determine how much Africans, Asians and Latin Americans are willing to pay for access to clean water. One of the earliest applications of this methodology in a poor country occurred in Haiti, where Whittington *et al.* (1990a) found that rural households were willing to pay US$1.78–US$2.22 per month for a connection to a potable water system. A somewhat wider range of values, from US$0.30 to US$6.15, was found in similar research carried out soon afterward in the Brazilian countryside (Briscoe *et al.*, 1990). CV studies in other parts of the developing world that address what rural households would pay for access to a potable water system have yielded the following results: US$3.93 per household per month in Kenya (Whittington *et al.*, 1990b); US$0.79–US$2.21 in Nigeria (Whittington *et al.*, 1991); and US$0.94–US$1.67 in the Philippines (Bohm *et al.*, 1993).

Something to note about these estimates is that the value that rural households place on water comprises an appreciable share of their modest earnings. Furthermore, differences among these estimates of WTP are not substantial, and certainly are not beyond what can be explained in terms of variations in earnings and other factors. These differences would have to be much larger to undermine confidence in the validity of CV. Indeed, this confidence is high enough that it is used routinely in economic appraisals of water projects (Whittington, 1998). For example, the Inter-American Development Bank recommends CV over alternative methodologies (Russell *et al.*, 2001).

Methodology

To assess the value that Cotacachi's rural population attaches to water and to analyse factors influencing these values, 80 households were interviewed about demographic, occupational, agricultural and other matters in September 2002. Half these households are in communities below 3200 metres above sea level (masl), which have irrigation, and the other half are in communities above this elevation level, where agriculture is rainfed. Along with questions relating to economic activities and income sources, the survey instrument contained referendum-style CV questions. In a typical question, a specific improvement in the local water system carrying a certain price was stipulated and the respondent was asked either to accept or reject the improvement and its price. Using the responses to CV questions as well as other survey data, we have undertaken econometric estimation of a model in which WTP for water quantity and quality improvements is the dependent variable and right-hand side variables include household earnings and the portion of income derived from off-farm employment.

Besides the CV analysis of potable water values, we have used data collected in the household survey to develop a linear programming (LP) model of a typical subsistence farm in the Cotacachi area. This model has been employed to assess the scarcity value of irrigation water.

LP Model

In the LP model, the maximand is agricultural gross margin, which is the difference between the value of agricultural output and the cost of variable agricultural inputs. Sumpsi *et al.* (1997) and Berbel and Gomez-Limón (2000) point out that this gross margin is a good measure of farm profits. In this model, the maximand amounts to:

$$NGM = f_{ocr} + f_{oli} + f_{oho} + a_{off}X_{off} - a_k X_k \quad (15.1)$$

which depends on objective function activities of the farm, off-farm work and capital:

f_{ocr} is the function of crop production; f_{oli} is the function of cattle and dairy production; f_{oho} is the function of hog production; $a_{off}\,X_{off}$ is the off-farming job of the household head, a_{off} is the farmer wage per day and X_{off} is the number of days dedicated to this activity; and $X_k - (1 + a_k)X_k$ is the capital availability for farmers.

The crop function can be defined as follows:

$$f_{ocr}\sum_{j=1}^{4} a_{sj}X_{sj}X_{Lj} - a_{bj}X_{bj} - Y_w - a_{Lj}X_{lj} - a_{rent}X_{Lrent} - la_j \qquad (15.2)$$

where a_{sj} is the price per quintal (q) of crop j, X_{sj} is the yield/ha of crop j; X_{Lj} is the amount of land dedicated to farming; a_{bj} is the price/q of product j that is bought for the minimum consumption requirement of the household; X_{bj} is the amount in quintals of product i that is bought for the minimum consumption requirement of the household; Y_w is the price per year of irrigation water that is paid by the farmer, Y_w is a fixed cost and it is reduced from NGM once the latest is estimated; a_{Lj} is the production cost per hectare of crop j; a_{rent} is the cost in dollars to rent 1 ha of land; X_{Lrent} is the amount of land that the farmer rents; and la_j is the cost in dollars of hiring extra labour. Four commodities (denoted by the subscript, j, in Equation 15.1) are included in the crop functions: (i) maize; (ii) beans; (iii) potatoes; and (iv) peas.

The part of the objective function related to cattle and dairy production can be defined as follows:

$$f_{oli} = a_{lis}X_{li}X_{lid} + a_{mi}a_yX_{li} - a_{lic}X_{li} \qquad (15.3)$$

where a_{lis} is the price of one unit of cattle sold at the second year; X_{li} is the amount of cattle units; X_{lid} is the cattle survival units estimated as a 0.85 survival rate; a_{mi} is the price for one unit of milk measures in cattle; a_y is the amount of milk per cattle unit; and a_{lic} is the cost to maintain one cattle unit.

The part of the objective function to do with hogs involves the farmer buying one suckling pig at the beginning of the year and selling it after fattening at the end of the year. The function can be defined as follows:

$$f_{oho} = a_{ohos}X_{ho}X_{hod} - a_{hoc}X_{ho} \qquad (15.4)$$

where a_{ohos} is the price per hog; X_{ho} is the number of hogs; X_{hod} is the hog survival rate, which is estimated as a 0.95 survival rate; a_{hoc} is the cost of raising a hog, which includes the price of buying the suckling pig, vaccines and other veterinary costs, and the cost of feeding them.

The head of the household can decide to work outside the farm, mainly as labour for other farms, in construction, or for local and government agencies. Earnings equal the daily salary, a_{off}, multiplied by the number of days worked, X_{off}.

Renting land is an option, and the cost of doing so is also included in the objective function: a_{Lrent} (annual per hectare payment) multiplied by hectares rented, X_{Lrent}. Likewise, interest (a_k per dollar) must be paid on outstanding loans (X_k).

The objective function, Equation 15.1, is maximized subject to various constraints. First, grain production cannot fall below a minimal level, which reflects subsistence farming in the study area. Second, production is bounded by available land, equal to the hectares owned by the household and those it is able to rent. Land use during the 4-month dry season is distinguished from land use during the wet season, which lasts 8 months. A rotation of maize and beans can be completed early in the former season, which allows enough time during the drier part of the year for a crop of peas. In contrast, a field planted to potatoes at the beginning of the rainy season is not harvested until well into the dry season, thereby precluding a complementary crop.

Land and the forage it provides is also required for livestock production, with 1 ha supporting three cows, six calves or a combination of the two. However, chickens are normally confined to small areas. So are hogs, which are fed hard maize, which must be purchased.

A third type of constraint in the LP model relates to water availability, both rainfall and what a household obtains by paying an annual fee: US$1.20 when the survey was conducted. During the wet season, each hectare used for the maize–bean rotation or planted to potatoes requires 1365 or 1869 m^3, respectively. A potato crop also needs 938 m^3/ha during the rest of the year. Alternatively, a dry season pea crop requires 727 m^3/ha.

Credit and labour are also constrained. Planting and other expenses cannot exceed the sum of cash available at the beginning of the year and up to US$1200 in loans taken out at that time and repaid soon after the harvest. The agricultural work force comprises the household head's spouse and dependent children over the age of 15, each of whom remains on the farm all week. The household head is assumed to work off-farm, which only allows him to tend to crops and livestock on one weekend day each week. Hiring labour is also an option.

CV regression model

An individual's consumption of a good or service is a function of its price, the prices of complementary and substitute goods, the individual's income and his or her individual preferences. Likewise, the most a respondent would pay for a good or service is a function of prices and his or her income and preferences.

In order to estimate the respondents' WTP for improvements of water systems in Cotacachi, this study used an approach featuring a dichotomous choice with follow-up. After the hypothetical project description, WTP questions began with one of three different bids: 1, 3 or 5 US$/month for a project with beneficial impacts on potable water supplies and 10, 30, or 50 US$/year as a fee for a project with irrigation benefits. These pre-determined bids were selected randomly for each household and a dichotomous answer (yes/no) was provided as alternatives for the household. If the household was willing to pay the initial bid, the monthly price was raised by US$1 and the household was asked again if it is willing or not to pay for the new bid. If the answer was still yes at the higher price, the respondent was asked to state the maximum amount that he is willing to pay. If the household rejected the project at the initial price, that value was reduced by half and the household was asked again to accept or reject. If the answer was still no, then the respondent was asked to state a maximum price for supporting the project.

As Whittington *et al.* (1990a) point out, the main biases that may affect CV regression model (CVM) estimates include strategic behaviour, hypothetical behaviour and starting-point biases. The first two biases did not arise because households were very familiar with water and its associated problems. Having different starting-point bids for each household that were selected randomly eliminated the starting-point bias. An important advantage in this region is the fact that all communities are familiar with water markets and very familiar with water prices.

According to Haab and McConnell (2002), the basic model for dichotomous choice responses is the random utility model. Accordingly, the indirect utility function for respondent *j* can be written as:

$$u_{ij} = u_i(y_j, z_j, \varepsilon_{ij}) \tag{15.5}$$

where $i = 1$ is the condition that prevails when a CVM is carried out, and the final state is $i = 0$. y_j is the income, z_j is an m-dimensional vector of the household characteristics, and ε_{ij} is the error term (Haab and McConnell, 2002).

Household annual income per capita can be defined as:

$$INC_j = f(WEALTH, DIST, FAMILY, EDUIND, PARTIND, \varepsilon) \tag{15.6}$$

where *WEALTH* is a proxy of household wealth (asset holding such as land ownership is seen as part of household wealth), *DIST* measures the distance in minutes from the household to the closest paved road, a proxy for the household access to the urban market and jobs, *FAMILY* is the number of members of the household that participate directly or indirectly in production activities, *EDUIND* is the stock of education of the household labour force employed during the year, *PARTIND* is community participation, and ε represents the error term and a proxy of risk.

An estimate of income was used as an exogenous variable in the equation for estimating maximum WTP for drinking water improvements, which is described below. A correlation exists between household income per capita (*INC*) and the ε of the second equation. To avoid this correlation problem, a predicted (fitted) value of income is used instead. Thus, Equation 15.6 was modified as follows:

$$INC_j^* = f(WEALTH, DIST, FAMILY, EDUIND, PARTIND, \varepsilon) \quad (15.7)$$

Land ownership is measured as a proxy of wealth *WEALTH*. It can be defined as:

$$WEALTH_i = q_i \varphi \quad (15.8)$$

where q_i is the quantity of land, measured in hectares, owned by household i, and φ is the shadow price of land in the area.

DIST is the measure of the time that takes for any member of the household to reach the closest paved road. This gives a proxy to measure household access to urban market where farmers can trade their yields and acquire the inputs needed, as well as the opportunity to access off-farming jobs.

Participation *PARTIND* is defined as all activities that members of each household get involved with in the community and is supposed to have an impact on the household annual income per capita *INC* (Robison *et al.*, 2000, 2002). Community participation is defined as:

$$PARTIND_i = \sum \left(\phi\delta_i \left[\left(\frac{t_i}{8} \right) \right], \phi\left(\frac{m_i}{T} \right), \phi\lambda_i \right) \quad (15.9)$$

where δ_i is defined as the leadership of household member i as elected officials (a proxy of voluntary participation); t_i is the time spent as elected officials of individual i; m_i is the total time of participation years of individual i; T is the time since the community organization was formed; λ_i is the participation perception rate of individual i; and ϕ is the participation weight.

This simple model of participation included a dichotomous option: voluntary participation or non-participation. The leadership and voluntary participation (δ_i) is weighted by multiplying the time (t_i) that an individual occupied an elected office. Thus, the model makes an important distinction between voluntary participation and non-participation. At the same time, the model does not give extra weight to the time spent on leadership activities. The time of leadership and voluntary participation is divided by 8 because Ecuadorian law allows two consecutive re-elections of 4 years term to the elected representative or executive office. In the area of study, the term in the elected representative or executive office is 1 year, thus 8 years of consecutive re-election terms seems reasonable enough as a time limit for public service.

In the simple model, non-voluntary participation is included. Non-voluntary participation is not weighted equally in the model. Neither are years of participation (m_i). In the simple model, the participation weight ϕ of voluntary/leadership participation (δ_i) is 2.3 times higher than the time of participation (m_i), and 3.5 times higher than non-voluntary participation (λ_i) (Table 15.1).

The second part of the regression model addresses households' WTP for improving drinking water systems. This value is a function of the expected household per capita square income INC_j^{*2} and the error term ε. The equation can be defined as follows:

$$MAX_j = f(INC_j^{*2}\ WEALTH, DIST, FAMILY, EDUIND, PARTIND, \varepsilon) \quad (15.10)$$

where MAX_j is the maximum amount of money that the respondent is willing to pay for improvements in the spring water systems, INC_j^{*2} is the predicted per capita square income and ε is the error term.

Additionally, the dichotomous with a follow-up approach gives 0 or 1 answers following with a maximum WTP. Therefore, it is necessary to know if there is a relationship between the maximum WTP and the randomly selected starting-point bid for each household. The hypothesis is that the

Table 15.1. Descriptive statistics of main variables.

Variable	Mean	SD	Skewness	Minimum	Maximum
WEALTH	114.7251	163.2254	2.858993	5.83	923.40
DIST	42.1875	18.03379	0.934561	15	90
FAMILY	5.5375	2.343554	–0.00915	1	11
EDUIND	0.173224	0.149436	1.479503	0.00	0.70
PARTIND	0.361997	0.125558	0.126105	0.02	0.75

starting-point bid may affect the final WTP bid. Thus, the initial dichotomous bid is a function of the starting-point bid and the other explanatory variables; the correspondent equation can be written as follows:

$$WTPINI_j = f(INC\hat{F}_j^2, BID, WEALTH, DIST, FAMILY, EDUIND, PARTIND, \varepsilon) \quad (15.11)$$

where INC_j^2 is the predicted per capita square income, *BID* is the initial randomly selected starting-point bid for each household (the bids are US$1, US$3 and US$5). Finally, ε is the error term.

Results and Discussion

One hundred and twenty households from ten different communities of the study area were originally included in the survey sample. However, one community, with approximately 30 households, refused to participate, although six of these households subsequently agreed to do so. Funding and time constraints also limited the sample. Consequently, data were collected from just 80 households.

From the survey results, on-farm and off-farm activities were recorded and factored into annual income. Income per capita comes from different production activities (Table 15.2). Only 35% of the sample population is dedicated exclusively to agriculture. The biggest source of income is off-farm labour; household heads, for example, typically work on urban construction sites or bigger farms.

Formal schooling in the communities of this study shows a pattern similar to that of other rural regions in Ecuador (Pichón and Bilsborrow, 1992; Rodríguez, 1996). Approximately 27% of the sample population does not have any education and another 35% did not go beyond elementary school since many children drop out during primary school to join the family labour pool. Together, these two categories include > 50% of the sample population. Only 3% of the sample population has a university diploma and approximately 7% has high or middle school education (Table 15.3).

The main crops are maize, beans, potatoes and peas in the highlands. Other important crops in the region are lima beans, lupine beans, wheat and quinoa. Recently, a small shift has been taking place on small farms to onions, cabbage and other annual crops, which can be sold for cash. Nevertheless, maize, beans and peas are still the main crops of Cotacachi communities. Approximately 93% of households plant and harvest maize, mainly for personal consumption. Beans are the second most important crop for these communities, planted by approximately 71% of total households. Peas are also important, with a little more than 50% of households harvesting this crop for their own consumption (Table 15.4).

The proportion of farming in the household annual income is compared with off-farming activities (Table 15.5). A summary of mean and standard deviation of the annual household income of the ten communities participating in this study is reported in the table, along with the percentage of that income from agricultural activities and off-farm.

The main objective of LP modelling and the CVM analysis in this study is to estimate the value of water. With estimates

Table 15.2. Sources of annual income (US$) in the Cotacachi communities.

	Mean	SD	Minimum	Maximum	Proportion of sample population
Farm production	125.8988	276.6928	0	1400.00	0.35
Herd production	110.9194	216.7663	0	1410.00	0.5125
Dairy products	215.3125	568.0837	0	3650.00	0.325
Textile sales	361.2	1407.547	0	9600.00	0.2125
External family assistance	4.65	29.23805	0	192.00	0.025
Off-farm labour	870.3	715.2119	0	3600.00	0.7875
Groceries store	57	276.0132	0	2160.00	0.0625

Table 15.3. Education level in Cotacachi communities.

	Proportion		
Education level	Sample population	Upland population	Lowland population
No school	0.271	0.328	0.175
Not attending	0.1869	0.216	0.137
Elementary school	0.3505	0.358	0.337
Junior high school	0.0701	0.045	0.113
High school	0.0794	0.03	0.162
University	0.0327	0.015	0.063
Professional or technical school	0.0093	0.008	0.013

Table 15.4. Crops yield and proportion of the sample raising crop.

Crops (in quintals)	Entire sample		Upland		Lowland		House-holds (*n*)	Proportion of sample raising crop
	Mean	SD	Mean	SD	Mean	SD		
Beans	1.3614	1.4793	1.4153	1.3301	1.2632	1.7765	57	0.7125
Barley	7.4875	12.2072	10.3269	14.4822	2.2143	1.584	20	0.25
Maize	13.2933	21.344	11.7979	11.871	15.8036	31.5934	75	0.9375
Potato	25.7734	51.8217	36.381	61.8563	5.5227	5.9775	32	0.4
Lima bean	6.5077	14.1439	9.2056	16.536	0.4375	0.2394	13	0.1625
Melloco	6.4375	4.4645	6.4375	4.4646	N/A	N/A	4	0.05
Oca	5.4375	3.7989	5.4375	3.799	N/A	N/A	4	0.05
Wheat	4.2143	7.8678	5.7188	10.3194	2.2083	1.8467	14	0.175
Lupin bean	2.1656	4.3174	1.376	1.8511	4.9857	8.4097	32	0.4
Peas	0.8274	1.0855	0.5542	0.5198	1.3306	1.5119	42	0.525
Lentils	2.2143	3.4953	0.25	0.00	3	3.9528	7	0.0875
Quinoa	0.6458	0.7794	0.45	0.2582	1.625	1.9445	12	0.15

N/A = not applicable.

Table 15.5. Household income and sources of income of the Cotacachi communities.

Community	Household income		Income from farming (%)	Income from non-farming activities (%)
	Mean	SD		
Upland without irrigation				
Arrayanes	777.42	485.0023	23.00	77.00
Italquí	1278.94	1473.891	11.00	89.00
Morochos	990.47	601.7152	10.00	90.00
Topo Grande	1018.84	770.857	15.00	85.00
Ugshapungo	3058.77	3036.075	69.00	31.00
Lowland with irrigation				
Chilcapamba	880.41	1199.674	29.00	71.00
Piavi San Pedro	2201.74	1728.358	34.00	57.00
Santa Barbara	1446.51	1439.641	34.00	66.00
Morales Chupa	2123.86	3286.92	9.00	91.00
Turucó	746.06	442.7867	46	54.00
Entire sample	1325.73	1612.085	24.28	75.18

of these values, policy proposals can be developed to improve living standards and to conserve natural resources. The shadow prices obtained from the LP model provide guidance for the development of water for agricultural uses. Likewise, CVM estimates of WTP are important criteria for the design of improvements in drinking water systems.

LP results

The LP model was run in two ways, one for an operation with irrigation water and the other for a farm without irrigation water. Additionally, one of the economic activities was modified in some runs of the model to make it more realistic, corresponding better to farming realities in Cotacachi. In particular, the LP initially had cattle and hogs competing for land with the other economic activities. However, farmers in Cotacachi generally use community land as a source of forage, along with their own fields. The model's land constraint was modified to reflect this practice, which obviously reduces the opportunity cost of livestock production from the household's perspective.

Runs of the LP models, both the standard version (in which the farmer is allowed to rent additional land for his livestock to graze) and the constrained version (in which cattle, hogs and other domesticated animals graze only on the farmer's own land), are compared with survey results (Table 15.6). LP analysis suggests that non-irrigated and irrigated farms should combine both farming and dairy because this would yield the highest net return.

When renting land is not an option, farmers dedicate all the land they possess to the maize–bean rotation during the rainy season and pea production during the dry season. With the standard LP, the maximum area (10 ha) is rented. The maize–bean rotation following pea production still features a higher net return than growing potatoes. The survey results show that the typical farmer with irrigated land sells part of his output: 10 q of maize and 4 q of beans and uses 1 ha for the maize–bean rotation, 1 ha for potatoes and 0.40 ha for peas. The typical farmer with non-irrigated land does not sell any crops, but uses 1.28 ha for the maize–bean rotation, 2.5 ha for potatoes and 0.305 ha for pea production. All the output from non-irrigated land is used for household consumption.

Since dairy farming is profitable, farmers should use community land, as allowed

Table 15.6. Linear programming solution and survey results.

	Non-irrigated land			Irrigated land		
	Survey results	Standard LP model	Constrained LP model	Survey results	Standard LP model	Constrained LP model
Net gross margin ($)	467.25	6,522.03	3,327.26	763.90	5,949.54	2,437.14
Maize sell (q)	0	280.31	90.85	10	235.08	53.66
Beans sell (q)	0	32.32	5.84	4	91.42	20
Potatoes sell (q)	0	0	0	0	0	0
Peas sell (q)	0	22.81	14.34	0	52.73	28.66
Land area maize–bean rotation (ha)	1.28	14.95	4.95	1	13.06	3.06
Land area peas (ha)	0.305	10.05	4.95	0.4	5.27	3.06
Maize purchase (q)	0	0	0	0	0	0
Beans purchase (q)	0	0	0	0	0	0
Potatoes purchase (q)	0	0	1.92	0	0	1.75
Peas purchase (q)	0	0	0	0	0	0
Hard maize purchase (q)	36	0	0	12	0	0
Hogs units	1	0	0	1	0	0
Livestock units	6	6	6	5	5	5
Dairy production (li)	607.5	3,645	3,645	547	2,733.75	2,733.75
Livestock sell units	1	5	5	1	5	5
Water needs, rain season (m^3)	669.8	19,432.29	6,085.41	669.8	17,153.13	3,506.15
Water needs, dry season (m^3)	–196.1	5,175.45	3,769.31	–196.1	4,004.48	2,394.68
Hired labour maize–bean rotation (days)	6	0	0	12	0	0.00
Hired labour potatoes (days)	6	0	0	12	0	0.00
Farm labour maize–bean rotation (days)	12	179.4	59.4	15	195.9	45.90
Farm labour potatoes (days)	12	0	0	15	0	0.00
Farm labour peas (days)	12	82.6	59.4	15	79.1	45.90
Farm labour livestock (days)	12	72	0	12	60	60.00
Head labour maize–beans rotation (days)	1	0	0	1	0	0
Head labour potatoes (days)	1	0	0	1	0	0
Head labour peas (days)	1	0	0	1	0	0
Off-farm work (days)	5	5	5	5	5	5
Land rent (ha)	0	9.999	0	0	10	0
Capital loan (US$)	0	0	0	0	0	0

in the LP model, for their herds. Differences between the findings of LP analysis and survey results are due to the fact that the LP model assumes that the entire herd can be dedicated to milk production, while the survey reveals that only one cow is used for this purpose. Survey results in both settings show that 12 days of labour should be dedicated to livestock, in contrast to what the LP models suggests: 60 days for an irrigated farm and 72 labour days for a non-irrigated operation.

All LP models suggest that the typical farmer should not hire anyone because family labour is sufficient for all farm requirements. The main reason for this is the fact that a household has to pay for the extra labour with crops. Contrary to the LP models, survey results show that the average irrigated farm hires workers for 12 days; for a non-irrigated farm, 6 labour days are hired. In both operations, pea production uses only family labour. The unconstrained LP models suggest using 179.4 and 195.9 days of labour (for non-irrigated and irrigated farms, respectively) for the maize–bean rotation, and the constrained LP models suggest using 59.4 and 45.9 days (non-irrigated and irrigated farms, respectively) for this activity. In unconstrained LP models, 82.6 and 79.1 days (non-irrigated and irrigated farms, respectively) are suggested. With the constrained models, 59.4 and 45.9 days, respectively, are used for this purpose.

There are some differences between the standard LP models and the constrained LP models mainly due to the limit on land rental incorporated in the latter. However, all LP models suggest that farmers are not using their resources efficiently. In particular, there is a substantial difference between the survey results and the LP models in terms of maximum net profits (Table 15.6).

The conditions of the two representative farms, with and without irrigation, illustrate the general conditions of the majority of farmers in the study area. Most of them engage in subsistence agriculture, with output used almost entirely for consumption. In this kind of subsistence agriculture, there is not much investment. Farmers do not use commercial fertilizer to enrich and protect the soil or insecticides to eliminate pests. Neither do they use commercial seeds that can produce more per area. Furthermore, there is little production of other crops, which could generate more revenues at a relatively low cost. Basically, farmers concentrate their efforts on their traditional crops for subsistence, such as maize, beans, potatoes and peas.

Another constraint can be added to the LP models, a constraint that makes the herd compete for the land available on the farm with crops. This would represent an alternative to the current practice of using community land to pasture livestock. With such a restriction, farm earnings would be reduced significantly, by approximately 70% on a non-irrigated farm and 80% on an irrigated holding. Based on information provided by the Ministry of Agriculture, it is necessary to have 2 ha of natural pasture per animal. In parts of the area of study, cattle compete with crops for available land. As a result, farmers dedicate part of their land just to have natural pasture. Yet, in other cases, such as the two representative farms analysed with the LP models, cattle graze on community land. This reduces the cost of raising cattle for the livestock owner. When cattle compete with crops for real estate, the model suggests that livestock production is unprofitable. Instead, raising crops is the most rewarding activity for farmers.

The shadow price for irrigated land is US$58.80/ha. In contrast, the shadow price of non-irrigated land is lower, at US$48.6/ha. This difference relates to the higher crop yields and agricultural net returns on irrigated fields. Another reason why irrigated land is worth more is that it is closer to urban areas than non-irrigated land. Also, there has been more subdivision of land in irrigated districts. Average farm size in communities with irrigated land is 2.58 ha, compared with an average of 3.29 ha in non-irrigated settings. Subdivision reflects population density, which correlates with demand for real estate.

An important reason for using the LP model in this study was to obtain a measure of the shadow price of water. In both representative farms, without and with irrigated land, the LP model yielded the values reported in Table 15.7. The shadow price in both representative farms is estimated after obtaining the amount of water required for each crop with the optimal solution.

Shadow prices of water show how undervalued the resource is in the region. The values obtained by the LP model are expressed in US$/m^3, which contrasts to how farmers are currently assessed for this resource. The values for non-irrigated and irrigated farms during the rainy season are very similar. Both reflect the amount of water available during the season as well as the productivity of the two farm categories. In contrast, land values during the dry season differ substantially. This reflects resource availability on both types of land when there is little precipitation.

The US$1.20 per growing season that each farmer now must pay reflects a history of heavy government subsidies. Farmers with irrigation should pay approximately US$0.83/m^3 of water/ha. With these subsidies, no conservation programme for water can be sustainable.

Capital is not binding in either farm, irrigated or non-irrigated. It is interesting that farmers in both settings do not request any loan from available sources, even with a line of credit of US$1200.00. Similarly, farmers in both versions of the LP model, without and with irrigation, do not apply for loans even when a line of credit is open. Of course, the absence of borrowing reflects modest use of purchased inputs.

Table 15.7. Shadow prices of land and water.

Constraints	Shadow price of land	
	Irrigated	Non-irrigated
Land wet season (ha) solution	58.8	48.6
Land dry season (ha) solution	0	0
Water wet season limit solution	0.273219	0.26
Water dry season limit solution	0.11	0.02

Land available during the dry season is also not binding. The model suggests using only 0.5 ha of the farmer's own land and the rest of the land should be rented to others. Limited availability of family labour is a binding constraint. However, fewer than the 40 days per year that the household head could dedicate to agriculture are actually used. This reflects the option that he has of working off-farm for a little less than US$3.00 per day.

Regression results

Survey responses from both non-irrigated land and irrigated land were combined and a discrete continuous model was estimated. Household income per capita was estimated using an ordinary least squares regression. For all farm and off-farm activities of the household, net revenues were computed from a detailed account of gross revenues and costs. Household consumption of its own production was carefully assigned for agricultural and non-agricultural activities. Various explanatory variables related to income were addressed in the survey. In Table 15.1, the means and standard deviations of these explanatory variables were presented. Coefficients, standard errors and *t*-statistics are summarized in Table 15.8. The results performed as expected with the main variables.

As expected, household wealth (*WEALTH*) has a positive and statistically significant coefficient, indicating that farmers who own more land were those who had higher incomes. However, two of the farmers with the highest income of the sample were dropped from the sample because they are outliers and their inclusion may have affected the analysis. In the surveyed communities, farmers with more land are considered the richest. Similarly, distance (*DIST*), which is measured in minutes of time that a household member takes to reach the nearest paved road and which serves as a proxy for access to urban markets and jobs, has the expected negative sign and is significant. Household family size (*FAMILY*) has a coefficient that is negative and statistically

Table 15.8. Estimation of income equation.

Dependent variable: income per capita
Method: ordinary least squares

Variable	Coefficient	SE	*t*-statistic	*P*
C	388.0754	109.5207	3.543397	0.0007
WEALTH	0.537994	0.160719	3.347415	0.0013
DIST	−2.980699	1.387526	−2.148212	0.0351
FAMILY	−23.70548	10.54953	−2.247065	0.0277
EDUIND	383.4886	174.6881	2.195276	0.0314
PARTIND	−33.63799	156.2893	−0.215229	0.8302
R^2	0.307529	Mean dependent variable		242.9099
Adjusted R^2	0.259440	SD dependent variable		248.0652
Log likelihood	−525.9099	*F*-statistic		6.395087
Durbin–Watson stat	2.334300	Probability (*F*-statistic)		0.000057

significant. Large families have lower per capita incomes, which is not a surprise because it was assumed that the incomes of the economic active members of the household are shared with the rest of the members. Education (*EDUIND*) also has the expected positive coefficient and shows significant levels in this study, as is the case in other studies.

Although it was expected to have an effect on income, individual participation in community activities (*PARTIND*) was not found to be statistically significant. One supposes that individuals would get involved in activities that provide them with benefits. The result of this study may be due to the likelihood that individuals could not see the potential benefits they can obtain from participation. Alternatively, these benefits are so far in the future that they do not have any effect on their current expected utility, which may be the reason for the negative sign of the coefficient.

Income estimation (Equation 15.6) was undertaken mainly to obtain the predicted (fitted) value of income (*INC**), which is used as explanatory variable for the maximum WTP (*MAX*) for improvements of the spring water systems. A censored TOBIT model was used to estimate *MAX*. A censored TOBIT model is used in an analysis when some information (called censored) is missing in the dependent variable, in this case the maximum WTP for the respondents. This missing information comes from the lack of information on those members of the communities that were not part of the survey. The sample of this study was small, and as a result could not record all community members' preferences. Ordinary least squares estimation of this censored regression model could generate biased and inconsistent parameters. Thus, a censored TOBIT model was used instead.

The maximum likelihood TOBIT model performed as expected with the main variables and indicated which factors tend to influence households' WTP. The *MAX* equation revealed that, on average, households would pay 50.5% more than what they are currently spending on water to improve the quality and reliability of drinking water. Respondents' maximum WTP was related positively to income; as expected, people with higher incomes were willing to pay more to improve the quality of their drinking water. The coefficient, standard errors and *t*-statistics are summarized in Table 15.9.

The size of the family presents a significant value and the expected positive sign in the coefficient. Households with more members are willing to pay more for improving the quality and availability of drinking water systems. This was also expected because the demand for water is higher in large households; as a result, they would

Table 15.9. Estimation of maximum willingness to pay (*MAX*) equation.

Dependent variable: maximum willingness to pay for improving water quality (*MAX*)
Method: ML-censored normal (TOBIT)

	Coefficient	SE	*z*-statistic	*P*
C	−0.670643	1.721370	−0.389598	0.6968
INCF^{*2}	2.05E-05	1.05E-05	1.947655	0.0515
WEALTH	−0.007866	0.004501	−1.747581	0.0805
DIST	−0.004118	0.019222	0.214210	0.8304
FAMILY	0.304212	0.149188	2.039116	0.0414
EDUIND	−3.558061	3.030550	−1.174064	0.2404
PARTIND	2.089149	1.426531	1.464496	0.1431
FERTI × *IRRI*	−0.336067	0.153692	−2.186634	0.0288
		Error distribution		
Scale: C(9)	1.715276	0.157091	10.91902	0.0000
R^2	0.136349	Mean dependent variable		1.845000
Adjusted R^2	0.039036	SD dependent variable		1.569237
Log likelihood	−142.1771	Hannan–Quinn criterion		3.886868
Average log likelihood	−1.777214			
Left censored observations	15	Right censored observations		0
Uncensored observations	65	Total observations		80

request increasing quality and reliability of the system and would be willing to pay to ensure it. The results also confirm how important water is for the members of Cotacachi communities and also how important family relationships are.

A variable that combines the effects of fertile soils and irrigation (*FERTI* × *IRRI*) is added to see if these factors have any effect on a household's WTP. Both variables estimated separately do not present significant levels, but as a product have a significant value and a negative coefficient. It seems that households with unfertile farms and no irrigation are willing to pay more for improving the spring water systems. Farmers with no irrigation may weigh their decisions more carefully to have better quality and reliable amounts of drinking water and are willing to pay more for improvements in the system.

The other variables of the maximum WTP equation did not have statistically significant coefficients. However, all but one of the coefficients had the expected signs. The only exception was for the coefficient for *WEALTH*, which was negative even though it had been expected that people with more assets would be willing to pay more for improvements in the water system.

As in the ordinary least squares estimation of income, participation (*PARTIND*) was statistically insignificant. Education (*EDUIND*) was not statistically significant either, even though it was expected that families with more education are more aware of the potential health problems that a spring water system can carry and thus may be willing to pay for dealing with this problem. Additionally, this study did not find any significant difference between households with non-irrigated land and those with irrigated land. Most of the households were willing to pay between US$1.00 and US$3.00. The maximum payments that the two groups are willing to make are reported in Figs 15.1 and 15.2.

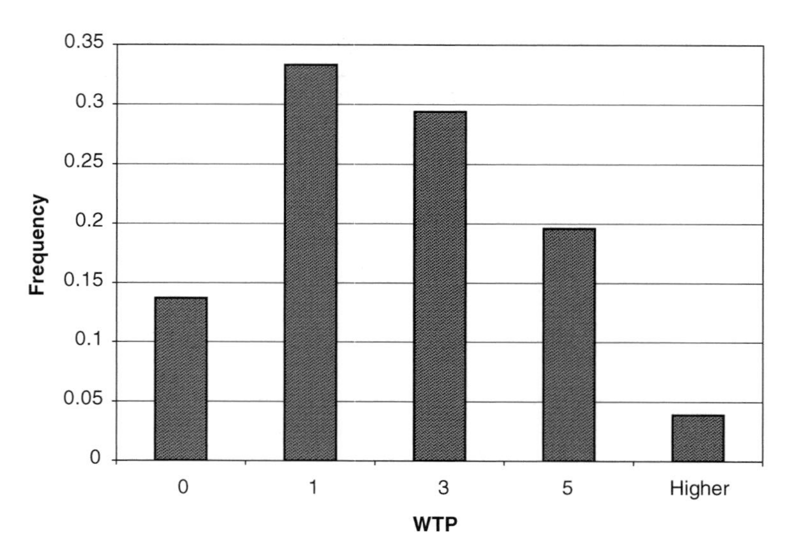

Fig. 15.1. Frequency of WTP bids (US$/month) in non-irrigated settings.

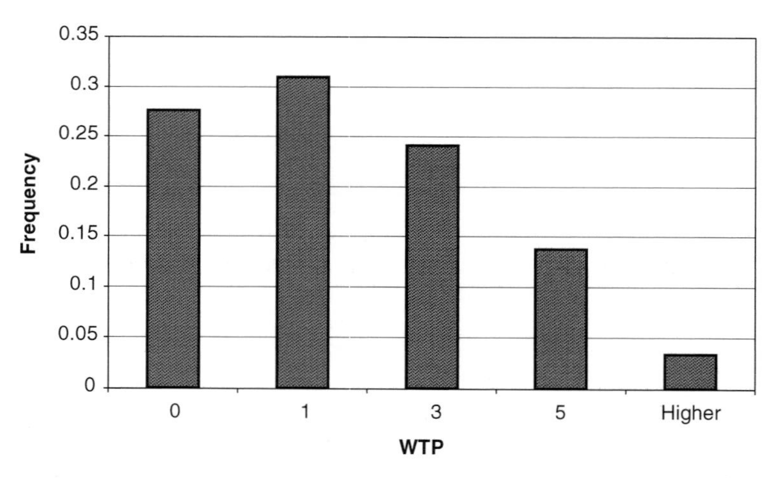

Fig. 15.2. Frequency of WTP bids (US$/month) in irrigated settings.

The zero responses may be due to, as the PROBIT model revealed, the income of the household, with families with lower incomes more likely to say no to the initial bid. As discussed earlier, additional analysis is required of the relationship between maximum WTP and the initial random bid selected for each household. A maximum likelihood PROBIT model was used to estimate the relationship between the randomly selected initial bid and maximum WTP. Potential payments for the improvement of spring water systems appear to have been influenced by the randomly selected initial bid (Table 15.10).

None of the explanatory variables has a statistically significant impact. However, most of the coefficients have the expected signs. As expected, the square of expected per capita income has a positive coefficient. Farmers with higher income are willing to say yes to participate and are also willing to pay more for improvements in their drinking water systems. Additionally, it

Table 15.10. Responses to a randomly selected initial bid.

Dependent variable: Initial dichotomous choice yes = 1, no = 2
Method: ML–binary probit

Variable	Coefficient	SE	z-statistic	P
C	−1.408094	1.192122	−1.181166	0.2375
INCF^{*2}	1.49E-06	7.82E-06	0.190461	0.8489
BID	0.028865	0.096229	0.299964	0.7642
WEALTH	−0.000112	0.003311	−0.033966	0.9729
DIST	0.011940	0.013069	0.913621	0.3609
FAMILY	0.079684	0.109169	0.729917	0.4654
EDUIND	−0.066220	2.234188	−0.029640	0.9764
PARTIND	−0.072276	1.094205	−0.066053	0.9473
Mean dependent variable	0.375000	SD dependent variable		0.487177
Log likelihood	−51.56118	Hannan–Quinn criterion		1.584532
Restr. log likelihood	−52.92506	Average–log likelihood		−0.644515
LR statistic (7 df)	2.727759	McFadden R^2		0.025770
Probability (LR stat)	0.908994			
Observations with Dep = 0	50	Total observations		80
Observations with Dep = 1	30			

seems that farmers are willing to say yes to a higher bid.

Conclusion and Recommendations

The main objective of this study was to determine if communities are willing to help pay for improving the quality and reliability of their spring drinking water systems. Failure of government policies has led communities of Cotacachi to seek outside assistance to build their own drinking water systems. They have received limited financial support from Switzerland to build the Cambugan system, which provides spring drinking water to 18 communities. Similar projects have been implemented during the last couple of years to reach more communities, and still more are being planned. Because of the shortage of funding, existing systems have limitations regarding quality of water. As a result, quality standards of the drinking water system have not matched those of urban areas. A study by Auburn University, undertaken for the SANREM–Andes project, found that all supply sources of drinking water are contaminated with *Escherichia coli* and other bacteria (Ruiz-Córdova *et al.*, Chapter 16, this volume). In light of the finding, our study sought to estimate how much more households would be willing to pay for a system that delivers water of a better quality. The results showed that respondents in the ten communities that participated in this study are willing to pay an average of US$ 1.84 per month (SD 1.5692, SEM 0.188) to improve the quality and reliability of their system, which is approximately 50% more than what they are currently paying.

This study reveals the problems related to drinking water that most indigenous communities in the Andes of Ecuador face. Most of these communities have enough water, but mismanagement has led to shortages and quality problems. Drinking water systems in the area of study tend to be under-dimensioned and have financial problems. Each system was not able to raise enough funds to be able to cover its maintenance and operating costs. There were problems collecting the water fees from consumers. Also, there seemed to be corruption in at least one of these community systems. Specifically, the individual who was hired to collect fees and fix any pipe problem kept those funds for himself. Fortunately, the problem was discovered early, and the collector was fired and replaced. However, these problems add to financial challenges. Regarding supply, there are constant problems with breaks in the pipe system. During the dry season, households complain about the quantity or lack of water. Along with these problems, a potable system that could provide high water quality would require heavy investments, including investment in water treatment capacity.

Economies of scale are an important dimension of these problems. To deal effectively with this factor, communities could organize and create a unified management system that could deal better with operating, maintenance and construction costs. Also, a unified system would be in a better position to access credit markets. In addition, cooperative action would allow the water management system to be more cost-effective and determine effectively its internal funding sources. Moreover, it would have more access to national or international assistance.

Finally, this study has a small sample and, because of that, the findings may not be consistent with the results of other studies. However, it shows that farmers in the area of Cotacachi are willing to pay something to improve the quality and reliability of their drinking water systems. Additional research is needed. Furthermore, it would be interesting to carry out a CV analysis addressing potential payments by farmers for improvements in irrigation systems. LP analysis suggests that the prices they currently pay are low relative to the value of these improvements.

References

Berbel, J. and Gómez-Limón, J.A. (2000) The impact of water pricing policy in Spain: an analysis of three irrigated areas. *Agricultural Water Management* 43, 219–238.

Bohm, R., Essenburg, T. and Fox, W. (1993) Sustainability of potable water services in the Philippines. *Water Resources Research* 29, 1955–1963.

Briscoe, J., Furtado de Castro, P., Griffin, C., North, J. and Olsen, O. (1990) Toward equitable and sustainable rural water supplies: a contingent valuation study in Brazil. *World Bank Economic Review* 4, 115–134.

Haab, T. and McConnell, K.E. (2002) Valuing environmental and natural resources: the econometrics of non-market valuation. In: *New Horizons in Environmental Economics Series.* Edward Elgar, Northampton, UK.

Pinchón, F.J. and Bilsborrow, R.E. (1992) Land-use System. Deforestation and Associated Demographic Factors in the Humid Tropics: Farm-level Evidence from Ecuador, In: *Seminar on Population and Deforestation in the Humid Tropics*; November 30 to December 3 (Draft Report), International Union for the Scientific Study of Population and Associação Brasileira de Estudos Populacionais, Campinas, Brazil.

Robison, L.J., Siles, M.E., Bokemeier, J.L., Beveridge, D., Fimmen, M., Grummon, P.T. and Fimmen, C. (2000) *Social Capital and Household Income Distribution: Evidence from Michigan and Illinois.* Social capital research report No. 12. Michigan State University, East Lansing, Michigan.

Robison, L.J., Siles, M.E. and Schmid, A.A. (2002) *Social Capital and Poverty Reduction: Toward a Mature Paradigm.* Social capital research report No. 13. Michigan State University, East Lansing, Michigan.

Rodríguez, F. (1996) *Economic Assessment of Timber and Non-timber Products Project at Yasuní National Park, Napo-Ecuador.* Tropical Research Institute, The Center for Tropical Forest Science Scientific Report, The Smithsonian Institute, The Smithsonian Tropical Research Institute, Washington, DC.

Russell, C., Vaughn, W., Clark, C., Rodríguez, D. and Darling, A. (2001) *Investing in Water Quality.* Inter-American Development Bank, Washington, DC.

Sumpsi, J.M., Amador, F. and Romero, C. (1997) On farmers' objectives: a multi-criteria approach. *European Journal of Operational Research* 96, 64–71.

Whittington, D. (1998) Administering contingent valuation surveys in developing countries. *World Development* 26, 21–30.

Whittington, D., Briscoe, J., Mu, X. and Barron, W. (1990a) Estimating the willingness to pay for water services in developing countries: a case study of the use of contingent valuation surveys in southern Haiti. *Economic Development and Cultural Change* 38, 293–311.

Whittington, D., Mu, X. and Roche, R. (1990b) Calculating the value of time spent collecting water: some estimates for Ukunda, Kenya. *World Development* 18, 296–280.

Whittington, D., Lauria, D.T. and Mu, X. (1991) A study of water vending and willingness to pay for water in Onitsha, Nigeria. *World Development* 19, 179–198.

16 Community-based Water Monitoring in Cotacachi

Sergio S. Ruiz-Córdova,[1] Bryan L. Duncan,[1] William Deutsch[1] and Nicolás Gómez[2]

[1]*Department of Fisheries and Allied Aquacultures, Auburn University, Auburn, AL 36849, USA*; [2]*SANREM–Andes, Cotacachi, Ecuador*

Introduction

Water quality and water quantity issues are critical and growing concerns of the 21st century, and are emerging as key issues for communities. In the Third World Water Forum held in Japan in 2003, representatives from > 170 countries agreed that worldwide > 1.2 billion people do not have access to adequate supplies of safe water, and 3 billion people do not have adequate sanitation (United Nations Non-Governmental Liason, 2003). Each year worldwide, an estimated 4 billion episodes of diarrhoea result in approximately 2 million deaths, mostly among children. Water-borne bacterial infections may account for as many as half of these episodes and deaths (Centers for Disease Control and Prevention, 2003). Eighty per cent of all diseases in developing countries are caused by contaminated water and sanitation-related problems. The poor pay the most for water while lacking piped water and sanitation systems, and suffer the greatest from impaired health and lost economic opportunities. Deutsch *et al.* (2003) noted that people's concerns often focus first on drinking water quality and public health (bacterial contamination), and second on soil erosion and sedimentation which directly affect their livelihoods. New approaches should emphasize prevention, rather than curative intervention, to overcome these failures, reduce poverty and conserve the environment. Effective water resources management must help alleviate poverty rather than making the poor the victims of bad decisions and policies.

Waterborne diseases in Latin America include epidemic cholera that re-emerged in 1991 after a 90-year absence. Since then, > 1 million cases of cholera have been reported in 20 countries in the region (Ampel, 1996). Other diseases commonly associated with water contaminated by faecal matter in tropical South America include: hepatitis A; trachoma blindness; typhoid fever; diarrhoeal diseases caused by *Escherichia coli* and other enteric bacteria such as *Salmonella*, *Shigella* and *Campylobacter*; amoebiasis; giardiasis; and cryptosporidiosis. Human health is further threatened by pollutants which enter drinking and bathing water from industry, mining and agricultural runoff, such as toxic chemicals, heavy metals, acids, pesticides and fertilizers. Environmental degradation, wildlife loss and reduced food production result. Effective water management must ensure involvement of all stakeholders, the best use and protection of available supplies, and should seek to limit conflicts over access to water.

In Ecuador, public health and the safety of drinking water is of greatest local concern.

Most public water supply systems depend upon conveyance of mountain spring water to community taps or homes, with intermittent or no treatment (chlorination, filtration, etc.). As a result, communities are reporting an increasing incidence of waterborne diseases. Increasing population and problems of waste disposal magnify the problem. These problems have been increasing in the last few decades, and are a cause for alarm among citizens (Duncan and Deutsch, 2001). Vulnerability to waterborne disease has been brought home by cholera outbreaks. From 1991 to 1993, the El Tor cholera pandemic in Ecuador resulted in 85,023 diagnosed cases and 977 fatalities (Whiteford *et al.*, 1996). After the pandemic had subsided in 1994, pockets of cholera remained. In 1996 a cholera outbreak occurred in Otavalo canton a few miles from Cotacachi. A total of 416 cases and four deaths were reported, the majority from indigenous communities surrounding the city of Otavalo (Pan American Health Organization, 1996).

On 11 September 2000, Cotacachi (Imbabura province, Ecuador) was declared an Ecological Canton and the 2003 Cantonal Assembly voted unanimously to place water as the number one priority for the canton. A municipal decree ('ordenanza') was approved in Cotacachi to establish exclusive zones along all water bodies (streams, springs, etc.) that should be left undisturbed by human activity. The municipal ordenanza calls for a 6 m wide buffer zone on both sides of streams known as stream management zones (SMZs). SMZs protect streams from direct solar radiation, maintaining stable diurnal and seasonal temperatures, are natural filters for elements that reach streams through runoff during rain events, and in general create more stable microhabitats for living aquatic communities. It is a challenge for canton, municipal and non-governmental groups to make people aware of the benefits of the SMZs, and to make them a reality. Public awareness and participation in the process of environmental management is required for lasting, positive impact. A lack of understanding of the biophysical alterations of water in Ecuador is as much a problem as the contamination itself. For that reason, this project strongly emphasizes the need for education and public involvement in water resource management to derive sustained solutions to environmental problems.

In Cotacachi canton, even the capital Cotacachi has no wastewater treatment. The Pichavi, Pichambiche and Yanayacu rivers all discharge into the Ambi river that is already degraded by sewage and wastewater from the neighbouring city of Otavalo. In additional to faecal contamination, land use changes are affecting water quality. Use of pesticides on agricultural and horticultural crops results in leaching and runoff of these chemicals into public water supplies. Mining activities in some areas threaten water quality and integrity of watersheds by altering natural stream flows and contaminating surface waters with metals and acid. Other industrial activities (leather, textiles, etc.) and the recent alarming increase of flower culture are introducing toxins to water in the area.

Watershed restoration and conservation, and its intimate link to water for human consumption, is a priority public issue in canton Cotacachi. Awareness and public interest is being created through the communication media. Local and national newspapers, radio talk programmes and television are presenting information related to ecology and water issues on a regular basis. The Community-based Water Monitoring (CBWM) project involves citizens in water monitoring activities, thus promoting community-based, participatory research, and empowering people dependent upon natural resources to have a role in their sustainable use and management.

The objectives of this research are as follows:

1. To foster the development of community-based water monitoring groups, to collect credible water quality/quantity data for environmental and policy improvements and to provide technical support to citizen monitoring groups for collection of data on water quantity and quality.

2. To ensure that the number and distribution of water testing sites is adequate to draw conclusions regarding water quality, and are strategically important for planning and conflict management.

3. To conduct bacteriological and water chemistry monitoring, develop quality assurance protocols and improve the water quality database.

4. To establish partnerships and linkages for research, outreach and training/ education activities in the Andean region and Latin America.

5. To develop a strategy for institutionalizing water monitoring and water resources management in targeted areas.

Site Description

Cotacachi is the principal city of the canton (political region) of Cotacachi located in the northern part of the Andes of Ecuador, about 2 h northeast of the country capital, Quito. It is the largest canton of the Province of Imbabura with an area of 1809 km^2 and a population in the year 2000 of approximately 37,000 people composed of mestizos, native Indians and Afro Ecuadorians. Elevations range from 200 metres above sea level (masl) in the subtropical area along the Intag River, to 4939 masl at the top of the snowcapped Cotacachi peak. Among the diverse landscapes in Cotacachi canton are some of the world's finest biological treasures, including threatened ecosystems that according to the world-renowned biologist E.O. Wilson include two of the world's 25 threatened hotspots of biodiversity. The 204,000 ha Cotacachi Cayapas Ecological Reserve is within the provinces of Imbabura and Esmeraldas, and contains one of the most diverse forests on Earth, and the 6000 ha Los Cedros Biological Reserve is virgin premontane wet tropical and cloud forest. It is well known that with just 0.2% of the Earth's landmass, Ecuador has 10% of the earth's plant species. Sadly, the rate of destruction of plant diversity and forests in canton Cotacachi far surpasses that of the Ecuadorian Amazon.

In Quichua tradition, Cotacachi has the gift of water. Traditional narratives state that the Cotacachi volcano is blessed with water in contrast to its close neighbour the Imbabura volcano. Three main rivers that run through Cotacachi (Pichaví, Pichambiche and Yanayacu), their locations and tributaries, are shown in Fig. 16.1, along with

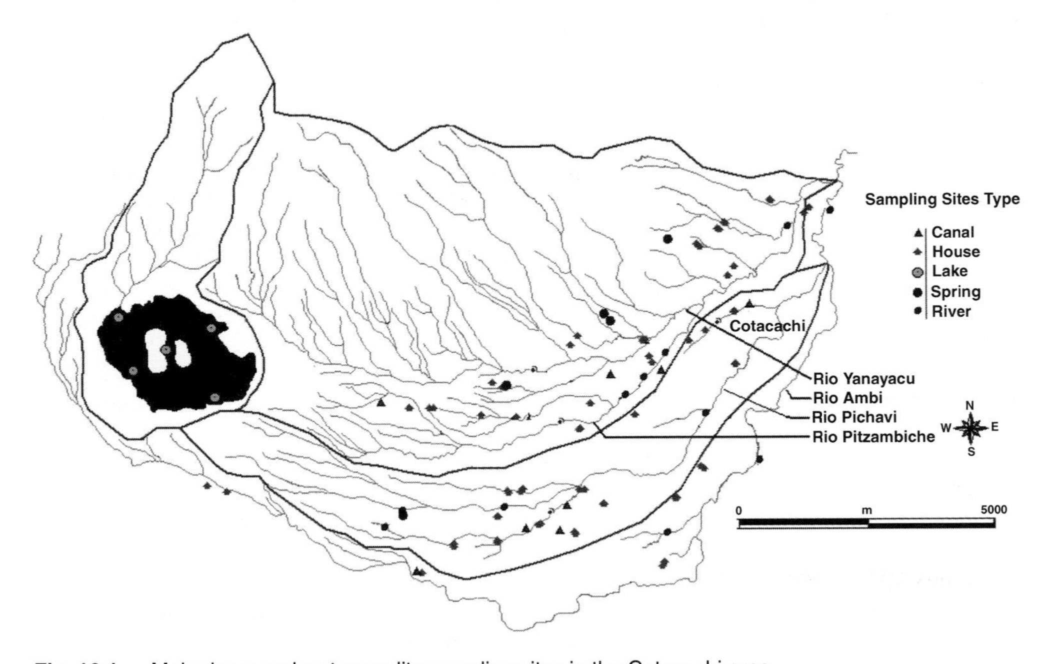

Fig. 16.1. Main rivers and water quality sampling sites in the Cotacachi area.

water quality monitoring sites. Water flow in all three rivers is subject to continuous change since many water intakes divert water to different 'haciendas' and communities mostly for agricultural purposes. All three rivers eventually enter the larger Ambi river.

Formation and Partners

On 9 December 1997, Auburn University was invited by leaders of the 'Union de Organizaciones Campesinas e Indigenas del Canton Cotacachi' (UNORCAC) to participate in a 'Water Quality Self-diagnostic Workshop'. UNORCAC is a strong organization of indigenous people. Associated with it are 43 community organizations and 17 Juntas de Agua, or water committees, charged with setting up and maintaining irrigation and drinking water systems in 40 of the 43 communities. In addition, women's committees at the municipal and community level work with culture and health.

Leaders of the 30 communities represented at that workshop reported flow reduction in springs and its possible relationship to deforestation, the opening of new agricultural fields, overgrazing and other resource management situations. In addition, they foresaw the need to improve and protect water quality in nature as a way to provide safe drinking water. Many participants remembered that in 1992, a cholera outbreak had shaken the entire country sending a powerful warning to people about the high risk of water-related diseases. It was the consensus of the participants that their problems could be summarized in the following seven statements:

- Their area lacked a Natural Resources Management Plan.
- Their traditional and cultural relationship with nature was changing.
- Many communities did not have legal access to water resources.
- Many springs were poorly managed.
- The Juntas de Agua (Water Boards) in the communities lacked institutionalization.
- Many drinking water systems were poorly designed.
- There was limited capacity to improve the problems described above.

On the occasion of the workshop, the CBWM programme of Alabama Water Watch was presented to the group, and physical–chemical and biological monitoring were demonstrated at the Ambi river. This was the beginning of Auburn University's involvement with the SANREM–Andes project in Cotacachi with UNORCAC as the main partner.

A proposal was born at the workshop to adapt the successful SANREM–Philippines model of CBWM to the Andean region to achieve research objectives (see Rhoades, Chapter 1, this volume).

Methods

Methods used in Ecuador are modelled after those developed and used by Alabama Water Watch, a citizen volunteer, water quality monitoring programme in the USA, and by the Tigbantay Wahig, a similar group of water monitors organized in the Philippines. US and Philippine citizen monitors, most of whom have little or no technical background, have been successfully using water quality test kits for 10 years. Manuals and other training materials were translated into Spanish for use in Ecuador, and adapted to the local situation.

Sampling sites are identified according to local concerns, and physical–chemical monitoring was conducted on a monthly basis. Six chemical parameters were tested using a specially designed field test kit. The tests are based on colorimetric techniques, and with adequate training allow ordinary citizens to measure air and water temperature, pH, hardness, alkalinity, dissolved oxygen and turbidity accurately. Water quality and quantity data were analysed, summarized and presented in appropriate formats for use by citizen groups, educators and local decision makers in Ecuador and the larger Andean region. The data were made available to other Andean SANREM researchers for incorporation into models, and for generating future scenarios.

Bacteriological monitoring of water for human consumption in participating communities and surface water (streams and springs) was conducted on a quarterly basis. The relatively new Coliscan Easygel method for coliform and faecal coliform testing is used, as approved for Alabama Water Watch by the US Environmental Protection Agency in December 1999. A 1 ml sample of water is collected using a sterile, plastic pipette and squirted into a 10 ml bottle of sterile, liquid medium. The medium (with colour indicators for coliforms) containing the water sample is poured onto a sterile, plastic dish, which is designed to induce the liquid to solidify. Incubation of sample plates at ambient tropical temperature was sufficient to grow the bacterial colonies for enumeration in about 36 h, though simple incubators are recommended to facilitate maintaining optimal temperatures of 29–37°C. No sterilizers or glassware are needed for this technique. Supplies, which cost about US$1.50 per sample, are easily transported to remote areas. Following incubation, colonies of *E. coli* and other coliform bacteria are enumerated and reported to the community. The equivalent number of colonies in 100 ml of water sample is reported.

Criteria suggested by US Environmental Protection Agency (EPA) were used for reporting *E. coli* results since most health organizations around the world follow this standard (US Environmental Protection Agency, 1986) (Table 16.1).

Table 16.1. US EPA upper limits of *E. coli* presence in water for diverse uses.

Water source	*E. coli* colonies/ 100 ml
Piped drinking water	0
Drinking water source before treatment	2000–4000
Designated beach area	235
Moderate swimming area	298
Light swimming area	406
Rarely used swimming area	576

All data were stored at the UNORCAC office in Cotacachi, as well as at the Catholic University in Quito, which serves as a major data collection point for the SANREM project. Most data are geo-referenced and reported in a form accessible to local communities, and policy and decision makers.

Results and Discussion

Seven Water Quality Monitoring Training Workshops conducted from March 1998 to April 2003 in canton Cotacachi involved 100 volunteers who were certified as Water Quality Monitors. The majority of trainees were UNORCAC members (Juntas de Agua members, health volunteers and residents representing 16 rural communities) and citizens from the urban areas of Quiroga and Cotacachi. Water quality monitoring was conducted with volunteers coordinated by local citizens, including Mr Nicolas Gomez, Mr Julian Pillaluisa, Ing Mario Landeta and Mr Horacio Narvaez, all of whom were trained and supervised. Intensive efforts were made in 2001 when 441 chemical and bacteriological samples were collected, accounting for 46% of the total.

From July 1998 to 2003, physical–chemical analyses from 52 sites were reported, accounting for 414 records. Eight bacteriological surveys tested 97 sites from drinking water sources as well as rivers and springs in Cotacachi (Table 16.2), yielding 545 records. Data were entered initially on data sheets stored in the SANREM computer at Jambi Mascaric.

More recently, a relational database was developed at Auburn University allowing easier management and reporting of

Table 16.2. Water sampling by year.

Year	Chemical	Bacteriological
1998	29	0
1999	12	49
2000	95	144
2001	126	315
2002	53	37
2003	99	–
Total	414	545

water quality data. Monthly data files were also sent via e-mail to Auburn University to be incorporated in the Global Water Watch database. Summaries of the data, graphs and site information are available from the database.

Greater emphasis was placed on surface water when conducting physical–chemical analyses, while bacteriological testing emphasized drinking water (Table 16.3). Fifty-four per cent of the sites tested for coliform bacteria were households. Sixty-six per cent of bacterial samples were from households. In most cases, sites tested for water chemistry were also tested for faecal coliforms. Almost 50% of sites tested for water chemistry were at rivers.

Sixteen per cent of the physical–chemical tests were conducted on water stored in concrete reservoirs and distribution tanks in ten communities. Physical–chemical measurements were within the expected ranges. Low dissolved oxygen values were expected from water stored in this type of structure, as it usually has been piped for long distances and has not had contact with air (Table 16.4). Lower values of pH, alkalinity and hardness were found closer to the slopes of the Cotacachi peak, while higher values were found in La Calera in the middle of the valley between Imbabura and Cotacachi. As suggested by its name, La Calera has abundant natural limestone, and hardness, alkalinity and pH in this area are consistent with those conditions.

Physical–chemical results from surface water were within expected values (Table 16.5). Air and water temperatures varied normally with altitude. Data for dissolved oxygen, air temperature and water temperature from July 2000 to August 2003 are shown for the three main rivers of Cotacachi (Fig. 16.2). Water temperatures follow air temperature closely, and all three rivers were within the range 10–25°C. Water temperature was more stable at the higher station in the Yanayacu River, staying between 11 and 17°C.

Dissolved oxygen followed expected patterns. It varied inversely to water temperature, as can be seen in Fig. 16.2. During some sampling events, dissolved oxygen

Table 16.3. Number and types of chemical and bacteriological sampling sites.

	Chemical sites	Chemical samples	Bacteria sites	Bacteria samples
Irrigation canals	7	52	6	29
Spring	10	80	9	44
River	25	217	17	85
Reservoir	10	65	11	17
Well	0	0	2	8
Household	0	0	52	362
Total	52	414	97	545

Table 16.4. Minimum and maximum physical–chemical parameters recorded from water tanks.

Parameter	Minimum	Site	Maximum	Site
Air temperature	7.0	San Nicolas	30	El Batan
Water temperature	6.0	San Nicolas	23	San Martin
pH	5.5	San Nicolas	8.0	La Calera
Dissolved oxygen	0.6	San Jose del Punge	9.6	Turuco
Hardness	20	Iltaqui	340	La Calera
Alkalinity	40	Iltaqui	480	La Calera

Table 16.5. Minimum and maximum physical–chemical parameters from surface water.

Parameter	Minimum	Site	Maximum	Site
Air temperature	10	Chilcapamba Spring	29	Rio Yanayacu Turuco
Water temperature	7.0	Chilcapamba Spring	24	Santa Barbara Acequia
pH	1.5	Rio Pichaví San Jose	8.5	Ten different sites
Dissolved oxygen	0	Rio Pichaví Estadio	9.2	Chilcapamba Acequia
Hardness	60	Cotacachi Spring	280	Rio Pichaví Estadio
Alkalinity	70	Chilcapamba Spring	375.0	Rio San Martin Calera

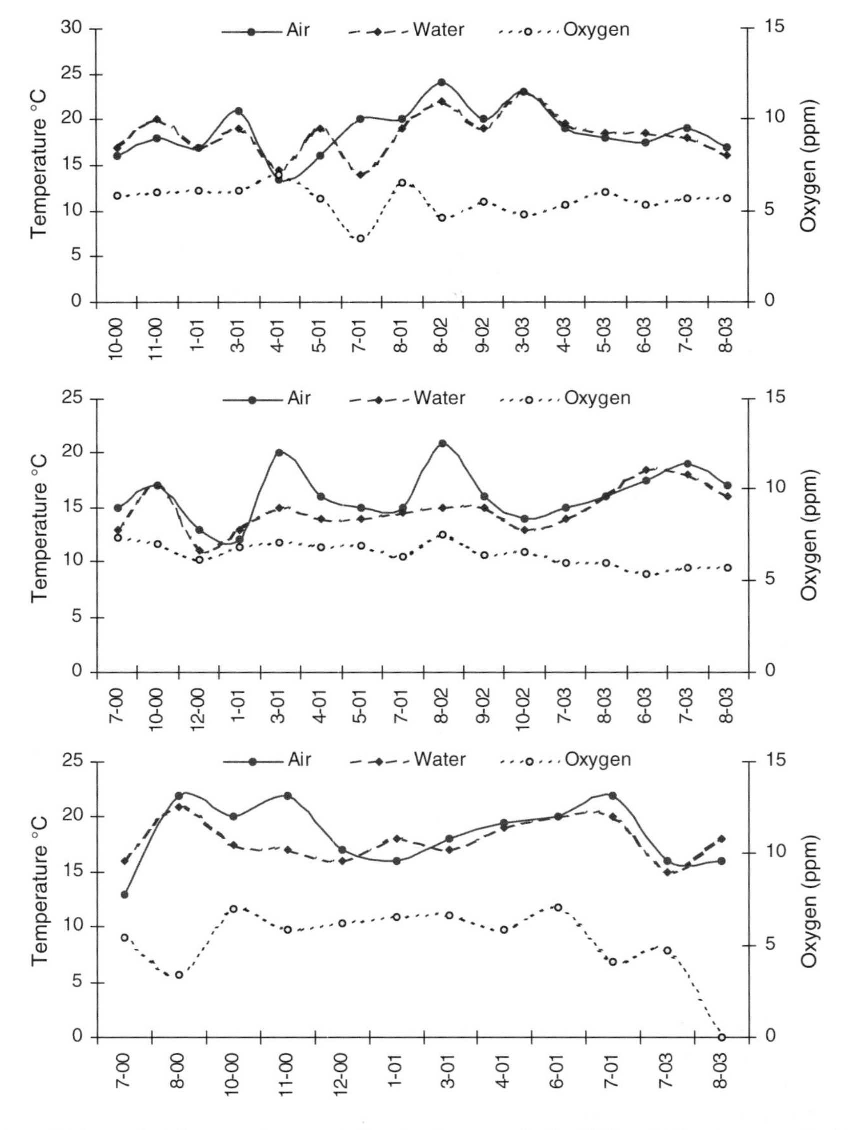

Fig. 16.2. Water and air temperatures and dissolved oxygen in the Pichambiche river near La Banda (top), Yanayacu river near Iltaqui (middle) and Pichaví river near the stadium (bottom).

measurements were below 1 ppm and under 5 ppm in 26% of all records. Of particular interest was the way dissolved oxygen seem to be decreasing in the Pichaví river and, more importantly, the decline to zero oxygen in August 2003. It is not surprising that there is little or no aquatic life in the Pichaví and the Yanayacu rivers near Cotacachi. Several other sites have similar conditions. Decreases in dissolved oxygen may be related to pollution that consumes oxygen such as municipal wastewater, livestock or other animal waste. Increasing water temperatures in the Yanayacu river could indicate the clearing of streamside vegetation which buffers and shades streams and moderates temperature fluctuations. Only after years of continuous and systematic monitoring can water quality trends be established for a particular site.

Data for pH, hardness and alkalinity from July 2000 to August 2003 are shown for the same three rivers (Fig. 16.3). Values for alkalinity and hardness are usually similar. Water from both rivers at those sampling points are in the category of moderately hard to very hard. Some measurements placed the water from the Pichaví River near the stadium as 'very hard'. In October 2002, alkalinity was 140 mg/l higher than hardness at the Pichaví river. This suggests that one or both of these parameters was affected by human activity. Observations upstream of the sampling site revealed leather processing activities that could have been disposing of wastewater containing chemicals that led to the separation between alkalinity and hardness values.

Several sewage pipes drain directly into this river, as they do to the other two rivers flowing near Cotacachi. Cotacachi and Quiroga lack sewage treatment plants. Citizens and local authorities have acknowledged the problem that > 24 municipal effluent outlets are coursing to the rivers. Low pH, like that observed on 8 May 2003 at Rio Pichaví San José when it was measured at 1.5, and low dissolved oxygen (Table 16.5) in streams are often an indication of heavy organic contamination and definitely detrimental to aquatic life. Hardness and alkalinity that usually are close in value to each other were 60 mg/l or higher at most sites, indicating a good source of bases coming in from the soils where the rivers are running, as could be the case with the Pichaví river flowing through locations such as San Martin and La Calera.

Drinking water, which is a major concern in Cotacachi, is not safe for human consumption in 75% of the communities assessed in this study. Eighty per cent of the sites tested positive for coliform bacteria and 47% revealed the presence of *E. coli*. Results from bacteriological surveys conducted from July 1999 to October 2002 identified potential problems with surface and drinking water in several UNORCAC communities in canton Cotacachi. Coliform bacteria were found in drinking water system tanks and households of 18 of 24 communities sampled. *Escherichia coli* was found in 45% of water samples with coliform bacteria (Table 16.6). This result may be related to the condition of the spring source (water from 16 spring sources was tested), the condition of pipes that convey water to community houses, and containers used for water storage. Water samples that were free of coliform contamination came from water treated with chlorine at distribution tanks or houses.

Bacteriological testing revealed that about 80% of the sites sampled had some degree of contamination with faecal coliforms (Fig. 16.4). Twenty-seven of 32 surface water sites (irrigation canals, rivers, creeks and springs) presented *E. coli*, and all had some form of contamination by coliforms.

Sites with higher faecal coliform contamination included the irrigation canals at El Batan, Chilcapamba, Piava San Pedro and Santa Barbara. At these sites, monitors often observed men, women and children using the water for domestic uses such as washing clothes and preparing food. The maximum values in both *E. coli* and total coliform contamination were found at the lower stations in the Yanayacu river (Piscina and Perafan) with about 20,000 bacteria colonies in 100 ml of water sample (Fig. 16.5). This last location is also used by locals to bathe and wash clothes. Samples from three springs (Cotacachi, Chilcapamba

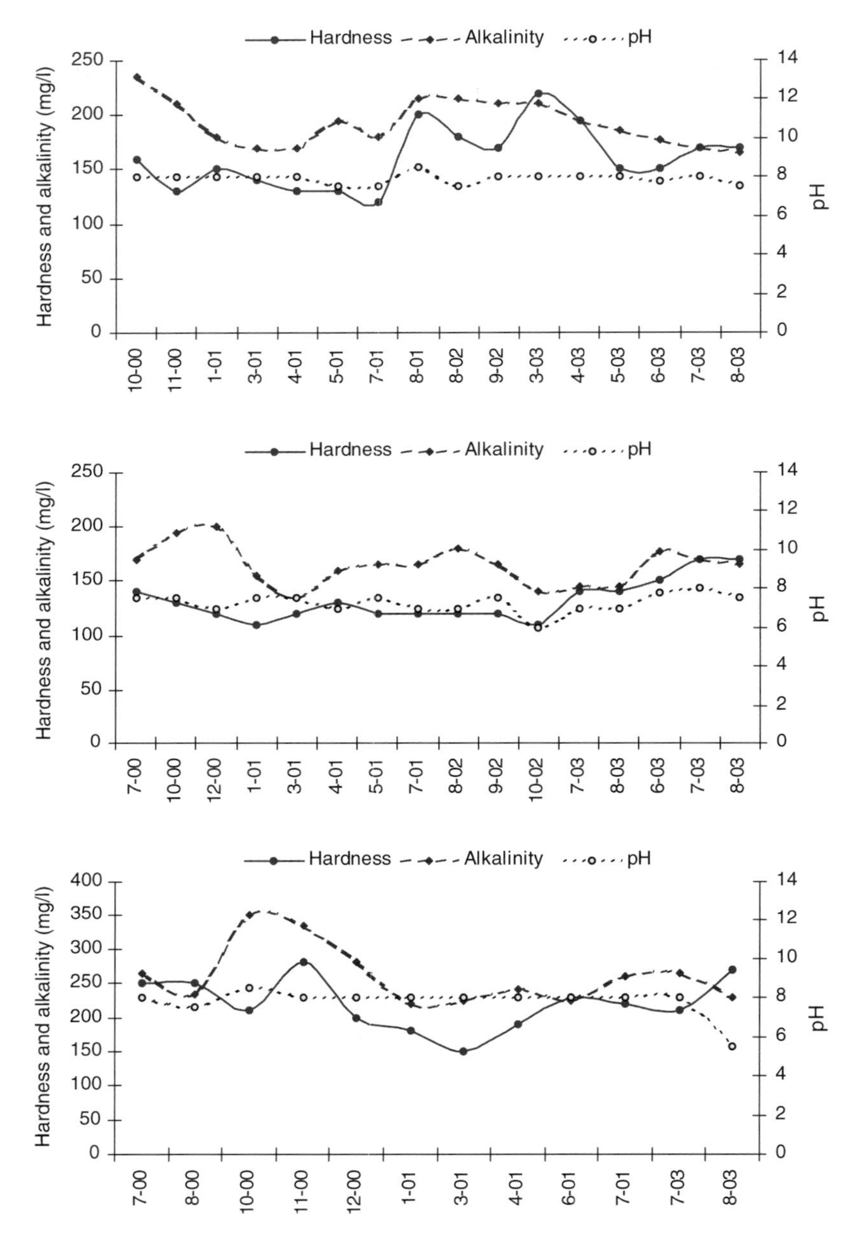

Fig. 16.3. Hardness, alkalinity and pH from the Pichambiche river near La Banda (top), Yanayacu river near Iltaqui (middle) and Pichaví river near the stadium (bottom).

and Quiroga), and the Alambuela Bajo creek were found not to be contaminated with *E. coli* during the surveys.

Bacteriological surveys in Cotacachi revealed communities with total coliforms ranging from 150 in 100 ml (once in the city of Quiroga) to a maximum of 23,350 colonies in 100 ml from a water faucet in a Cumbas Conde home. Homes in 13 communities had counts of > 10,000 coliform bacterial colonies in 100 ml of water (Fig. 16.6). In addition, water samples from 56% of these communities tested positive for *E. coli*, with counts ranging from 100 (San Martin) to

Table 16.6. Number of colonies of *E. coli* and other faecal coliforms found in water samples from households in the communities of the Cotacachi area 1999–2002.

Community	*E. coli*	Total coliforms
Ashambuela	000	000
Cotacachi	000	000
Cuicocha Centro	000	3350
El Ejido	000	000
Perafan	000	20000
Piava Chupa	000	000
Piava San Pedro	000	000
Pilchibuela	000	000
Quiroga	000	150
San José del Punge	000	19300
San Miguel	000	12200
San Nicolás	000	20000
Santa Barbara	000	20000
Turuco	000	20000
San Martin	100	4950
Iltaqui	450	20000
La Calera	700	4450
Chilcapamba	750	12250
Cumbas Conde	3350	23350
Domingo Sabio	3350	10000
San Antonio del Punge	3350	22650
El Batan	4200	18200
Arrayanes	4650	6750
Guitarra Urco	10000	16700

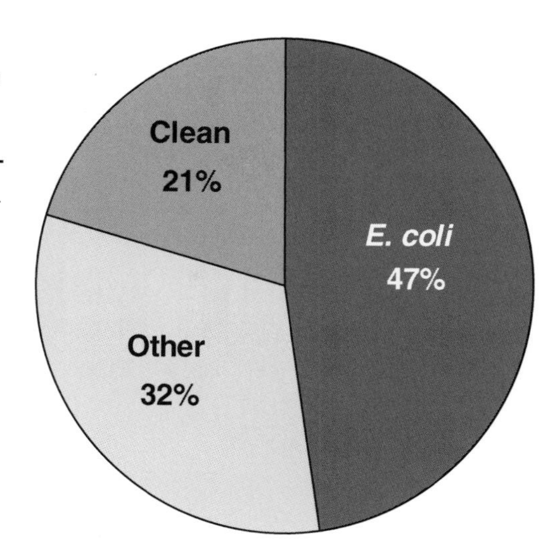

Fig. 16.4. Percentages of sites with bacterial contamination in the Cotacachi area from July 2000 to April 2002 ($n = 97$).

10,000 (Guitarra Uco) colonies in 100 ml. Some of these concentrations are far above that considered 'safe for whole body contact' by the US-EPA (Table 16.1). Drinking water from the city of Cotacachi, Barrio El Ejido and four other communities was completely free from faecal contamination. These communities are all supplied by the Cotacachi water system that comes from a large spring.

All surface water samples contained some degree of coliform contamination except the Chilcapamba spring, and 85% of surface water samples had *E. coli* (Table 16.7). Water from the two wells tested had coliform bacteria. Sixty-four per cent of water storage and distribution tanks were also contaminated, although only one of them had evidence of *E. coli*.

Water in 71% of households was positive for coliforms and 30% for *E. coli*. Dissemination of these results increased awareness of bacteriological quality of water. Repairs were made where broken pipes were related to high bacterial counts in drinking water supplies. The results of these bacteriological surveys conducted in Cotacachi revealed that the drinking water supply system of several communities have sections that are contaminated with coliform bacteria in concentrations that are probably dangerous for human consumption. People in these locations are aware of the potential for water-borne diseases, but seeing results of bacteriological analyses has prompted them to do something to solve the problems or to prevent them. Community members who attended water meetings and workshops indicated that bacteriological and chemical testing of drinking water is the primary motivator for their participation in the programme.

Bacteriological tests of water samples from the communities of Santa Barbara and Turuco in 1999 and 2000 showed high contamination with coliforms. However, similar tests conducted in 2001 showed an absence

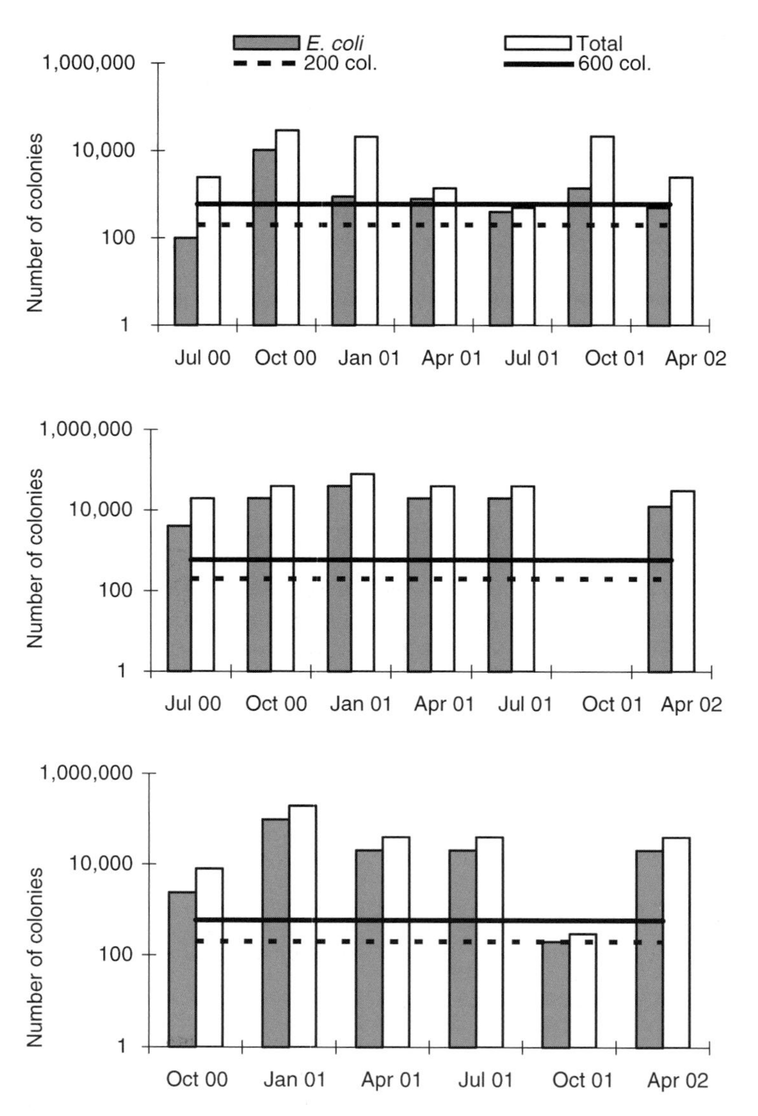

Fig. 16.5. Colonies of *E. coli* and total coliform bacteria collected during bacteriological surveys at random sites in the Pichambiche river near San Francisco (top), the Pichaví river near the stadium (middle) and the Yanayacu river near the piscine (bottom).

of coliforms. This improvement is attributed to remediation work of community members and Juntas de Agua operators, who cleaned and fenced water collecting tanks. Families were happily able to discontinue chlorination of their drinking water. There is resistance to chlorination even when drinking water sources are known to be contaminated. This is one of more than ten cases of remediation of water systems after CBWM revealed faecal contamination.

Conclusion

Auburn University water quality workshops and fieldwork revealed that local Cotacachi citizens had a strong interest in testing drinking and surface water quality. Contamination of household water generally is associated with poor sanitation at the collecting tanks and springs, as well as in water storage containers in the houses. There have been cases of improvement in the infrastructure of

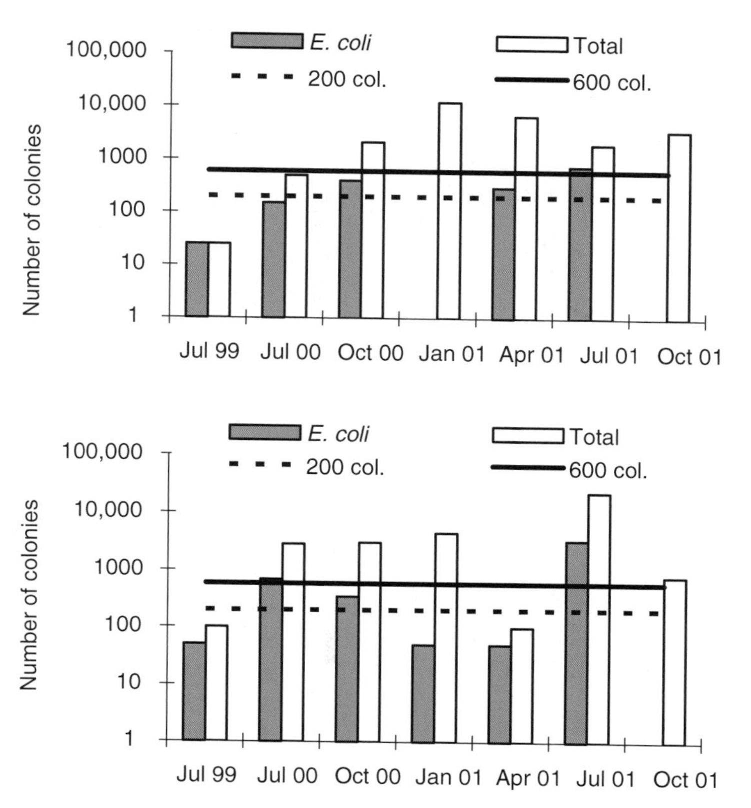

Fig. 16.6. Colonies of *E. coli* and total coliform bacteria collected during bacteriological surveys at random homes in Chilcapamba (top) and San Antonio del Punge (bottom).

Table 16.7. Total sites that tested positive for *E. coli* during 1999–2002.

Site type	*E. coli* positive	Total sites
Homes	17	52
Irrigation canals	6	6
Tanks	1	11
Springs	4	8
Rivers	17	18
Wells	1	2
Total	46	97

the water system following SANREM water quality monitoring activities in canton Cotacachi. UNORCAC and community members have used bacteriological test results to identify and eliminate faecal contamination of drinking water sources.

Water quality data collected by citizens are more meaningful to them if integrated with other activities such as soils and agroecological projects, biodiversity projects and environmental education, and development of watershed and natural resources management plans. A good example to follow is in the Philippines where results of community-based water monitoring have been incorporated into Natural Resource Management Plans (Deutsch, 2003). In Ecuador, water is very important in Quichua culture and traditions. Many times each year, hundreds of people take sacred baths and have rituals at several water bodies during festivities. As Deutsch and Neely (1997, unpublished manuscript) have discussed, there is an urgent need to think about watersheds holistically with attention to culture. Watershed education should be conducted by merging the spiritual and cultural traditions of a place with biophysical research.

The partnership established with UNORCAC in Cotacachi has been fruitful.

Frequent interactions with the Mayor, Water Commissioner and other Municipality officials, and representatives of the Cantonal Assembly, have created common ground and purpose for CBWM in the canton of Cotacachi, and generated support for CBWM. Desires for expansion and intensification of CBWM have been expressed: a group of organic coffee growers are willing to start testing the quality of the water they use for watering their coffee plantations; an official proposal is being developed, in collaboration with the city of Cotacachi Water Commissioner, to establish more CBWM groups in the canton.

The Municipality is implementing sewage treatment plants for some communities in the Cotacachi canton and requested SANREM assistance for establishing the capacity for regular testing for biochemical oxygen demand (BOD). The Municipality is considering adding more technicians to their personnel in the environmental commission (now only one person), particularly in the area of water quality, an opportunity that will be open to some of the monitors certified by this programme. This new area of knowledge by local citizens is resulting in their taking an active role in environmental issues, especially those affecting water quality. Local citizen involvement will result in community and Municipality policies aimed at conserving and protecting water resources.

Experiences in Ecuador have contributed to creation of interest and perceptions of usefulness of water monitoring in institutions in Peru and Brazil. Programmes in these countries are in various stages of implementation. Brazil is most advanced, with site-based activities since October 1999. The Brazil programme is largely funded by a non-governmental organization (NGO) partner, which is making significant progress to institutionalize the programme. An integrated relational database has been developed at Auburn for data compilation that greatly assists the programme by facilitating quality assurance measures, correction of data, and reducing the risk of error in data entry. The database allows special analysis and reporting of data. It has the capability for sharing of data, information and experiences across sites and countries.

Monitoring and describing physicochemical and biological characteristics of surface waters should be a goal of Watershed Management Plans. Deutsch (1997) foresaw the potential of CBWM to complement what government agencies, universities and other research organizations do. Citizen volunteer groups can reach greater numbers of sites, collect greater numbers of samples, sample with greater frequency and do so at a much lower cost. The result is comprehensive baseline water resource data, water quality trends, education of the citizenry, and powerful participation of citizenry in the development of public policy, and the remediation and protection of water resources.

Other Accomplishments

During the journey of SANREM in Cotacachi, scores of children have been introduced to the importance of water quality monitoring through talks, demonstrations in the classroom, field trips and festivals. CBWM activities have influenced decision making of community leaders in Cotacachi. Water quality and quantity plays an important role during political campaigns in Cotacachi. The importance and impact of SANREM's goal of providing local decision makers with appropriate data, decision support tools and methods is evident in political processes, including election of municipal leadership. CBWM has strengthened the unity of indigenous and mestizo people, as they express common cause in water resource issues. It is hoped that increased interest in citizen monitoring, and increased knowledge about water issues, will have the effect of reducing confrontation and conflict over water.

The SANREM water quality monitoring programme has improved the basis for cooperation between municipal and the indigenous organizations in Cotacachi by providing common ground. Regular meetings with the Mayor of the city of Cotacachi have had positive results. The Mayor supports most activities that UNORCAC proposes, and has offered to mediate relationships with other governmental organizations and NGOs working in the area in similar or related projects. With SANREM assistance, the president of UNORCAC and the Mayor of Cotacachi signed a cooperation agreement, on 24 January 2002, to conduct water quality monitoring in the watersheds of the canton as part of a general plan for natural resources management. The Municipality of Cotacachi requested SANREM to introduce an inexpensive method for BOD testing for monitoring stream water quality at 24 sites where municipal waste is discharged. Strengthening institutional partnerships and linkages was further accomplished through joint activities, including training, workshops, technical support and database development. The Cotacachi Cantonal Assembly has demonstrated strong interest in CBWM. They have provided entry into several communities, transportation for UNORCAC monitors, funds to print workshop manuals and involvement of cantonal youth groups in CBWM.

Envisaged in the beginning as an 'offspring' of the SANREM–Philippines experience, the Water Resources Management and Environmental Education project in the Andes approached problems of water through community-based, participatory research. Successful implementation of this work in Cotacachi has resulted in 'scaling up' activities to other areas of Ecuador and to Peru. Visits to neighbouring sites in Ecuador, such as San Pablo El Lago, Cayambe and even as far as Cotopaxi province, have revealed potential, local interest and the need to expand CBWM activities. In late 2003, an agreement was reached between Auburn University and Heifer Project in Ecuador to provide facilitation and support for continued CBWM at existing sites, and expansion of CBWM to other sites in Ecuador where Heifer Project is active. SANREM was also invited to assist with CBWM and environmental management in the Amazon region of Ecuador. The organization for Eco-Development of the Ecuadorian Amazon Region (ECORAE) approached SANREM partners to explore the possibility of starting water quality monitoring activities in this region that covers 50% of the country, and from where 90% of Ecuador's oil is extracted. Over the past decade, poverty and environmental degradation have dramatically increased in the Amazon. ECORAE and Auburn University signed a memorandum of understanding to collaborate in water quality monitoring and environmental management. The presence of headwaters of the Amazon river system gives this region considerable importance. Auburn University personnel were conducted on a 2-day tour of the region by ECORAE, and the director visited Auburn University for discussions. Leveraging of oil company funding is a possible source of support. These types of relationships have the potential to outlive SANREM and make a water resource management programme sustainable.

An important outgrowth of the SANREM water monitoring experience has been the conception and implementation of a Global Water Watch Network, which extends beyond the 'SANREM countries'. Organizations in Brazil, Thailand, and China are the most recent Auburn partners. An exciting development is Global Water Watch, a worldwide network of community-based water monitoring groups interconnected through a global database that is being implemented.

References

Ampel, N.M. (1996) *Emerging Infectious Diseases* 2, 109–116.

Centers for Disease Control and Prevention (CDC) (2003) US Department of Health and Human Services. Available at: http://www.cdc.gov/ncidod/dbmd/diseaseinfo/waterbornediseases_t.htm

Deutsch, W.G. (1997) Volunteer water quality monitoring – the Alabama water watch. In: *Highlights of State and Tribal Nonpoint Source Programs.* Section 319 Success Stories: Vol. II. US-EPA 841-R-97-001, pp. 8–10.

Deutsch, W.G. (2003) Community-based water monitoring in Alabama and the Philippines. *Volunteer Monitor* pp. 14–15.

Deutsch, W.G., Duncan, B.L. and RuizCórdova, S.S. (2003) *Manual de Certificación Básica, Monitoreo Físico–Químico del Agua.* Centro Internacional de Acuacultura y Ambientes Acuáticos, Universidad de Auburn, Auburn, Alabama.

Duncan, B.L. and Deutsch, W.G. (2001) Water resources and environmental education in two Andean watersheds. In: *Innovative Research for Sustainable Agriculture and Natural Resources.* SANREM CRSP 2000 Annual Report. Watkinsville, Georgia.

Pan American Health Organization Country Office Official Fax, 8 February, (1996) Quito, Ecuador.

United Nations Non-Governmental Liason (UNNGLS) (2003) *NGLS Roundup.* Third World Water Forum and Ministerial Conference held in Japan. Geneva, Switzerland.

US Environmental Protection Agency (1986) *Ambient Water Quality Criteria for Bacteria.* US Environmental Protection Agency. US-EPA-440-5-84-002.

Whiteford, L.M., Laspina, C. and Torres, M. (1996) *Environmental Health Project. Project No. 936-5994.* Office of Health and Nutrition, USAID, Washington, DC.

Part IV

Negotiating 'Development with Identity'

Robert E. Rhoades confers with two past presidents of UNORCAC: Cornelio Orbes (centre) and Rafael Guitarra (right) (photo: Virginia D. Nazarea).

The chapters in this concluding section deal with how indigenous Cotacachi is engaging and negotiating development at local, national and global scales. This book shows that Cotacachi is an innovative experiment in seeking strategies to sustainable development without losing group identity and ethnicity. This, as the authors point out, is neither a predetermined outcome nor an easy, straightforward process. The future is uncertain while both powerful internal and external forces operate to undermine the maintenance of traditional values and social systems. Nevertheless, the following chapters document a process of creativity and synergy in striving for indigenous development as defined by Cotacacheños themselves.

Despite enormous transformations in agriculture over the past 50 years and the growth of non-farm employment, Cotacachi remains today in heart and mind predominantly an agricultural society. Young Cotacacheños who take up employment in Ibarra, Quito and increasingly in other countries often dream of returning to purchase land and pursue part-time farming. Return to the land is a sacred and civil calling for indigenous people. It is painfully clear, however, that under present market conditions, farming is not a profitable undertaking in Cotacachi. Our research attempts to understand this decline of farming as well as seek new alternatives. Anthropologist Brian Campbell in Chapter 17 presents his analysis of indigenous farmers' perceptions of agricultural change in Cotacachi communities. He uses in-depth histories in three communities to compare their encounters with external development agencies which produced unique responses by each community. He notes that while the Green Revolution was a significant force in change, multiple other factors such as climate change, diminishing landholdings, loss of farm animals and labour constraints fostered a declining agriculture. Despite these negative forces, he finds that indigenous systems of collective action and knowledge could be combined with ecologically sound, sustainable science and development alternatives such as agrotourism and organic production.

Changes in the landscape and lifescape are two sides of the same reality. Gabriela Flora in Chapter 18 presents her research on circular migration and the maintenance of a sense of place and ethnicity in Cotacachi. Although a classical migration 'push area' where people leave for better opportunity, the indigenous people of Cotacachi utilize circular migration for work in nearby areas and Quito as a way to anchor their local roots. They work away from home but maintain a connection to Coctacachi and invest savings in land. Migration, while typically a mechanism for change, is a way the indigenous people preserve their ethnic identity and build a land base. In recent years, Cotacacheños have emulated their Otavalo neighbours to the east by engaging in transnational migration and the marketing of artesian goods and Andean music. This form of migration, however, has not yet become a major force in Cotacachi social change.

Drawing on their extensive backgrounds in the study of social capital, Jan and Cornelia Flora along with Florencia Campana, Mary García Bravo and Edith Fernández-Baca examine in Chapter 19 how local Cotacachi groups form advocacy coalitions to influence environmental policy at local, national and even international levels. Using an advocacy coalition framework (ACF), they look at how civil, government and commercial groups sought and formed alliances around two environmental issues: governance of the Cotacachi Cayapas Ecological Reserve and open-pit mining in Intag. Analysis of the outcomes of these two cases points to important lessons on how stakeholders can overcome differences in desired future conditions to work towards sustainable development.

Chapter 20 by Robert Rhoades and Xavier Zapata Ríos presents the results of the future visioning methodology developed by the SANREM–Andes programme. The objective of community-based future visioning is to provide local people with scientifically derived future scenarios based on a trajectory of past trends and actions. These scenarios, in turn, will be utilized as a dialogue platform to discuss Cotacacheños own culturally created desired scenarios of the future. Using LUC simulation and photo-manipulation techniques, panoramic scenarios of Cotacachi were created for 3 years: 1963, 2000 (an actual photograph) and a projection to 2030. In a community workshop, these scenarios were discussed and local scenarios created by different age and gender groups. In addition to affirming that Cotacachi wants 'development with identity', this future scenario-building method also highlights differences between scientists' and local people's perceptions of what matters in the environment.

The final chapter of this volume by Robert Rhoades (Chapter 21) discusses how to connect sustainability research and global scientific questions with indigenous communities' priorities and needs. It is no longer possible to do sustainability science to establish general principles without also returning something of value to local people. This chapter describes the ethical and contractual conditions under which SANREM–Andes operated in Cotacachi and the creative ways both scientists and local people achieved mutual and distinct goals. Rhoades presents six case studies where research was combined with priorities requests of either the communities, UNORCAC, cantonal government or water associations. As communities around the world become more aware of their rights and role in sustainable development, these experiences of SANREM can provide insights and methods for reconciling differences in agendas in a positive fashion for both researcher and researched.

17 Why is the Earth Tired? A Comparative Analysis of Agricultural Change and Intervention in Northern Ecuador

B.C. Campbell
University of Georgia, Department of Anthropology, 250 Baldwin Hall, Athens, GA 30605, USA

> The ground will no longer produce without the chemicals. We are obligated to use them. The Earth is tired.
> (Indigenous farmer, Cotacachi, Ecuador)

Introduction

Traditional Andean societies function within a cultural logic of complementary socioeconomic and ecological relationships that emphasizes solidarity and cooperation (Masuda *et al.*, 1985; Stanish, 1992). The indigenous peoples of the Andes endured the violent subjugations of the Inca and Spanish empires and exploitation throughout the colonial and national periods primarily because of these collective solutions (Oberem, 1976; Murra, 1985; Campbell, 2002; Moates and Campbell, Chapter 3, this volume). Now, however, with the penetration of the capitalist economic system and multinational corporations, the foundation of indigenous Andean society, the 'embeddedness' of Andean economy within their distinct complementary social relationships, has begun to succumb to the individualistic approach of Western capitalism. As Stern (1987, p. 5) explains:

> Scholars tend to agree that the penetration of capitalism accentuates the internal differentiation of peasant society into rich and poor strata. More precisely, capitalism breaks down institutional constraints that pressured wealthy peasants and villagers to channel their resources into 'redistributive' or prestige-earning paths that blocked the free conversion of wealth into investment capital.

Yet, in the northern Andes of Ecuador, indigenous communities are consciously attempting to curb this trend, to rejuvenate the social and agricultural traditions that have allowed them collectively to survive. Contact with various exogenous institutions, ideologies and organizations within the last century has affected the indigenous communities of northern Ecuador in diverse ways in terms of land tenure, land management, worldview, agroecology and sociopolitical organization. In Cotacachi, indigenous communities realize that the cultural fabric of their communities might continue to unravel as they become more enmeshed in the capitalist economy. They have responded in a variety of ways; however, the most evident reaction has been the formation and pro-active stance of

 Development with Identity: Community, Culture and Sustainability in the Andes (ed. R.E. Rhoades)

UNORCAC, which originated in order to combat peasant and indigenous exploitation and now serves as a second-level institution that facilitates local development on local terms. UNORCAC represents 43 campesino and indigenous communities within the Cotacachi canton, the majority of which farm for their livelihood. These communities have identified decreases in agricultural production and the loss of traditional cultivars, varieties and practices as the most significant agricultural changes in their communities (UNORCAC, 1996; Campbell, 2002). They recognize that the sustainability of their land is synonymous with the maintenance of their communities and they are concerned that an influx of modern agricultural practices may endanger that sustainability. They requested that the SANREM–Andes research programme collaborate with them in an effort to examine and better understand the causes of these changes and to propose strategies to combat them.

The objectives of this historical agroecology analysis are twofold. First, to better understand local conceptualizations of the agricultural decline, I examine indigenous farmers' perceptions of agricultural change in five Cotacachi communities where surveys, participant observation and interviews were conducted. Second, to elucidate the possible causes of decreased agricultural productivity, I present detailed summaries of the social and agricultural histories and contemporary characteristics of three of these five indigenous communities and compare their distinct encounters with exogenous institutions over the last century and the concomitant changes in their agricultural practices. I selected for comparison these three communities in particular because they have each experienced, almost exclusively, the distinct forms of intervention that have occurred in the region during the recent past. Their comparative analysis provides heuristic insights into the agricultural tragectories and social effects of distinct development initiatives and sociopolitical experiences and allows for an assessment of their respective roles in contemporary agricultural change.

Methods

Participant observation, semi-structured interviews, surveys and agroecological zoning were employed in five primarily indigenous communities in the Cotacachi province. They were conducted with either the head female or male of 12 households in four of the communities, and with six households in the dispersed community of Ugshapungo. As Brush (1977, p. 136) and other Andean researchers indicate, the household serves as the ideal research subject in the study of Andean political economy because it 'is the basic economic unit of production and consumption' (Mayer, 1985; Stanish, 1992). Agroecological zoning methods, including field plotting and agroecological transects, were conducted in each community with an indigenous assistant, who was a member of UNORCAC, and served as a Quichua translator. Interviews and participatory field plots were conducted in both Spanish and Quichua depending on the level of Spanish fluency of the respondent. Archival research, including UNORCAC's diagnostic reports and census data, filled in gaps in our survey data on specific community demographics.

The Research Area

The five communities under examination are located in the Andean zone 2200 metres above sea level (masl) where indigenous communities have traditionally settled in attempts to avoid exploitation at the hands of Inca conquerors or Spanish *hacendados* (hacienda owners) (Campbell, 2002). The indigenous peoples of this region of Ecuador refer to themselves and their language as Quichua.

The indigenous communities of Cotacachi employ a variety of representative Andean social mechanisms that aim to resolve internal conflicts and equitably distribute communal resources. These mechanisms include reciprocity as the base of group solidarity, the *minga* as organized mandatory communal labour, and celebrations as means for reducing socioeconomic differentiation and ritual resolutions of interpersonal, communal and familial conflicts. Traditional subsistence consisted of various strategies designed for the vertical exploitation of multiple ecological zones and a reliance on relatively lush volcanic soils for crop production with short fallow cycles (Campbell, 2002; see Moates and Campbell, Chapter 3, this volume). Currently, however, the synergistic effects of degraded local soils and insufficient animal manure, tenuous land tenure and Green Revolution introductions preclude the traditional vertical complementarity in most cases (Murra, 1985; Campbell, 2002; see Moates and Campbell, Chapter 3, this volume). The indigenous communities share geographic, ethnic and sociocultural characteristics, in addition to a tumultuously exploitative history of mistreatment. The three communities discussed in detail and compared, however, differ to certain degrees in land management approaches, land tenure, agricultural and social history, and biophysical characteristics. This research explores the proposition that, while traditional Andean sociocultural and agricultural practices have endured subjugation for 400 years, albeit in muted or modified forms, contemporary capitalist enterprises masked in development initiatives have subverted fundamental Andean lifestyles because dependence precludes the tradition of subsistence farming.

Perceptions of Agricultural Change

UNORCAC expressed general concern regarding soil fertility, reduced agricultural productivity and the loss of traditional varieties, knowledge and practices. They identified decreased soil fertility as the major agricultural change in the region that needed to be addressed. Yet, when I administered surveys and interviews in the communities, asking indigenous farmers what they perceived as the most significant agricultural change in the region, a more complex picture developed. While decreased agricultural production emerges as the general problem, the causes of this agricultural decline are multifarious; historical, cultural and ecological (see Skarbø, Chapter 9, this volume). The explanation resides both in the exploitative history of the region and in the climatic changes experienced and observed first hand by the closest residents to the disappearing glacier atop *Mama Cotacachi*. As Chapter 5 (Rhoades *et al.*, this volume) corroborates, and Table 17.1 and Fig. 17.1 demonstrate, local farmers perceive climate change as the most significant variable in their decreased production. Compared with the < 4% of interviewed farmers who ranked soil impoverishment as the most significant agricultural change in the region, approximately 30% stated that climate change was the most significant. As the glacier melted over the past century, it produced abundant waters for the farmers; however, now, since the glacier is no longer melting and producing water in the form of rainfall or rushing rivers, farmers

Table 17.1. Perceptions of agricultural change by households in the respective communities.

Community	Climate change	Reduced production	Lost crops	Loss of traditional agricultural practices	Pests	Impoverished soils	Total
Chilcapamba	6	0	1	1	4	0	12
Iltaqui	3	4	2	3	0	0	12
Morochos	2	6	2	1	0	1	12
Topo Grande	4	2	4	1	0	1	12
Ugshapungo	1	0	2	3	0	0	6
Total	16	12	11	9	4	2	54

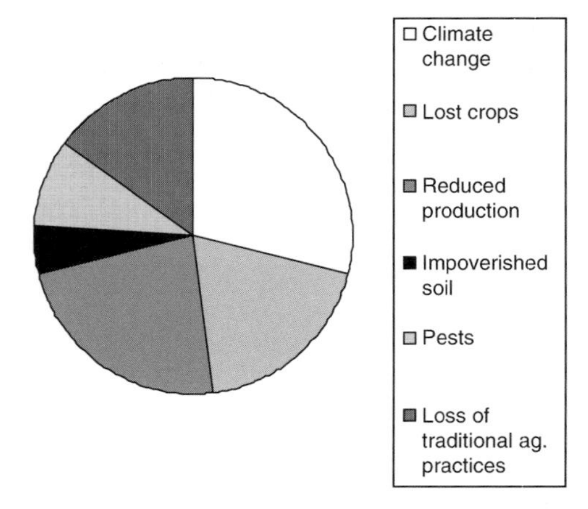

Fig. 17.1. Graph of most significant agricultural changes.

are having significant difficulties with their agricultural pursuits.

These climatic changes have not only complicated agricultural practices, they have also affected perceptions of traditional knowledge. Indigenous farmers invariably commented on the abundant, plentiful production that they experienced in the past. They believe that they are losing their ancestors' sacred agricultural knowledge; their awareness of the relationship between lunar phases and agricultural tasks that allowed them to realize such wonderful production. With climate fluctuation, traditional planting dates no longer correspond with contemporary climatic and agroecological conditions. Traditional crops that grew in their particular vertical zone no longer produce, and the farmers believe that they are to blame, because they have mismanaged or neglected the traditional knowledge of their ancestors.

Besides these climatic factors, cultural and historical variables also have significantly affected contemporary agricultural strategies and soil productivity. Although farmers traditionally kept their animals in harvested fields, allowing the animals simultaneously to graze on stalks and other crop residues and organically fertilize their fields, farmers are no longer able to engage in this practice. The theft of livestock has reached such levels in these remote, indigenous communities that farmers can no longer leave their animals tied up in their fields or anywhere outside of their view or earshot at night without having them stolen. Non-indigenous livestock rustlers from nearby cities frequently sweep into indigenous villages at night and steal any unmonitored livestock that is not locked up.

In addition to theft, indigenous farmers also have other barriers to the implementation of traditional agricultural practices. As Chapter 4 (Zapata Ríos *et al.*, this volume) shows in the aerial photos of the region, when agrarian reform was implemented in Ecuador, the 'breaking up' of the haciendas in Cotacachi resulted in nothing more than a minor allocation by the *hacendados* of small, marginal parcels of land to the *huasipungueros*, rather than the lofty stated intent of agrarian reform, which was to redistribute the hacienda land among the poor, exploited farmers. The result has been that farmers do not own, nor have access, to enough land to keep sufficient animals to produce organic fertilizer for their fields. Also, in many cases, they no longer engage in the traditional use of multiple agroecological zones for production (see Moates and Campbell, Chapter 3, this volume). These historical social factors have resulted in decreased production and a precarious existence that force indigenous farmers to accept exogenous interventions as possible solutions. While indigenous farmers perceive themselves at least partially to blame, because of their loss or neglect of traditional knowledge, the culprit appears to be exogenous in nature. The next section presents comparative histories of three communities and their distinct intervention experiences and analyses the impacts on agricultural strategies. These histories are derived largely from interviews with indigenous members of each community and represent the voices not included in the 'official' annals of history (Wolf, 1982).

Community One: Chilcapamba

The community of Chilcapamba is located at an elevation between 2500 and 2700 masl.

The Pichambiche river passes along one side of the community, and the Suarez wash, a human-manipulated offshoot of the Yanayacu river, passes through the heart of the community. The water from the Yanayacu has traditionally been used for washing, irrigation and consumption; however, recently it has dried up considerably. The community has especially sandy soils that require extensive organic fertilizer for decent agricultural production. Migration has increased significantly as a result of decreased production and a concomitant inability to secure abundant harvests (see Flora, Chapter 18, this volume).

History of exploitation

Before the 1940s, the community of Chilcapamba did not exist. It was subsumed within the confines of the 'La Compania' hacienda, where the relatives and ancestors of current Chilcapamba residents worked as virtual serfs. Besides the indigenous serf inhabitants of the hacienda, there were also indigenous families that lived on the fringe of the hacienda. The hacienda encompassed the prime agricultural land in the area, including water access and grazing lands. Therefore, local indigenous people were forced to sneak on to the hacienda in order to get water from the Yanayacu. If the hacienda owner caught the indigenous people using water from the Yanayacu, they were beaten and forced to work without pay for 3–5 days.

Eventually, four serf families coordinated with some local Quichua to revolt against the *hacendado*. After a lengthy struggle, the indigenous people achieved their freedom, but afterwards they were denied admittance to the hacienda lands. After this revolt, the *hacendado* invited poor mestizos (non-indigenous) from neighbouring communities to work the land, yet they were never paid. Some poor mestizos worked with the indigenous Quichua to organize and gain control of some of the hacienda land. The hacienda owners then immigrated to Quito and abandoned the hacienda to the state. Ancestors of present-day Chilcapamba residents walked to the capital of Ecuador and petitioned the government for the lands. Finally, the government declared Chilcapamba an official community with full right to their lands. Yet, Chilcapamba indigenous peoples have continuously struggled with the mestizos in the region. Some mestizos manipulated the Quichua and took control of the best agricultural lands. Also, the cantonal government of Cotacachi exploited them until the 1980s through threats and violence. They forced the indigenous peoples to work on city projects such as roads and buildings, using them as beasts of burden to carry materials.

Land tenure and management

Chilcapamba's total area is approximately 100 ha divided relatively evenly among 62 households. While the average number of hectares per household works out as 1.25, interviews demonstrated that the few mestizos living in the community, who are recent in-migrants, have larger land holdings than the indigenous inhabitants. Of the 600 inhabitants of Chilcapamba, 200, or 33.3% are mestizo. The average land holding for Quichua residents is < 1 ha.

Due to the scarcity of land and the frequent inability of elders to pass on a substantial plot of land to their numerous heirs, young Quichua males tend to emigrate to Quito or other cities in search of a paying job. Sometimes, after earning money in the city, or ending up on the streets, descendants return to Chilcapamba to take care of their family's land. Their familiarity with farming, however, is relatively limited, because instead of learning to cultivate while they were young, they were working in the city.

The agricultural practices in the community are primarily organic and traditional. The main approaches for tilling the land are the use of the *yunta*, or yoke (see Fig. 17.2), and on small plots, locally fashioned steel hoes with wooden handles called *azadones*.

Fig. 17.2. Tilling with the *yunta* (Photo: B.C. Campbell).

The primary fertilization technique is the grazing of domesticated animals, cows, pigs, sheep, goats and chickens, in the fields while they are fallow. Also, the manure of guinea pigs, known locally as *cuy*, which they typically house in the corner of the cooking area, is collected and applied to the fields. Many of the older inhabitants of Chilcapamba plant their cultigens according to the phases of the moon or according to Catholic Saints' days. The major crops grown in Chilcapamba are quinoa, potatoes, maize, fava beans, broad beans, wheat, barley, peas and several species of cucurbits (Campbell, 2002). Intercropping is practised, especially the widely employed traditional Native American 'three sisters' approach, with maize, beans and squash all intercropped.

Exogenous interactions

The majority of the indigenous population currently professes faith in evangelical Protestant religions. They have a Protestant church in the centre of their community. While they profess allegiance to various forms of Protestant Christianity, they maintain their traditional Andean cosmovision alongside their Protestant beliefs. North American church volunteers assisted in the construction of the church and other various structures in Chilcapamba. The traditional style of home construction in Chilcapamba consists of either wood or earthen walls and a thatched roof. The varied assistance of missionaries over the last couple of decades, however, has led to the emergence of homes with concrete and cement block bases, foundations and walls. The more prominent members of the community now have cement homes, while those with less resources still reside in traditional thatch roof homes.

The church distributed pamphlets to the indigenous people of Chilcapamba, which advocate a strictly organic agricultural approach, comparing it with their traditional religion that places the earth, *Pachamama*, in a sacred realm that should not be polluted with foreign substances. Interestingly, the interweaving of Western and Quichua religious components with land management and agriculture emerged as a common theme in interviews with inhabitants of Chilcapamba. These beliefs remain ubiquitous because they are tied up in the lingering Catholicism implanted by Spanish *hacendados* and priests and also by more recent encounters with visiting missionaries. A 64-year-old Quichua-speaking man from Chilcapamba recounted this story demonstrating the intermixing of indigenous and Western religious systems and their influence on agricultural practices:

> On Saint John's Day, like on the 24th of June, which we know as *Inti Raymi*, a white priest came and gave a blessing and all the people there thanked him for blessing their crops. Therefore, they didn't lose the harvest and the ground continued

> producing more. But we haven't had the white priests here and that's why we're having lower production. Where the priests go they have better production.
>
> Before planting, we always carry seeds . . . to the religious idols of the Saints and Virgin Mary on the mountainside to have them blessed by the Saints' and Virgins' statues. Then we take the blessed seeds back and mix the blessed seeds with the rest of our seeds at the house and then we plant them.

Besides encounters with exogenous religious groups, the people of Chilcapamba have also had Ecuadorian institutions such as the Ministry of Agriculture advise them on land management issues. Segundo Morales, the president of Chilcapamba at the time of this study, succinctly explained his critical stance of the Ministry of Agriculture;

> A change that has occurred because of the Ministry of Agriculture is that they have told us that we need to use the 'improved' seeds because they are better. Then, we have confided in them because they tell us that they are better. But, the truth is that the ground here is very sandy and the seeds cannot adapt and they need more fertilizers. If we blindly follow their advice, then we lose. That's happened to us. Clearly, they do not want us to lose, rather they want to help us produce more.

Segundo and other inhabitants of Chilcapamba know that the unique soil structure of their community precludes the cultivation of any random seeds; the infertile and extremely sandy soils of Chilcapamba require landraces that have adapted to such conditions. Therefore, they are hesitant to jump into sweeping changes in their land management without experimentation. Also, contact with the Ministry of Agriculture in Chilcapamba was much more ephemeral and superficial than the relationship established by the Ministry with other communities. Chilcapamba has established a greenhouse where they are growing organic tomatoes, tree tomatoes, strawberries and numerous other marketable fruits and vegetables. Despite these recent innovations, they have had limited success lately growing their staple crop, maize, because of their infertile soils, lack of animals and pasture land, and unpredictable rains.

Community Two: Morochos

Morochos stands out as an extremely conservative indigenous community. All of the 750 members of the community are indigenous. They adhere closely to the traditional Andean socioeconomic framework, employing communal decision making, intracommunal exchange networks, a relatively strong degree of ecological complementarity and frequent *mingas*. The community is located at approximately 2800 masl; however, the communal land extends up to > 4000 masl, because of the traditional links between ecozones. Morochos communally cultivates a variety of Andean tubers and grains in their *páramo* territory (Campbell, 2002).

History of exploitation

The Spanish conquest of northern Ecuador required little effort on the part of the conquistadors because the Inca, only a number of years previously, had devastated the local populations in the region in a 20–30 year battle (Salomon, 1986; see Moates and Campbell, Chapter 3, this volume). As Spaniards appeared and quickly appropriated all the lush valleys, some indigenous peoples attempted to escape to higher elevations to avoid the requisite *encomienda* duties the Spaniards forced on the indigenous people. In addition to Spanish conquistador *hacendados*, the Catholic Church also appropriated vast tracts of land throughout Ecuador, including much of the Cotacachi mountainside. Ancestors of the contemporary community of Morochos consisted of *huasipungueros*, a term that refers to indebted tenant farmers who lived within the confines of an hacienda, and fugitives who managed to escape enslavement. The hacienda was an enormous livestock operation that spanned multiple ecological zones and dominated the landscape. The Quichua who evaded enslavement remained obligated by Catholic priests to pay a tithe to the

Church, the best products from each harvest, and labour to the *hacendado* when they were caught trespassing. Similar to Chilcapamba, after they obtained recognition of their official status as the owners of their community through struggle and legal petitions, they continued to be harassed, mistreated and forced by the local Cotacachi government to contribute uncompensated labour to infrastructure projects for the city.

Land tenure and management

The community consists of a total of 300 ha, 120 households and 750 community members. The mean average per household works out as 2.5 ha; however, the survey data showed a more consistent median figure of approximately 1 ha per household. The community has extensive communal land in multiple ecological zones and remains conceptually divided according to the sectors that were established when the community was a hacienda livestock operation. The names of the various sectors of the community correspond to the hacienda use for them; the lower altitude region named *Chichu Vaca* (pregnant cows) was used for breeding and calving, the higher altitude regions of *Conrayano* (pasture land) and *Pinllu Corral* (bulls and oxen) were used as grazing land and feed and stable yard, respectively.

The people of Morochos require that all able-bodied members of their community participate in *mingas* on communal land. They take advantage of the fact that they own communal land in distinct ecological zones by cultivating distinct crops in each zone. They cultivate the 'three sisters', maize, beans and squash, along with broad beans and other cucurbits. In the *páramo*, they cultivate a variety of Andean tubers; oca, melloco, mashua and potato, along with grains such as quinua, wheat, barley and rye. Due to the different climates of these zones, cultivation and harvest are staggered so that they plant the lower territory first, then the *páramo* zone, and they harvest the crops in the same respective order. The multizone agriculture of Morochos provides a complementary array of food; with the staggered cultivation and harvests, the Quichua are able to allocate communal labour without conflicting cultivation or harvest schedules. The community has very strong social cohesion, reflected in part by their strict employment of the *minga* multiple times each week and their inflexible reliance on communal consensus for decision making.

Despite the conservatism and cohesion of the community, there remains much intra-communal variation. Many elders of the community continue to employ traditional approaches to fertilization, land management and local resource use, exemplified by a *tayta* from Morochos, which translates from Quichua into 'a respected or revered elder of Morochos'. He continues to practise the traditional use of local plants for medicinals, clothing, fertilizers and other varied purposes (Mendizabal, 1999). One plant in particular, *Cabuya Blanca* (*Fourcroya andina* Trel), a succulent that he processes into fibres for ropes, bags, clothing and shoes, produces an enormous amount of moist, green waste material when processed, which he applies to his fields as a fertilizer. The *tayta* from Morochos describes his land management and economic practices (see Fig. 17.3):

> Sometimes I work in the fields, with yuntas, tilling, I prepare the lands to plant some products, either here where I live or also in the paramo. Up on the mountain I plant potatoes, mellocos, ocas, wheat, barley, and here below, I plant corn, beans, sambo (*Cucurbita pepo* L.; ficifolia), just a little of everything. I work only with organic manure; nothing of chemicals – I don't use them. For potatoes and other crops I use the waste from crops and the waste from the Cabuya serves as an organic fertilizer, and it lasts three or four years in the soil. We use it, also the animal waste as fertilizer.
>
> The Cabuyas we sell in the cities, like Quiroga, Otavalo, and also within the community here. We sell it as rope, and bags, depending on who wants it. The ropes are used to tie animals, like cows, sheep, and pigs.

While the *tayta* of Morochos represents most elders and some of the general population in his land use approach, there

Fig. 17.3. The *tayta* from Morochos with Cabuya waste material and a burro ready for market (Photo: B.C. Campbell).

Table 17.2. Tillage method by households in the respective communities.

Community	Yunta	Tractor	Azadon	Tractor and yunta	Yunta and azadon	Tractor, yunta and azadon	Total
Chilcapamba	8	0	0	2	2	0	12
Iltaqui	3	1	0	6	1	1	12
Morochos	1	2	1	7	1	0	12
Topo Grande	6	0	0	4	2	0	12
Ugshapungo	0	0	0	6	0	0	6
Total	18	3	1	25	6	1	54

exists another faction of Morochos, primarily younger residents, that have come in contact with more industrialized agriculture approaches and have adopted the use of tractors in land preparation. As demonstrated in Table 17.2, the amount of households utilizing a tractor for tillage outnumbers those using the traditional methods of *yunta* and *azadon*.

The use of tractors demonstrates a break with tradition. Elders claim that the young reject traditional methods because they are lazy. Those that utilize tractors are in agreement. They frequently criticize themselves for applying the tractor on their land because it goes against their traditional views that the land needs to feel the human and animal tracks, breath and faeces upon it in order to become fertile. The allure of rapid land preparation, with the tractor requiring less than a full day for several hectares, while the *yunta* necessitates at least a full day of tiring, back-breaking work, has meant that many have succumbed to abandon tradition and opt for modern technology. The financial cost of a tractor rental, based on the reduced time for preparation, ends up being comparable with the rental of bulls for a *yunta*. Also, the recent thefts of cattle have forced inhabitants to weigh their options, and more often than not, when livestock is absent, the Quichua of Morochos choose the easier, but non-traditional method (Fig. 17.4).

Exogenous interactions

Morochos did not engage in agricultural modernization projects as early as other

Fig. 17.4. Quichua child in Morochos learning tractor basics (Photo: B.C. Campbell).

communities of the region; only within the last 10–20 years has Morochos interacted with and implemented 'Green Revolution' style agricultural packages. They were encouraged by international non-governmental organizations (NGOs) to adopt agrochemical and improved seed approaches to agriculture if they expected to receive loans from external sources. In response to interaction with lending agencies and foreign NGOs, Morochos began to apply agrochemicals for specific crops on communal lands, potatoes and vegetables more than grains. Recently, however, community members have spoken against chemical use at community meetings, and consensus has decided to halt the application of agrochemicals on communal lands. The Quichua of Morochos only engage in relationships with exogenous agencies if there is communal consensus. In 2001, they collaborated with UNORCAC and the Heifer International programme to establish an alpaca production operation that they believe to be more in line with their ancestors' practices and beliefs. Other recent interactions with international development agencies have focused less on agriculture and more on basic sanitation and education issues. Since 1980, they have been assisted financially by several NGOs in the construction of a communal meeting house (*Casa Comunal*) and in improvements to their potable water systems.

Community Three: Ugshapungo

Ugshapungo is situated at 3000–3200 masl, a remote highland community overlooking the impressively grand crater lake, *Laguna Cuicocha*. The community consists of 150 residents that are primarily indigenous, with only ten mestizo community members that constitute 6.6% of the population. The community has experienced enormous changes over the last century. The members currently engage in market-oriented intensive agriculture. They are obligated to purchase and apply modern agrichemicals to produce even the most menial harvests. Korovkin (1997) has demonstrated that the high zones in Ecuador, such as Ugshapungo, have been most affected by Green Revolution introductions and most susceptible to the disintegration of the traditional verticality approach to agricultural production.

History of exploitation

At the turn of the 20th century, Ugshapungo was a remote hacienda owned by a family that hired local indigenous people to round up other Quichua and form *mingas* to work the hacienda land for negligible pay. In the 1940s, the *hacendados* offered to sell the land. Indigenous peoples began networking and formed a cooperative, consisting of indigenous people from the Ugshapungo region and well beyond, in order to purchase the hacienda. Many of my older respondents were children at the time and remember a drastic transition, from their parents working in the hacienda to them becoming owners of their own land. Stories abound of bags and bags of tubers being grown in the dark volcanic soils. Local Quichua recall abundant harvests with virtually no manure inputs; usually the plant stalks were left to decompose as a green manure and that was the only fertilization used.

Shortly after the land entered the hands of the indigenous cooperative, however, engineers from the Ministry of Agriculture and the *Misión Andina* (Andean Mission) appeared in Ugshapungo. They promised to help the indigenous farmers to improve their harvests. They brought in bulldozers and land movers to flatten the steep mountain terrain to facilitate large-scale cultivation. Local Quichua were reluctant to apply the Green Revolution packages being pushed by the engineers to their land. They were content with their harvests. Therefore, agricultural engineers coerced them with experimental plots of chemical fertilizers, pesticides and input-responsive 'high-yielding' varieties of potatoes and free handouts to try on their own land. The result, in the minds of local Quichua, was mind-bogglingly enormous potatoes. The Quichua already had > 20 different varieties of potatoes, all that served distinct culinary and agricultural purposes and risk aversion, but those produced on the test plot were bigger and more uniform than any they had ever seen. They accepted the free handouts; 'Green Revolution' packages with agrochemicals and 'high-yielding', input-responsive varieties of potatoes. Convinced by the scientific experiments, they began to devote large expanses of their recently acquired land to the new varieties and less to their traditional varieties. They extensively applied the chemical fertilizers and pesticides. Since that first demonstration, the Quichua and their land in Ugshapungo have become dependent upon agrochemicals and input-responsive varieties. Over the years, they have conducted their own experiments; they planted 'improved' and traditional seeds without agrochemicals and the plants simply would not mature. The engineers never explained to the unsuspecting Quichua that the varieties required agrochemicals to produce and that their soils would no longer self-replenish their nutrient content because the chemicals had destroyed some of the requisite microorganisms. Nor did they explain that if the indigenous people expected to cultivate the next season, that they would be forced to travel long distances to cities to buy more agrochemicals and more 'improved' seeds.

Land tenure and management

Ugshapungo is a disparate community, with approximately 20 households spread over 100 ha. There are approximately 150 members of the community and each household owns an average of 5 ha. As the aforementioned history of the community demonstrates, land management has been severely altered by contact with exogenous groups. Land management practices still include the use of the *yunta* and *azadon*, and animal manure is still applied to the land; however, the cultigens will not produce without the application of agrochemicals. The potatoes especially require large amounts of pesticides because of their susceptibility to late blight. As one local man explains; 'Now we always have to fumigate three, four, five times during the development but when we do that we take the risk that the fruit will be contaminated and be worth nothing.' Tables 17.3 and 17.4 demonstrate the difference in chemical

Table 17.3. Pesticide use by community.

Community	Chemical pesticide	No chemical pesticide	Chemical pesticide sometimes	Total
Chilcapamba	1	10	1	12
Iltaqui	1	11	0	12
Morochos	1	9	2	12
Topo Grande	0	11	1	12
Ugshapungo	6	0	0	6
Total	9	41	4	54

Table 17.4. Fertilizer use by households in the respective communities.

Community	Organic	Chemical	Organic and chemical	Total
Chilcapamba	10	1	1	12
Iltaqui	10	1	1	12
Morochos	11	0	1	12
Topo Grande	12	0	0	12
Ugshapungo	0	1	5	6
Total	43	3	8	54

pesticide and fertilizer use between the communities being explored.

As Table 17.3 indicates, household pesticide use in Ugshapungo is a serious outlier. While the other Cotacachi communities use relatively few pesticides, households in Ugshapungo consistently apply them. As can be seen from Table 17.4, most households in Ugshapungo continue to utilize organic fertilizers; however, they all also apply chemical fertilizers.

While the Quichua of Ugshapungo have lost almost all of their traditional potato varieties, having replaced them with 'high-yielding' varieties, they still produce various landraces of other tubers such as *melloco*, *oca* and *mashua*. As one local man explains, another major change that has occurred in the land management since 'Green Revolution' introductions were made is the quantity of land they cultivate:

> Before we planted only a little, but it gave us great production and in that time we didn't sell any of the harvest. It was only for consumption. Already now, after the arrival of the Andean Mission, they showed us how to use chemicals and produce more and with the chemicals we continue. We have to buy them and sell our harvest to buy even more.

Exogenous interactions

The landscape and land management approaches in Ugshapungo have been most affected by development groups, such as the Andean Mission and government agencies encouraging Green Revolution technologies. Local residents became distraught when discussing their naïve acceptance of Green Revolution packages; their present-day dependence on agrochemicals and 'certified' seeds frustrates them enormously. An excerpt from an interview with a 64-year-old Quichua man from Ugshapungo poignantly illustrates the nature of exogenous interactions there:

> Researcher: Do you have any stories about how your ancestors worked the land?
>
> Well, the Andean Mission came to tell us that we should use chemicals and tractors. It was a support project, but before they came the land was already producing very well.
>
> Researcher: Can you explain when the Andean Mission came and what they did here?
>
> Yes, people . . . who had been studying in the colleges in Quito and Latacunga, were brought here from below as a help for this region. At first, they helped with the tractors and chemical fertilizers and we eventually began to rent the tractor. We bought chemical fertilizers because at that time they were very cheap. I remember that the cost to rent a tractor was about 100 sucres per hectare and the chemical fertilizers about five sucres per bag. Now these things are very expensive. We buy a bag for about 250,000 to 300,000 sucres and to rent a tractor is 100,000 an hour. On one hand, they did us well because all this land was mountains; but on the other hand it's bad because we have lost the use of organic fertilizer for the application of chemicals.

> Now the seeds come certified and we are totally obliged to use chemicals with them. I did an experiment planting with and without chemicals. I see that what I planted with chemical fertilizer grows well, while the plants without chemicals grow small and sickly. Therefore, we have to use them, chemical fertilizers and pesticides. Also, we plant different cultivars and we take them to the market in Cotacachi to sell them; like I said, I plant continuous crops to have the most security to obtain economic resources and food.

Comparison

The historical mistreatment of the indigenous peoples of these communities represents a constant variable. The manner in which they emerged from this exploitation, the land holdings that they were able to secure as a community, and the forms of exogenous and mestizo interactions and interventions designate the salient differences between the communities that help explain their contemporary agriculture (see Table 17.5). The over-riding factor in contemporary agricultural difficulties in Chilcapamba has less to do with exogenous interventions, and more to do with historical exploitation and the concomitant lack of land. Table 17.5 illustrates that Chilcapamba indigenous farmers have comparatively smaller land parcels and significantly more mestizos in their communities. The mestizo presence in Chilcapamba not only resulted in small land holdings among indigenous inhabitants, it also inhibits traditional communal solidarity and collective decision making because the mestizos have different belief systems and agricultural goals from the indigenous inhabitants. While the Quichua strive to survive collectively, mestizos, for the most part, perceive the world through an individualistic, capitalist mindset of wealth accumulation. If Chilcapamba farmers had access to sufficient land for animals and the simultaneous use of multiple ecozones, as in Morochos, their agricultural opportunities would be heightened. This is especially true, because in Chilcapamba, where exogenous interventions were primarily religious groups, advocating not the adoption of modern technologies, but rather Christian faith, farmers do not depend on foreign inputs.

The development agencies that arrived in Ugshapungo intended to help the farmers modernize their agricultural practices. Most probably, they saw what they considered poverty and felt they could alleviate it through the introduction of more advanced agricultural technology. Yet, we are finding that sometimes such interventions have different effects (Peet and Watts, 1996). The juxtaposition of Ugshapungo pesticide and chemical fertilizer use with that of other Cotacachi communities, as seen in Tables 17.3 and 17.4, respectively, reveals a consequential difference. Contrary to the other Cotacachi communities, Ugshapungo farmers consistently apply chemical fertilizers and pesticides. Ugshapungo experienced relatively early agricultural development interventions, deliberate attempts to convert Ugshapungo farmers into 'Green Revolution' proteges, whereas in other local communities such an agricultural development presence was minimal. The agricultural decline, loss of traditional cultivars and varieties, and dependence on agrochemicals must be positively correlated with exogenous interventions in Ugshapungo.

In the Morochos case, their cultural solidarity and conservatism, combined with limited interactions with 'modern' agricultural development agencies, has resulted in a deliberately reserved adoption of modern agricultural practices. They have minimal dependency on agrochemicals, but have adopted the use and ownership, when feasible, of tractors. The absence of development interventions that encourage full-scale modernization in these communities demonstrates that they have the ability to pick and choose, experiment with and/or reject modern technologies, rather than having the technologies and dependency forced upon them with no alternatives. The fact that they have obtained ownership and access to multiple ecological zones has allowed them to retain traditional agro-ecological strategies and less dependence on exogenous methods.

Table 17.5. Explanatory variables in contemporary agricultural decline.

Community	Exploitative social history	Emancipation	Mestizo population	Landholdings	Exogenous interactions
Chilcapamba	*Pre-1990* • Hacienda • Slavery • Mestizos • Cotacachi government	Indigenous serfs obtained small parcels of hacienda and mestizos appropriated the best agricultural lands	33.3% (200/600)	Minifundio < 1 ha per indigenous household • Negligible communal land	Non-agricultural • Evangelical protestant missionaries Agricultural • Ministry of Agriculture (minimal contact)
Morochos	*Pre-1990* • Hacienda • Huasipungo • Catholic Church • Cotacachi government	Indigenous community obtained legal right to entire hacienda, constituting multiple ecological zones	> 1% (0/750)	Varied > 1 ha per indigenous family • Extensive communal land in multiple zones	Non-agricultural • Swiss Aid Agricultural • Heifer International Green Revolution • (DRI) Desarrollo Rural International
Ugshapungo	*Pre-1990* • Hacienda • Huasipungo • Cotacachi government	Indigenous leaders purchased the hacienda for future generations	6.6% (10/150)	Larger tracts ~5 ha per indigenous household	Green Revolution • Misión Andina • Ministry of Agriculture

Conclusion

This research has found that besides development initiatives of the Green Revolution variety, other explanatory factors must be addressed. Climatic change, inaccessibility to sufficient land and animals, and historic subjugation all play significant roles in contemporary agricultural struggles. This is not to say that capitalist enterprises infiltrating indigenous communities under the guise of 'development' have not impacted traditional land management substantially. The reported impacts of 'Green Revolution' packages in Cotacachi and throughout the developing world range from severe environmental degradation to social and cultural disruption to severe health deterioration (Ponting, 1991; Sherwood, 1999; Frolich *et al.*, 2000). As Frolich *et al.* (2000, p. 5) explain:

> Studies have shown that overuse and careless management of pesticides contribute to severe human health effects, including poisonings (171/100,000), dermatitis (48% of applicators), pigmentation disorders (25% of applicators), and severe neuropsychological effects (peripheral nerve damage, abnormal deep tendon reflexes and coordination). Mortality due to pesticides was among the highest reported anywhere. These health impacts were found to be about twice as severe in rural settings.

Increased dependence on inputs requires income. Therefore, with limited jobs in the region, another effect has been that migration rates have dramatically increased (Painter, 1995). Bebbington (1996) examined similar problems in Ecuador and came to the conclusion that indigenous peoples could embrace 'Green Revolution' technologies in order to curb these escalating problems. He states, 'In short, the incorporation of modern technologies can be a sign of being liberated from a past of domination, even if this may imply new dependencies. It may be that incorporating modern techniques may be politically empowering rather than culturally disempowering' (p. 92). Yet, considering the enormous problems and dependence associated with these practices historically, and the concomitants – forced migration and reliance on exogenous corporations – how can an embracing of 'modern technologies' politically empower?

While Bebbington (1996) criticizes the universal applicability of Farmer-First models, he proposes that modernization initiatives could further indigenous revitalization and increase market options for poor peasant farmers in marginal Andean zones. This study has shown that socioeconomic situations differ based on agricultural, biophysical and sociopolitical historical processes and contacts, and that increased collaboration with development and government organizations that encourage unsustainable, exogenous inputs without education on alternatives only furthers dependence and precludes local control. Therefore, just as Bebbington (1996) denounced Farmer-First models for their blindness to local realities, the same holds true for the adoption or continuation of degradatory land management in the name of cultural rejuvenation, especially when cultural identity and sustainable, traditional agricultural practices are far from mutually exclusive. Local peoples of Cotacachi do not want to continue to deny the environmental wisdom of their ancestors. While their past was one of domination and subjugation, the current state of affairs, with dependency continuing at the hand of outside corporate interests, needs to be changed. UNORCAC, in conjunction with SANREM, advocates, an 'ancestral future' that maintains the cultural identity and integrity of Cotacachi's indigenous people by fusing traditional practices and worldview with ecologically sound, sustainable development alternatives such as ecotourism, alternative energy and organic agricultural exports. UNORCAC's search for sustainable development alternatives may be a slight essentialism of a difficult past, but political empowerment lies not in foreign control of decision making, but in the 'embeddedness' of the local economy in the traditional Andean framework of collective solutions and, most of all, in the local control of resources and land management.

References

Bebbington, A. (1996) Movements, modernizations, and markets: indigenous organizations and agrarian strategies in Ecuador. In: Peet, R. and Watts, M. (eds) *Liberation Ecologies: Environment, Development, Social Movements*. Routledge, London, pp. 86–109.

Brush, S. (1977) *Mountain, Field, and Family: the Economy and Human Ecology of an Andean Valley*. University of Pennsylvania Press, Philadelphia, Pennsylvania.

Campbell, B. (2002) Ancestral futures? Historical ecology in the north Ecuadorian sierra. In: Stepp, J.R., Wyndham, F.S. and Zarger, R.K. (eds) *Ethnobiology and Biocultural Diversity*. University of Georgia Press, Athens, Georgia, pp. 442–463.

Frolich, L., Sherwood, S., Hemphill, A. and Guevara, E. (2000) *Eco-Papas: Through Potato Conservation Towards Agroecology*. ILEA, Ibarra, Ecuador.

Korovkin, T. (1997) Indigenous peasant struggles and the capitalist modernization of agriculture: Chimborazo, 1964–1991. *Latin American Perspectives* 24, 28.

Masuda, S., Shimada, I. and Morris, C. (eds) (1985) *Andean Ecology and Civilization*. University of Tokyo Press, Tokyo.

Mayer, E. (1985) Production zones. In: Masuda, S., Shimada, I. and Morris, C. (eds) *Andean Ecology and Civilization*. University of Tokyo Press, Tokyo, pp. 45–84.

Mendizabal, T. (1999) *Medicina Tradicional e Interaccion de Sistemas Medicos en las Comunidades Andinas del Canton Cotacachi*. Medicos Sin Fronteras, Quito, Ecuador.

Murra, J. (1985) The limits and limitations of the 'vertical archipelago' in the Andes. In: Masuda, S., Shimada, I. and Morris, C. (eds) *Andean Ecology and Civilization*. University of Tokyo Press, Tokyo, pp. 3–15.

Oberem, U. (1976) El acceso a recursos naturales de diferentes ecologias en la sierra ecuatoriana. Siglo XVI. In: Oberem, S.M.a.U. (ed.) *Contribucion a la Etnohistoria Ecuatoriana. Coleccion Pendoneros*, Vol. 20. Instituto Otavaleno de Antropologia Otavalo, Ecuador.

Painter, M. (1995) Upland–lowland production linkages and land degradation in Bolivia. In: Durham, M.P.a.W. (ed.) *The Social Causes of Environmental Degradation in Latin America*. University of Michigan Press, Ann Arbor, Michigan, pp. 133–168.

Peet, R. and Watts, M. (1996) *Liberation Ecologies: Environment, Development, Social Movements*. Routledge, London.

Ponting, C. (1991) *A Green History of the World*. Penguin Books, New York.

Salomon, F. (1986) *Native Lords of Quito in the Age of the Incas*. Cambridge University Press, New York.

Sherwood, S. (1999) *Reporte sobre Cultivos de Cobertura*. Centro Internacional de la Papa (CIP), Quito, Ecuador.

Stanish, C. (1992) *Ancient Andean Political Economy*. University of Texas Press, Austin, Texas.

Stern, S.J. (ed.) (1987) *Resistance, Rebellion, and Consciousness in the Andean Peasant World, 18th to 20th Centuries*. University of Wisconsin Press, Madison, Wisconsin.

UNORCAC (1996) *Memoria del Taller de Autodiagnostico en la UNORCAC*. UNORCAC, Cotacachi, Ecuador.

Wolf, E. (1982) *Europe and the People Without History*. University of California Press, Berkeley, California.

18 Circular Migration and Community Identity: Their Relationship to the Land

Gabriela Flora
American Friends Service Committee, Central Region Project Voice Organizer, 901 W. 14th Avenue, Suite 7, Denver, CO 80204, USA

Introduction

The rural highland indigenous communities of Cotacachi, Ecuador, have many of the classic 'push' characteristics of a migrant sending area. Yet the desire of its indigenous inhabitants to leave these communities is low. While permanent out-migration from the rural communities of Cotacachi does occur, many residents leave for varying lengths of time, from overnight to most of the year, but still consider themselves culturally active members of their rural community. This chapter examines how an indigenous population manages to stay in a locale where *campesinos* (peasants) have very small land holdings, access to water is limited and there are few possibilities for employment in their own community. *Circulation*, short-term migration for work, is one of the primary strategies to remain rooted to place. Circulation broadens the economic options available but maintains cultural capital.

The chapter begins with a literature review of circulation and its long-standing role in highland Ecuador. This is followed by a description of the methodology utilized to understand economic activities, attachment to place, and circulation and migration in rural Cotacachi. The data analysis provides insight into the economic, social and cultural realities that allow people to stay tied to their lands in rural communities of the canton of Cotacachi.

Literature Review

Circulation

Circulation is described as repetitive, short-term or cyclical population movement. The common feature of this type of mobility is that the actors do not intend to relocate permanently. Jones calls it a 'flexible form of migration' (1990, p. 222). Chapman and Prothero point out that circulation centres on the 'territorial separation of obligations, activities, and goods' (1985, p. 1). Migration theorists have traditionally viewed this 'bi-local' (Cohen, 1996, p. xvi) lifestyle as occurring between a rural and urban area where there is 'dual dependence on city and village' (Jones, 1990, p. 222). These dependencies go far beyond the focus of neo-classical migration theorists. They encompass social and cultural, as well as economic, relationships.

Chapman and Prothero cite three overlapping perspectives on circulation. First, circulation is an 'integration of distinct places and circumstances'. Second, it occurs because of 'socioeconomic disequilibrium'. Third, it 'involves the interchange of labour

 Development with Identity: Community, Culture and Sustainability in the Andes (ed. R.E. Rhoades)

between one mode of production and another' (1985, p. 2). Studies of circulation by economists, geographers and demographers have focused more on the 'interchange only of people between complementary places or situation' (Chapman and Prothero, 1985, p. 4). Other social scientists, anthropologists in particular, often view circulation as an exchange of ideas, customs and goods, in addition to people and labour (Collins, 1988; Werbner, 1990; Gardner, 1995; Paerregaard, 1997).

Circulation usually occurs because not all of an individual's or family's needs are able to be met in one place. Hoops and Whiteford comment that the rural poor in Latin America 'are forced to combine a variety of strategies over the year to maintain themselves' (1983, p. 261). These strategies often involve circulation (Waters, 1997).

History of circulation in highland Ecuador

The rural indigenous communities of the eastern part of the canton of Cotacachi have not lived only from subsistence agriculture, but rather have had mixed survival strategies for at least 400 years. Circulation became a central component of these mixed strategies after independence from Spain, but even during the colonial period circulation was important for some families. In his analysis of Ecuadorian market-place trade, Bromley concludes, 'periodic and daily markets have formed a major focus for circulation in the Central Highlands for at least four hundred years' (1985, p. 348).

During the colonial period in Ecuador, the Spanish crown allotted land grants, which along with a large parcel of land included access to the labour of the people that were living on the land. In Ecuador, this feudal sharecropping system was called *huasipungo*. All adult *huasipungueros* (i.e. serfs) were required to dedicate a portion of their work to the *hacendado's* (i.e. lord's) agricultural production and household maintenance. In exchange for this, the *huasipungueros* were given usufruct rights to a plot of land where the family could raise crops for personal consumption, meaning they had the right to use and enjoy the profits and advantages of the land, as long as the property was not damaged or altered in any way. In addition, access to *hacienda* (large farm) resources was allowed (Hidrobo, 1992; Zamosc, 1994; Korovkin, 1997; Lentz, 1997). The *huasipungueros* thus engaged in mixed survival strategies by dividing their time between work for the *hacendado* and subsistence agriculture, along with other household tasks.

There were many *campesinos* whose economic survival was not linked to *haciendas*. Additionally, there were others that worked for pay rather than having a feudal relationship with the *haciendas*. Some *campesinos* had access to communal land shared by others in their community and some 'worked within the economy of the hacienda, but without a *huasipungo* relationship' (Crissman, 2003, author's translation).

The *huasipungo* system dramatically affected the mechanisms of economic survival in the Andes highlands. Because of their ties to the *hacendado*, the *huasipungueros*' survival strategies were restricted to the hacienda and involved little mobility. Those who were independent of the *huasipungo* system and gained ownership of small plots of land engaged in circulation to obtain money in exchange for their labour. Those people who were able to accumulate some capital before the end of the *huasipungo* system were at a great advantage when *hacienda* land was being sold, as they were able to purchase land beyond their government allotment (Lentz, 1997).

The *huasipungo* system continued long after independence until national land reforms of 1964 and 1973 were enacted. The land reforms claimed officially to divide the *hacienda* lands into parcels to be distributed to the former *huasipungueros*. However, many of the *haciendas* maintained large tracks of land (Brown *et al.*, 1988; Lawson, 1990; Zapata Ríos *et al.*, Chapter 4, this volume).

Brown *et al.* (1988) and Lentz (1997) describe why agriculture no longer provided the base for family subsistence after the end of the *huasipungo*. Their description fits into four basic themes. First, although the former *huasipungueros* (considered *campesinos* after the end of the *huasipungo* system) were usually given title to the same amount of land they had been cultivating for their subsistence crops on the *hacienda*, the land they were given was often of poorer quality. Second, the *campesinos* no longer had access to *hacienda* resources. These resources included pastureland, firewood, water and roads. Thus resources that had previously been available were no longer easily accessible. Third, there has been a continual parcelization of land as the population increased. Fourth, there was a decrease in agricultural productivity due to deterioration of soil quality and other factors discussed in this volume.

This decrease in soil fertility and increase in soil erosion were both exacerbated and caused by two interacting factors. First, the land the *campesinos* were given after the land reform was marginal (of low fertility and steeper and higher elevation than that kept by the *hacendados*, who generally occupied the valley floors). Second, *campesinos* with *minifundios* could not leave the land fallow for the necessary period of time for organic matter and nutrients to be replaced.

The combination of these four factors cited by Brown *et al.* (1988) and Lentz (1997) meant *campesino* families had to engage in labour off their home plots to meet all their needs even after they were officially landowners. With the end of the *huasipungo* system, there was a shift from work in kind to paid labour (Lentz, 1997). Although large *haciendas* that hired agricultural labourers existed in the highland communities, many *campesinos* had to go outside their community to obtain income needed for family survival. Although they were working outside their communities, they returned to their community frequently and considered it home.

Farrell *et al.* (1988) and Waters (1997) stress that in highland Ecuador, circulation to urban areas for jobs is a means of maintaining rural identity in a context where not all needs can be met in one's own community, rather than being a rejection of rural ideals. When speaking about highland Ecuadorian *campesinos* working in the informal sector in Quito, Waters provides a variety of factors that prevent families from meeting all their survival needs in the rural area. These include 'limited access to land, the paucity of rural employment opportunities, unstable tenancy patterns, low levels of productivity, low incomes, and lack of credit and technical assistance' (1997, p. 56).

Farrell *et al.* (1988) stress that Ecuadorian indigenous highland *campesinos* circulate to urban areas not just for material reproduction, but also to reproduce their rural indigenous lifestyle, e.g. maintain their cultural heritage. Circulation is at the same time an integration into the capitalist world system and an act of resistance (Martinez, 1985; Farrell *et al.*, 1988). As Lentz emphasizes, capitalist modes of production both 'perpetuate and deform the traditional modes of production and lifestyles' in highland Ecuador (1997, p. 10).

In sum, multiple survival strategies – of which circulation is a central part – are not new in Cotacachi, but the scope of circulation has expanded and the distance travelled to engage in multiple activities has increased. Research in the rural highland, indigenous communities of Cotacachi in 1997 and 1998 examined the role of circulation in family survival strategies. This research was part of a larger migration project which also included SANREM's sites of Nanegal and Las Golindrinas (Rhoades *et al.*, 2001).

Methodology

For 4 weeks in June and July of 1997, the author conducted exploratory research on water issues and migration in the eastern part of the canton of Cotacachi. A short migration interview schedule, a daily labour calendar and an agricultural calendar were completed in 12 households in five rural communities. This research served as a base for a return to Cotacachi in March and April of 1998 to administer migration interview schedules.

The 1998 interview schedule elicited basic demographic information and details on the location of the respondent's occupation(s) at the time of the interview. It contained questions related to views on the area and other questions related to attachment to place. Additionally, the interview schedule elicited information on all the places the respondent had lived, the occupations held in each of these locations and the motivations for leaving each place where they had lived. Other information asked for included land holdings and crop and animal production.

Quantitative data from 276 migration interviews from 17 communities were analysed using the Statistical Program for the Social Sciences (SPSS). The quantitative data were supplemented with qualitative data collected through participant observation and conversations with UNORCAC leaders and community members. In addition, interviews by the author and her interactions with interviewers through the interview schedule monitoring process provided ethnographic information that complemented the quantitative data.

Establishing a sampling frame

Of the 43 communities that are members of UNORCAC, 17 were selected in which to conduct interviews. (Four of the five communities involved in the 1997 research were included in this sample.) The formula for a stratified random sample with a 95% confidence level and heterogeneity of 25/75 was utilized to determine that 276 interviews needed to be completed.

The number of interviews in each of the 17 communities was based on the number of households in each community. The calculation meant that roughly every fifth house in each of the communities was to be sampled (Table 18.1).

Table 18.1. The 17 communities where interviews were conducted.

Community	Population	Number of households	Number of interviews	% of households interviewed
Alambuela[a]	169	37	8	22
Ambi Grande	548	117	26	22
Azaya	250	54	14	26
El Batán	212	45	10	22
La Calera	626	158	35	22
Chilcapamba[a]	430	90	19	21
Colimbuela	430	90	20	22
Cumbas Conde	543	120	26	22
Morales Chupa	422	72	11	15[b]
Morochos	595	127	28	22
Perafán	237	51	12	24
Piava Chupa	144	30	8	27
Piava San Pedro	180	40	14	35
Quitugo	343	73	16	22
San Miguel	104	22	5	23
San Pedro[a]	400	82	18	22
Santa Barbara[a]	105	27	6	22

Source for population data: 1997 *Medicos Sin Fronteras*/UNORCAC *Jambi Mascaric Project.*
[a]Communities where agriculture and daily labour and short migration interviews were conducted in July 1997.
[b]Morales Chupa is under-represented because the interviewer assigned to that community did not complete all his interviews.

Slightly more males than females were interviewed (Table 18.2). This parallels the 1990 Ecuador census, which showed that in the canton of Cotacachi the gender breakdown of the 27,225 rural population was 51.7% males and 48.3% females (Instituto Nacional de Estadística y Censos, 1991, p. 19). The 276 interviews were divided up utilizing five age categories, 15–25, 25–34, 35–44, 45–54 and > 54 (Table 18.3). The migration literature consistently shows that young adults are more inclined to migrate compared with older adults (Lee, 1966; Valesco, 1985; Jones, 1990). Because of this and the fact that out-migration of the young was of concern to UNORCAC (Andrango, Ecuador, 1997, personal communication; Calapi, Ecuador, 1997, personal communication; Guitarra, Ecuador, 1997, 1998, personal communication), the younger age categories were oversampled along a continuum, with the smallest sampling rate being applied to those in the > 54 category.

Access to communities

Community organization is particularly high in the highland indigenous communities of Cotacachi, which have a legacy of organization and where community cohesion and communication have been central to survival. Access to the local power structure was facilitated by previous SANREM links with UNORCAC and the canton of Cotacachi government via COMUNIDEC (an Ecuadorian non-governmental organization (NGO) that was a SANREM Ecuador partner). In addition to the indigenous organization and canton government approval, local community level approval was also important. The *cabildo* is the lowest level of formal communal organization in Ecuador and serves as the local authority in the communities. Direct approval by a *cabildo* member was obtained in one-third of all the communities where interviews were conducted.

Table 18.2. General description of the sample: interviewees by gender and language of interview (%).

Language of interview	Males ($n = 145$)	Females ($n = 131$)	Total ($n = 276$)
Quichua ($n = 144$)	52.4	51.9	52.2
Quichua and Spanish ($n = 47$)	18.6	15.3	17.0
Spanish ($n = 85$)[a]	29.0	32.8	30.8

[a]Twenty-two of the interviews were conducted with *mestizo*, non-indigenous, people (8% of total interviews) and all of these were in Spanish.

Table 18.3. General description of the sample: education level of respondents by age (%).

Level of education	15–24	25–34	35–44	45–54	≥ 55	Total
None	12.7	22.1	49.2	41.7	52.0	30.4
Some primary	34.2	41.6	42.4	44.4	44.0	40.2
Attended 6th grade	24.1	23.4	3.4	11.1	4.0	15.9
Some secondary	26.6	6.5	1.7	2.8	0	10.1
Attended 12th grade[a]	2.5	6.5	3.4	0	0	3.3
Total	100 ($n = 79$)	100 ($n = 77$)	100 ($n = 59$)	100 ($n = 36$)	100 ($n = 25$)	100 ($n = 276$)

χ^2 significant at 0.00000.
[a]Three respondents (all of whom were males) had some post-secondary education.

Data collection and monitoring

Ten local indigenous, bilingual (Quichua and Spanish) youth took part in a day and a half of training. During the training, the objectives of the migration study and reason for each question were explained. Trainees underwent an intensive series of mock interviews. The 276 interviews were carried out in two and a half days during Holy Week, the days leading up to Easter Sunday where people return to their community of origin to participate in parades and parties. On average, an interview took 30–45 min, although a few extended to 1 h. Two days were spent reviewing the interview schedules with the interviewers.

Results and Analysis

Principal activities

Each interviewee was asked to provide a migration history. In each of the places lived, the respondent provided up to two economic/survival activities engaged in while living there. These were open-ended questions and were coded later. The data on activities carried out do not include all the activities ever engaged in throughout the life of each individual, but rather the most prominent or recent activities in each place of residence.

Most household members engaged in a variety of survival activities. If the household owned land or had access to cultivated land, agriculture and animal husbandry were an important part of the household survival strategy. Sixty-six per cent owned their own land (while not necessarily having title to it), with the average land holding for those reporting it being 0.90 ha. Thirty-eight per cent of respondents worked land belonging to extended family members with an average size of 1.07 ha. Twenty per cent sharecropped the land *al partir* with an average size of 0.90 ha. Three per cent rented land with an average size of 1.04 ha. (These categories are not mutually exclusive; some respondents worked land from more than one source.)

Agriculture production was principally for subsistence and not for commercial sale. Only 22.5% of the respondents' households sold food crops, while 95.7% grew them. Somewhat fewer households (84.4%) engaged in animal husbandry. However, more sold animals (33.3%) than was the case with agricultural produce.

Although agriculture was important in the mix of survival activities, the small land holdings of the rural populace make it virtually impossible to live off the land without an outside source of income. A permanent job is the exception for most residents in the rural communities of the eastern part of the canton of Cotacachi. Thus, most individuals engage in multiple activities. Although some people have year-round jobs at *haciendas* as agricultural labourers, most *hacienda* labour requirements are seasonal. Thus males often switch between agricultural labour and construction work, while helping out with home plot agriculture. Thirty-five per cent of males reported having worked as agricultural labourers on *haciendas*. Fifteen per cent of females reported having worked as agricultural labourers. There are *haciendas* scattered throughout the eastern part of the canton of Cotacachi, so some agricultural labourers work in their own community. Older children in school sometimes work on *haciendas* on the weekends and/or during their vacations. As in most cultures, the activities an individual engages in are shaped by one's gender.

Thirty-seven per cent of the males interviewed reported having worked in construction. Many of the males that work in construction go to Quito for the week, where they rent a room (which is usually shared with a number of friends or relatives), and return to their community on the weekends. Others find construction work in the urban centre of Cotacachi or in the nearby towns of Otavalo and Ibarra (Table 18.4).

Most of the females engaged in agriculture on home plots. Sixty per cent reported engaging in housework and agriculture and/or animal husbandry on home plots, and 46% reported engaging in agriculture and/or animal husbandry on home plots. The majority of

Table 18.4. Most common activities of informants (%).

Occupation	Males	Females	Total
Agriculture on home plot	53.8	45.8	50.0
Housework and agriculture on home plot	2.8	65.6	32.6
Handicrafts in home	20.7	41.2	30.4
Agricultural labourer	34.5	14.5	25.0
Construction worker	37.2	0	19.6
Domestic help	0	26.7	12.7
Handicrafts in workshop	13.8	2.3	8.3
Total	$n = 145$	$n = 131$	$n = 276$

These are not mutually exclusive categories, thus percentages exceed 100. Each respondent was encouraged to indicate two activities.

males helped out with home production. Fifty-four per cent of men reported engaging in agriculture on home plots. Females were generally the primary agriculturists, as males, in most cases, were working away from the home.

The agricultural calendars conducted in July of 1997 showed that home plot agricultural productivity is not impacted by circulation. Small land holdings mean that labour is not a constraint for home agricultural production. At the time of the interviews, over half of the males were gone during the day and another 18% were gone during the week, but both males and females adjust what they do to fit the agricultural calendar. Although females do the majority of the farm work around the house, the 13 agricultural calendars gathered during June and July of 1997 showed that the traditional gender division of labour related to land preparation that Boserup (1970) found in her cross-cultural studies was still adhered to in the rural communities of Cotacachi. Females may do much of the agricultural work, but males still do all the ploughing, with rare exception.

Most of the females interviewed engaged in activities around the house, housekeeping, agriculture and animal care. Making handicrafts is an income-generating activity that can be performed in the home around other household obligations.

Forty-one per cent of the females interviewed reported having done handicrafts in their home. Males also engage in handicrafts, as it is something they can do for money while they are in-between wage jobs. Twenty-one per cent of males interviewed reported having made handicrafts in their home. The principal handicraft activity of males is weaving on looms. More males than females make handicrafts in a workshop, with 13.8% of the males and 2.3% of the females having worked in a handicraft workshop. Those workshops are all located in the urban centre of Cotacachi and in Otavalo.

Female handicrafts work includes weaving on a loom, embroidering, knitting and sewing on a machine. Women often do contract embroidering of blouses. In contract work, sewers are provided with the shirts and the thread. They embroider designs on the shirts with their own sewing machine or by hand, and return the finished blouses to the intermediary. The craft that takes the least amount of inputs and is easiest to do is making bracelets out of coloured string. Children often make these bracelets.

Many of the home handicrafts involve making items that are part of the traditional dress of the indigenous females, sewing the material which is wrapped around as a skirt (called an *anaco*) and weaving the belts that secure women's skirts. More recently, females purchase pieces of leather and yarn and knit them together into vests that are sold to tourists. Some women produce handicrafts on a piece work basis, while others buy the materials and sell the resulting crafts to their neighbours or go to Otavalo on Saturdays and sell at the market.

Twenty-seven per cent of all females reported having worked as domestic help at some point in their life. Domestic help includes working as an *empleada*, or housekeeper, which is most commonly done by younger, single females. They often live with a family and engage in housekeeping duties. In addition, such jobs include washing clothes, cleaning, cooking or caring for small children for private homes on a part-time basis.

Live-in domestic help is the principal activity of females that work in Quito. Younger females sometimes combine working as live-in domestic help and going to school. The town of Cotacachi is a common place where young females who have finished their primary education in their rural community of the canton of Cotacachi go to live and work for a family. Younger *empleadas* often go to school at night. There are only grade schools in the rural communities of Cotacachi. In order to attend a secondary school, a student must go to the urban centre of Cotacachi or another town farther away. Classes are much shorter in night school and thus keep young women working as *empleadas* away from the household for less time. Not surprisingly, they obtain a lower quality of education than if they had attended day school. Generally, females going to school, and working as domestics, are paid little as the employer helps pay school expenses. It is very difficult to work as live-in domestic help and care for one's own family (which usually includes home plot agriculture), so married women who engage in domestic paid work generally do it on a part-time basis.

Community Attachment

Attachment to area

Attachment to the area was analysed because those that are more attached to a specific region are less likely to want to move out of the area. The interview schedule asked if the respondents want to stay and live in the area for the rest of their lives. Nearly nine out of ten respondents said they wish to stay. Even among the age group with the highest propensity to migrate, those aged 15–24, over three-quarters desired to live in the area for the rest of their lives (Table 18.5). The language in which the interview was conducted was more significant than age in explaining community attachment.

All the responses that provided insight into attachment to the area were included in a factor analysis. These responses included all the positive and negative characteristics named by the respondents, whether the respondent thought it was better to work in the area or a city, whether the respondent thought it was better to live in the area or a city, whether the respondent wanted to stay in the area to live the rest of her/his life, whether the respondent would like to buy (more) land in the area and whether the respondent had previously moved.

There were four primary variables that factor analysis showed to be highly correlated: (i) desire to live in the area; (ii) preference to work in the area over a large city; (iii) preference to live in the area over a large city; and (iv) a desire to buy land in the area. These four factors were significantly linked to the language in which the interview was conducted. The language in which the interview was conducted was used as an indicator of acculturation. Table 18.8 shows that Quichua speakers had a higher attachment to the area. In all but the preference to work in the area, Spanish speakers had a higher attachment to the area than the bilingual speakers.

The vast majority of rural residents have strong ties to place. The language in

Table 18.5. Desire to live in area for rest of life by age (%).

Age group	Males[a]	Females[b]	Total
15–24	76.9	75.0	75.9
25–34	100.0	92.1	96.1
35–44	93.9	100.0	96.6
45–54	90.5	86.7	88.9
$\geq$ 55	100.0	91.7	96.0
N	145	131	276

[a]χ^2 significant at 0.00473.
[b]χ^2 significant at 0.03193.

Table 18.6. Attachment to area by gender (%).

Question	Male	Female	Total
Prefer to live in the area rather than the city	95.1	91.5	93.4 ($n = 273$)
Want to live rest of life in the area	91.0	87.8	89.5 ($n = 276$)
Would like to buy land in the area	86.5	91.3	88.8 ($n = 268$)
Prefer to work in the area rather than the city	74.8	82.8	78.6 ($n = 271$)

Note: no statistically significant differences.

Table 18.7. Attachment to area by age (%).

Question	15–24	25–34	35–44	45–54	≥ 55	Total
Prefer to live in the area rather than the city	91.0	92.2	98.3	91.7	95.8	93.4 ($n = 273$)
Want to live rest of life in the area	75.9	96.1	96.6	88.9	96.0	89.5 ($n = 276$)
Would like to buy land in the area	86.5	93.4	94.8	80.0	80.0	88.8 ($n = 268$)
Prefer to work in the area rather than the city	74.7	80.3	75.4	88.6	79.2	78.6 ($n = 271$)

Note: the results for 'live here all your life' were significant, $\chi^2 = 0.00011$; others not significant.

Table 18.8. Attachment to area by language of interview (%).

Question	Quichua	Bilingual	Spanish	Total
Prefer to live in the area rather than the city	97.2	87.0	90.5	93.4 ($n = 273$)
Want to live rest of life in the area	97.9	76.6	82.4	89.5 ($n = 276$)
Would like to buy land in the area	90.6	80.9	90.4	88.8 ($n = 268$)
Prefer to work in the area rather than the city	86.5	76.1	66.7	78.6 ($n = 271$)

Note: all statistically significant except 'buy land' variable. Live rest of life $\chi^2 = 0.00001$; work $\chi^2 = 0.00188$; live in area $\chi^2 = 0.02209$. When these questions were disaggregated by education level, there were no significant relationships.

which the interview was conducted proved to be more significant than either gender or age in predicted attachment to place, but it was high for all age groups (Tables 18.6, 18.7 and 18.8.).

Positive and Negative Characteristics of the Area

Positive characteristics

To gain insight into perceived positive and negative characteristics of the area, the respondents were asked what they would tell someone who was thinking about moving to the area, what were the good things about the area and what were the bad things about the area. Although the questions were asked about the area, most respondents' answers reflect their views on their particular community. Open-ended questions were utilized to avoid pre-coded categories that impose the researcher's view of potential strengths and weaknesses. Interviewers were taught to probe without suggesting particular answers. Twenty-one per cent of respondents did not provide a codeable answer when asked to cite two good characteristics of the area. Neither gender, age nor the language in which the interview was conducted was related to naming specific positive characteristics (Table 18.9). 'Infrastructure', i.e. basic services, was the most common positive characteristic mentioned by respondents; 33.5% of those that responded gave it.

The second most frequently mentioned category was 'agriculture'. This category encompasses all positive comments related to agriculture and animal husbandry. It includes the mention of good productivity in the area and reference to good soil quality. In addition, it includes comments about the fact that in this area one can raise one's own food (this implies a contrast to urban areas or less open spaces) and that you can have animals. For example, one 40-year-old woman said a good thing about living where she does is that '*puedo tener pollitos sueltos*' ('I can have free-range chickens').

A quarter of respondents mentioned something that fits into the category 'natural environment'. This included statements such as 'fresh air', 'beautiful landscape', 'good climate', 'open space' and 'there are trees'.

The fourth most frequent category, 'relations with people', encompasses all statements that have to do with positive social interactions. Examples of this are 'people are friendly', '*amistad*' ('friendship'), 'there are not thieves', 'the youth are enthusiastic' and 'we do not have delinquency'.

The fifth most frequently mentioned category was 'tranquility'. Seventeen per cent of respondents who answered the question said that tranquility was something good about the area; thus it was given it its own category.

'Organization', the sixth most commonly mentioned category, included all the statements relating to positive social organization. Examples are: 'we have good leaders', 'the community is moving forward', 'we have a communal store (workshop, dance group)' and 'we have participatory people'.

'Other' includes all other characteristics. The following categories were collapsed into 'other' because none of them had > 3%: family well being, economic vitality, everything, and support from government organizations and/or NGOs.

Table 18.9. Positive characteristics of area named by interviewees by gender (%).

Characteristic	Males	Females	Total
Infrastructure	32.8	34.3	33.5
Agriculture	31.1	20.2	26.2
Natural environment	21.0	29.3	24.8
Relations with people	21.8	17.2	19.7
Tranquility	17.6	15.2	16.5
Organization	16.8	9.1	13.3
Other	6.7	13.1	9.6
Nothing	3.4	6.1	4.6
Total	$n = 119$	$n = 99$	$n = 218$

Table 18.10. Negative characteristics of area named by interviewees by gender (%).

Characteristic	Males	Females	Total
Infrastructure	24.8	27.5	26.1
Social problems	17.9	16.8	17.4
Robbery	12.4	17.6	14.9
Nothing			12.0
Lack of work	13.8	9.2	11.6
Organization	13.8	9.2	11.6
Other	13.1	8.4	10.9
Land[a]	6.2	5.3	5.8
Natural environment	4.8	4.6	4.7
Total	$n = 145$	$n = 131$	$n = 276$

[a]This category was split evenly between lack of land and poor quality of land.

Negative characteristics

Twelve per cent of the respondents said there is nothing bad about the area. Over a quarter of respondents mentioned 'infrastructure' as a negative characteristic of the area. The mentioning of the lack or poor quality of one or more characteristics such as the following was coded as infrastructure: a school, irrigation, piped water, electricity, roads, etc. (Table 18.10).

Seventeen per cent of all respondents mentioned 'social problems' as a negative characteristic of the area. Under the category 'social problems', alcoholism was included (which 4.3% of all respondents mentioned), youth being disrespectful to adults (which 3.6% of all respondents mentioned), gangs (which 1.4% of all respondents mentioned), in addition to other conflictive social environmental characteristics. The general conflictive social environmental category included such statements as 'some people are bad', 'there is division within the community', 'there are jealous people', 'people gossip a lot', 'there is not equality' and 'people aren't friendly'. Robbery was separated from other social problems because robbery was mentioned so often. Theft of animals appears to be the most common form of robbery.

Over one in ten respondents mentioned 'lack of work' as a negative characteristic of the area. 'Organization' was mentioned as a negative characteristic by 11.6% of respondents. Comments such as 'we don't have solidarity in the community', 'people don't cooperate' and 'we lack leadership' were coded under this category.

The categories 'lack of support from government and/or non-governmental organizations', 'poor communication', 'poverty' and 'everything' were collapsed into the category 'other' because none of them had more than 2.2%. In addition, 'other' included such statements as 'very far from Cotacachi', 'the males are abusive', 'the people don't work', 'there is not mass', 'the residents are leaving', 'a house nearby is not clean' and 'there are *mestizos*' (a *mestizo* is a term used for someone considered to be part European and part Indian, but who has adapted a Western lifestyle and dress).

Although small land holdings dominate among the *campesinos* in highland Cotacachi, only 5.8% mentioned lack of land and poor quality of land as a negative characteristic of the area. Five per cent mentioned 'natural environment' as a negative characteristic. The vast majority of the responses coded under this category were negative comments related to the climate (i.e. 'it is cold', 'we don't get enough rain', etc.).

Migration and Circulation

Place of birth

In-migration into the indigenous rural communities of Cotacachi is very low. When it does occur, people move to the area from nearby. Ninety-six per cent of respondents were born in the canton of Cotacachi. Ninety-eight per cent of all respondents were born in the Province of Imbabura, while 1.8% (five individuals) were born in the neighbouring Province of Pichincha. Four of these five were born in Quito.

Although age is not statistically significant in relation to place of birth, all those born outside of Imbabura Province, (in Pichincha Province) were under 45 years of age. They were probably born there when their parents were temporarily working in Pichincha. The fact that no one over the age of 44 was born outside the canton of Cotacachi suggests that the pattern of temporary migration to Quito is relatively recent.

Places lived

Aside from circulation for work, there is very little migration in the rural communities of Cotacachi. The migration that has occurred was primarily within the rural areas in the eastern part of the canton of Cotacachi. Fifty-five per cent of all interviewees have lived in the same community in Cotacachi all their lives. There was no difference between males and females.

Motives for moving

Of the 276 interviewed, 45.3% had lived somewhere besides their place of birth. These respondents were asked for two motives for leaving a place. This was an open-ended question (Table 18.11).

The most commonly cited reason for leaving a place of residence by females was family reasons, with 85.2% mentioning it. Family reasons are those that are not individually based; rather decisions are made by family members or influenced by a family

Table 18.11. Motives for leaving a place of residence by gender (%).

Motives	Males	Females	Total
Family	48.4	85.2	66.4[a]
Economic/work	59.4	37.7	46.4
Personal	18.8	13.1	16.0
Land	9.4	14.8	12.0
Education	3.1	4.9	4.0
Other	3.1	4.9	4.0
Total	$n = 64$	$n = 61$	$n = 125$

[a]The sum of the percentages exceed 100% since respondent were allowed two responses per move and respondents had up to eight moves.
Males and females were significantly different on 'family' ($\chi^2 = 0.00001$) and 'Economic/work' ($\chi^2 = 0.00289$). None of the other motives was significantly different at the 0.05 level.

situation. For males, family was the second most frequently listed motive for leaving a place of residence after economic and work-related reasons, with 48.4% citing it. Relocation due to marriage to someone from another community was one of the most common descriptions of familial motivations for both males and females.

Economic/work reasons constituted the primary motivation for males to leave a place of residence. Over half of males cited this factor. For females, economic/work reasons fell in second place, with 37.7% of females citing it. In addition to 'motives pertaining to lack of money or other resources for surviving' (Martínez and Rhoades, 1997, unpublished manuscript), the category of economic/work reasons included statements about jobs and things such as 'to progress', 'to buy myself clothes' and 'to send money to my family'. (The latter was also coded as a family reason.) Economic/work was the only other motive besides family reasons that had a statistically significant relationship with gender.

Most people who moved for a specific job had either a family (if this was mentioned, it was also coded as a family reason) or friendship connection in the new location. Social networks play a very significant role in migration streams and job procurement among the people of the rural communities of Cotacachi.

'Land reasons' was the third most frequent motive for females and the fifth for men. Fifteen per cent of females and 9.4% of males cited it. In addition to statements related to lack of land and poor quality of land, 'land reasons' included moving back to a community in Cotacachi because the respondent or a family member had land there. The importance of land ownership as a tie to place, even when that land is poor and not large enough to ensure self-sufficiency, is noteworthy.

Circulation

Location of current work

Among the 276 respondents, gender is the only salient feature that had a significant relationship with location of work (Table 18.12). Forty-three per cent of females worked outside their home, compared with 71.0% of males. Rural communities in Cotacachi were the primary location of work outside the home for both females (15.3%) and males (15.9%). The primary activity in the rural communities of Cotacachi was as agricultural labourers on *haciendas*. The town of Cotacachi was the third most frequent location of work. Males principally worked in construction and females as domestic workers, but some worked as private employees in businesses including hotels and handicraft workshops.

Eight per cent of the females and 12.4% of the males worked in Otavalo. Those who worked in Otavalo either sold handicrafts in the market once a week (principally females), did construction (only males), carried out domestic work (only females) or worked in the handicraft workshops (principally males).

Quito is the fifth most common place the respondents worked. In Quito, males principally worked in construction and females as domestic workers. La Molienda, a sugarcane plantation where *panela* (brown sugar) is made, employed everyone that worked in Salinas. Activities of those

Table 18.12. Location of work at time of interview by gender (%).

Location	Males	Females	Total
At home in own community	29.0	57.3	42.4
Rural community in Cotacachi	15.9	15.3	15.6
Town of Cotacachi	15.2	10.7	13.0
Otavalo	12.4	8.4	10.5
Quito	9.0	6.9	8.0
Salinas	8.3	0.8	4.7
Other cantons of Imbabura[a]	4.8	0.8	2.9
Colombia	2.8	0	1.4
Other places in sierra and Amazon	2.8	0	1.4
Total	100	100	100
	(n = 145)	(n = 131)	(n = 276)

χ^2 significant at 0.00003.
[a]This category excludes Otavalo, which is in Imbabura. It had a high enough frequency of respondents working there to merit its own category. Thus this category includes the following cantons in Imbabura: Ibarra, Antonio Ante, Pimampiro and San Miguel de Urcuqui.

working in Salinas ranged from working in the field harvesting and planting sugarcane to operating the ovens that heat the cauldrons to boil down the sugarcane juice, to cleaning the moulds where the hot syrup dries into brown sugar.

Surprisingly, only 1.4% worked in Ibarra (all males) despite the fact that it has an urban population of 80,477 (Instituto Nacional de Estadística y Censos, 1991) and is only 13 km away from the town of Cotacachi. Because it was such a small percentage, Ibarra was collapsed into the category 'other cantons of Imbabura', which was the seventh most common place for respondents to work. The other cantons in Imbabura that were included in this category are Antonio Ante, Pimampiro and San Miguel de Urcuqui.

Three per cent of respondents (all male) worked in other parts of Ecuador outside of Imbabura. Another 3% worked in Colombia. Those that work in Colombia sold handicrafts made in Cotacachi or Otavalo or sold Ecuadorian clothes, as clothes prices were lower in Ecuador than in Colombia prior to Ecuador's dollarization. People tended to go to Colombia for several months at a time. It is interesting to note that the locations of work farthest from Cotacachi (Colombia and other places in the sierra and Amazon) were only frequented by those in the two youngest age categories (Table 8.13). This reflects the fact that an individual's life cycle stage shapes the distance of circulation. The young are less likely to have established families of their own and, if they have, often have fewer obligations to those families.

Eighteen per cent of the males interviewed working at the time of the interview a far enough distance from their home (this includes Quito; other cantons in Imbabura, excluding Ibarra; Colombia, and other places in the sierra and Amazon) stated that they only returned to their community on weekends, holidays or in some cases every few months. The language in which the interview was conducted did not have a significant relationship with location of work.

Conclusion

Although the rural communities in the highlands of Cotacachi have many of the typical characteristics of a sending area, analysis of a stratified random survey of rural residents reveals a strong attachment to place. A surprising number of individuals want to live in rural communities in Cotacachi, although fewer (but still a majority of the survey respondents) desire to

Table 18.13. Location of work at time of interview by age (%).

Location	15–25	25–34	35–44	45–54	≥ 55	Total
At home in own community	43.0	40.3	47.5	30.6	52.0	42.4
Rural community in Cotacachi	11.4	14.3	20.3	22.2	12.0	15.6
Town of Cotacachi	5.1	16.9	13.6	19.4	16.0	13.0
Otavalo	15.2	14.3	1.7	11.1	4.0	10.5
Quito	12.7	7.8	3.4	8.3	4.0	8.0
Salinas (Imbabura)	1.3	1.3	10.2	8.3	8.0	4.7
Other cantons of Imbabura	5.1	1.3	3.4	0	4.0	2.9
Colombia	3.8	1.3	0	0	0	1.4
Other places in sierra and Amazon	2.5	2.6	0	0	0	1.4
Total	100 ($n = 79$)	100 ($n = 77$)	100 ($n = 59$)	100 ($n = 36$)	100 ($n = 25$)	100 ($n = 276$)

χ^2 not significant.

work there. The ability to separate place of work from place of residence and identity is critical to the continuance of rural communities and rural culture.

UNORCAC leaders have been concerned about the out-migration of young people, which they voiced to SANREM. However, the application of 276 migration interview schedules in 1998 found that the migration of youth was only a small part of a system that employs circulation as a component of a complex of survival strategies. Those that define themselves as residents of the rural communities engage in a range of economic activities to survive, have strong attachment to their community and have little desire to leave the area permanently. Yet they will spend extended periods away from the community to earn cash to support their family and their position in the community.

In the rural highland indigenous communities of Cotacachi, attachment to place does not provide economic security, but is essential to ethnic, cultural and social identity and their reproduction. Very few people have moved into the area, and those that are there have a strong desire to stay connected, despite lack of access to land, water and employment. Circular migration is a major mechanism of cultural and economic survival for rural households in Cotacachi.

Lack of jobs and limited access to land and water are significant economic challenges for the rural residents of Cotacachi. Yet, when asked to cite negative aspects of the area, these issues were not among the top characteristics mentioned. It could be that jobs, water and land were not mentioned as frequently because people feel that there is little that can be done about them. Yet, when directly asked whether they would like to buy more land in the area, almost nine out of ten people said, 'yes.' Thus, when directly asked about land access, most respondents felt it was an important issue, but when asked to characterize the area, respondents did not focus on land, water and jobs. Circulation mitigates the lack of access to land, water and jobs.

Infrastructure was mentioned as both a positive and a negative characteristic more frequently than any other. During the first decade of UNORCAC, the organization focused on infrastructural issues and utilized demonstrations and rallies to motivate the government of the canton to invest, at least minimally, in schools, water systems and other infrastructural necessities of the rural communities. The legacy of these fights for basic necessities left people with a sense either that a lot had been gained (infrastructure a positive characteristic) or that more was needed (infrastructure a negative characteristic).

Contrary to expectations, working outside of the community through circulation does not increase the desire to move. Qualitative research suggests that circulation encases individuals in the community, rather than separating them from it. Circulation is usually based on close networks that stem from community ties, which it serves to maintain. While younger individuals were more likely to be less attached to place and more likely to want to live elsewhere than older respondents, even among the young attachment to place is high. Migration is often the vehicle for cultural change, but for the indigenous communities of Cotacachi, circular migration provides an important mechanism by which ethnic identity and economic viability can be maintained.

References

Bosrup, E. (1970) *Woman's Role in Economic Development.* John Wiley and Sons, New York.

Bromley, R. (1985) Circulation within systems of periodic and daily markets: the case of central highland Ecuador. In: Prothero, R.M. and Chapman, M. (eds) *Circulation in Third World Countries.* Routledge and Kegan Paul, London, pp. 325–349.

Brown, L.A., Brea, J. and Andrew, R.G. (1988) Policy aspects of development and individual mobility: migration and circulation from Ecuador's rural sierra. *Economic Geography* 64, 147–170.

Chapman, M. and Prothero, R.M. (1985) Themes on circulation in the Third World. In: Prothero, R.M and Chapman, M. (eds) *Circulation in Third World Countries.* Routledge and Kegan Paul, London, pp. 1–26.

Cohen, R. (1996) Introduction. In: Cohen, R. (ed.) *Theories of Migration.* Edward Elgar Publishing Company, Cheltenham, UK, pp. xi–xvii.

Collins, J.L. (1988) *Unseasonal Migrations: the Effects of Rural Labor Scarcity in Peru.* Princeton University Press, Princeton, New Jersey.

Crissman, C.C. (2003) La Agricultura en los Páramos: Estrategias para el Uso del Espacio. Available at: http://www/cipotato.org/market/PDFdocs/CondesanWP_full.pdf

Farrell, G., Pachano, S. and Carrasco, H. (1988) *Caminatas y Retornos.* Instituto de Estudios Ecuatorianos, Quito, Ecuador.

Gardner, K. (1995) *Global Migrants Local Lives: Travel and Transformation in Rural Bangladesh.* Oxford University Press, Oxford, UK.

Hidrobo, J.A. (1992) *Power and Industrialization in Ecuador.* Westview Press, Boulder, Colorado.

Hoops, T. and Whiteford, S. (1983) Transcending rural–urban boundaries: a comparative view of two labor reserves and family strategies. In: Hunter, J.M. *et al.* (eds) *Population Growth and Urbanization in Latin America: the Rural–Urban Interface.* Schenkman Publishing Company, Inc., Cambridge, Massachusetts, pp. 261–280.

Instituto Nacional de Estadística y Censos (1991) V Censo de Población y IV.

Jones, H. (1990) *Population Geography.* Paul Chapman Publishing Ltd, London.

Korovkin, T. (1997) Indigenous peasant struggles and the capitalist modernization of agriculture: Chimborazo, 1964–1991. *Latin American Perspectives* 24, 25–49.

Lawson, V.A. (1990) Workforce fragmentation in Latin America and its empirical manifestations in Ecuador. *World Development* 18, 641–657.

Lee, E.S. (1966) A theory of migration. *Demography* 3, 47–57.

Lentz, C. (1997) *Migración e Identidad Etnica: la Transformación Histórico de una Comunidad Indígena en la Sierra Ecuatoriana.* Abya-Yala, Quito, Ecuador.

Martínez, L. (1985) Migración y cambios en las estrategias familiares de las comunidades indígenas de la sierra. *Ecuador Debate* 8, 110–128.

Paerregaard, K. (1997) *Linking Separate Worlds: Urban Migrants and Rural Lives in Peru.* Berg, New York.

Rhoades, R.E., Martínez, A. and Jones, E. (2001) Migration and the landscape of Nanegal. In: Rhoades, R.E. (ed.) *Bridging Human and Ecological Landscapes.* Kendall/Hunt Publishing Co., Dubuque, Iowa.

Velasco, J.L. (1985) Las migraciones unternas en el Ecuador: una aproximación geográfica. Migraciones y migrantes. *Ecuador Debate*, 8.

de Vivienda (1990) Población del Ecuador por Area y Sexo, según Provincias, Cantones y Parroquias (Datos Provisionales), Quito, Ecuador.

Waters, W.F. (1997) The road of many returns: rural bases of the informal urban economy in Ecuador. *Latin American Perspectives* 94, 50–64.

Werbner, P. (1990) *The Migration Process: Capital, Gifts and Offerings Among British Pakistanis.* St Martin's Press, New York.

Zamosc, L. (1994) Agrarian protest and the Indian movement in the Ecuadorian highlands. *Latin American Research Review* 29, 37–68.

19 Social Capital and Advocacy Coalitions: Examples of Environmental Issues from Ecuador

Jan L. Flora,[1] Cornelia B. Flora,[1] Florencia Campana,[2] Mary García Bravo[2] and Edith Fernández-Baca[3]
[1]*Iowa State University, 317 D. East Hall, Ames, IA 50011, USA*; [2]*Heifer Project-Ecuador, Quito, Ecuador*; [3]*Grupo Yanapai, Peru and Iowa State University, 107 Curtiss Hall, Ames, IA 50011, USA*

Introduction to Advocacy Coalitions

Policies – their formation and their implementation – depend on power. Grassroots groups that come together around a perceived threat or opportunity that is influenced by policy or its implementation have little power compared with groups and organizations that are regional, national or international in scope. Furthermore, the fact that groups and organizations come from different sectors – market, state and civil society – makes having a voice that will be listened to even more difficult. Thus local groups generally use one of two tactics: protest (violent or non-violent) or supplication. One is based on power to mobilize locally and the other is based on ability to show abject need. Neither tactic is particularly effective in the development process.

In this chapter, we examine how local groups in a rural region of Ecuador formed and reformed advocacy coalitions to influence policy of immediate concern in their daily lives. We build on a tradition of analysis that has focused first on policies, and then on impacts. We start with locally identified threats and opportunities to determine the degree to which locally generated advocacy coalitions can influence policy not only at the local, but also at the national and even the international level.

Theoretical Approach

Local groups both depend on and distrust the three sectors (market, state and civil society) that are outside their local domain. The market provides incentives for production and ensures efficient distribution of traded goods and services. However, in many cases, market firms have extracted wealth from the locality and left environmental destruction. A well-functioning state provides the rules under which the market functions and enforces those rules. Yet states have been corrupt and made rules in ways that profit only the wealthy or those who hold political power. A robust and diverse civil society reduces transaction costs in the other two sectors by building trust and diversifying social networks, and, in the best of circumstances, provides social values that legitimize the state's regulation of the market. However, civil society organizations can also use localities to further their larger goals, without attention to long-term local consequences.

Thus, local groups seeking to influence the larger events that impact them directly are of two minds when seeking allies to help them in their cause. They must weigh the added power of an international non-governmental organization (NGO) or green firm against the possibility of being only a tool for gaining a market share among donors or buyers.

We have chosen to conduct policy research by examining relationships among grassroots groups and these three sectors using an *advocacy coalition framework* (ACF). Sabatier and Jenkins-Smith (1993), who developed the framework, argue that organizations, agencies and firms form alliances, or advocacy coalitions, around concrete issues in order to achieve common desired futures. These public and private institutional acts at various geographic scales thus share: (i) certain basic beliefs that anchor common desired futures (ends); (ii) mental causal models, implicit or explicit means for reaching those futures (means); and (iii) rules of evidence that allow for members of the coalition mutually to ascertain progress towards the goals. By determining from each potential ally where it wants to go (their declared and implicit missions) and how they think they will get there (the means they see as viable and effective), local groups can seek appropriate alliances for varying periods of time – advocacy coalitions – in order to work towards their desired future in light of the specific threat or opportunity.

It follows, then, that effective advocacy coalitions share common desired futures and *mental causal models* (perception of relationship between ends and means) but are also sufficiently diverse in their contacts and external linkages to garner a diversity of resources and information/knowledge. They are effective in combining *bridging* and *bonding* social capital (see Narayan, 1999). In a group with high bonding capital, members know one another in multiple settings or roles.[1] Bridging social capital connects diverse groups within the community to one another and to groups outside the community. Bridging social capital, like Granovetter's (1973) *weak ties*, tends to involve instrumental single-purpose linkages between two groups or individuals.

A coalition in which the members share common goals and a degree of consensus on appropriate evidence for showing whether they are progressing towards those goals is more likely to endure than is a marriage of convenience. The richer a coalition is in bridging social capital, the more likely it will be able to increase its access to and appropriately to combine the human, financial and natural capital that are necessary for it to prevail or negotiate effectively with an opposing coalition. Advocacy coalitions incorporating a diversity of institutions – from multiple sectors and representing various geographic scales – are likely to have access to more diverse information and resources than a less diverse or more geographically isolated coalition. Furthermore, success by such a coalition in reaching its goal is likely to strengthen bridging and bonding social capital further.

Having common desired futures does not ensure that different entities will cooperate to achieve those ends. If groups have different mental causal models, it may matter little that they share similar desired futures. Alternatively, desired futures may differ sharply, making any sort of cooperation or compromise quite unlikely. It either case, two or more oppositional advocacy coalitions may form, resulting in gridlock, or in the triumph of a powerful coalition over a socially excluded or poorly organized one.

Unlike Sabatier and Jenkins-Smith (1993), we believe that economic interests, although introducing methodological challenges, must be taken into account in a model that seeks to understand policy formation. Historical analysis of previous issues, alliances and movements is helpful in this regard. Bourdieu's (see Bourdieu and Wacquant, 1992) perspective on the unequal ability of different social classes to wield social and cultural capital is more helpful to us than is Coleman's (1988) view of social capital, because Bourdieu *is* explicit about economic interests and about political power (see Sharp *et al.*,

2003, for a way to link community power to network analysis).

In the ACF, policymaking is not unilinear. It is contested and manoeuvred by different sectors from different levels (Münch *et al.*, 2000). Nor can policymaking be captured in a series of prescribed steps to be taken by decision makers, which if appropriately executed would almost automatically lead to optimum decisions. Such an overly rationalist element often creeps into decision-making models that take a technical rather than a political approach.

ACF can be understood as a specification of stakeholder analysis (SA) (Clarkson Centre for Business Ethics, 1999), which is much better known and widely practised than is the ACF. SA was initially developed within management science to deal with the political problem of externalities – firms and agencies were often blindsided by civil society groups that were negatively affected by their projects and policies, often derailing those efforts. An example is the failure of a state–market coalition, which included McNamara's Defense Department and large firms in the aircraft industry, to build a Supersonic Transport during the Johnson Administration, due in part to their inability to anticipate the strong opposition of environmentalists (cited in Freeman, 1984, pp. 136–139). They needed a way to identify these less obvious stakeholders, who were most likely to wreck the implementation of their plans because they failed to negotiate with them up front. Mason and Mitroff (1981, pp. 98–99) used the example of the snail darter and the environmentalists, who, fearing its extinction, nearly stopped the construction of a dam project. The focus on a single firm or a state–market coalition of powerful players meant that SA initially was rather vertical in its orientation (Grimble and Chan, 1995).

If a firm or a government agency wanted to implement a particular project or change a particular policy, which groups might it attract as allies and which would be likely to oppose it? A first cut in identifying these individual or institutional *stakeholders* (those likely to be affected by or to affect the project or policy) could result from looking at the relevant *interests* of different entities (Freeman, 1984, p. 135). SA has evolved and broadened so that horizontal and vertical relationships can be assessed. Indeed, analysis of coalitions has been included as part of the methodology. Participatory approaches are used to identify stakeholders and to identify and shape the projects or policies so that few stakeholders will be hurt by the project/policy (Grimble and Chan, 1995).

The ACF methodology focuses explicitly on institutional actors and the coalitions that develop among those actors or stakeholders. The methodology mimics the policymaking or decision-making process by identifying existing issue-specific coalitions (and potential or emerging coalitions) and bringing coalition members physically together, for the purposes of gathering data (if a strict research project) and/or for cementing and strengthening such coalitions (if the project is more applied).

The building of coalitions is based first on identification of common desired futures and similar rationalities (mental causal models) for achieving those goals (Sabatier and Jenkins-Smith, 1993). Information becomes a tool in coalition formation. The ACF assumes that institutional actors consider information to be relevant and useful if it is congruent with their experience and interests. This is quite different from the policy analysis approach that assumes 'sound science' will resolve conflicts (Münch *et al.*, 2000).

The ACF uses multiple methods tentatively to identify inter-institutional coalitions; this is not unique among stakeholder approaches. What is different is the sorting of the entities by their desired future states and their rules of evidence for judging what actions move them toward those states. The coalitions thus identified have emergent qualities: the commonalties in goals and mental causal models gives them the characteristic of a social group in which the whole is greater than the sum of its parts. It may also mean that the coalition is only *emerging* in another sense rather than existing full-blown. It is this characteristic that gives the theory an inherent applied element. The analyst becomes more than a casual observer but must take a stance vis-à-vis the different coalitions, for the method of gathering data in groups can either foster or fail to foster the solidification of different coalitions, depending on how the focus group is organized and directed.

Advocacy coalitions arise when institutional groups of actors see problems and their solutions in an integrated fashion and seek appropriate collective action. Thus, a typical advocacy coalition can be characterized as 'emerging' rather than institutionalized. They are ever changing as certain groups are incorporated and others drop out. This is not to say that advocacy coalitions do not have structure. They tell us a great deal about patterns of relationships. Even though particular entities may enter and exit a coalition with alacrity, the aggregation of interest that an advocacy coalition represents has greater permanence than the participation of any single institutional actor. Advocacy coalitions serve as structures that allow us to identify the social location of institutional actors and to analyse the discourse that is carried forward to support their positions, often in juxtaposition to another coalition that is taking a contrary (or at least a contrasting) position. These coalitions are submerged within a larger universe of discourse in which the actors attempt to persuade others that a particular decision could benefit a greater number of interests than would the alternative decision. Such persuasion depends on building bridging social capital.

Our approach to advocacy coalitions is to begin with a local issue and the groups that have formed around them. With those groups, we identify potential institutional actors at various levels, and policies (existing or potential) that are relevant to that coalition of actors. In the course of the research, policies of various state, market and civil society actors are identified to modify or leverage. Because policy formation and implementation is a dialectic and dynamic process, we monitor how those coalitions change over time.

Examples of Advocacy Coalitions from Ecuador

In the following section, we illustrate the advocacy coalition approach with concrete examples. In northern Ecuador, we examined advocacy coalitions around the issues of: (i) governance of the Cotacachi Cayapas Ecological Reserve on which the canton of Cotacachi abuts; and (ii) whether what would probably be open-pit copper mining will occur in the semi-tropical part of Cotacachi, called Intag. The two examples involve one issue in which desired futures were (and probably still are) reconcilable under a compromise solution (the reserve), but immediate interests vary. In the case of the mining controversy in Intag, a compromise solution is difficult because the desired futures of the actors are quite divergent, and because of the size of the proposed project. With the power of certain external actors, a win–win solution is not readily imaginable. We examine bridging social capital in opposing coalitions in terms of the incorporation of market, state and civil society organizations and in the way in which ties to different levels are utilized, i.e. coalitions that include market, state and civil society, and embrace entities at all levels – local, regional, national and international – can be said to have considerable bridging social capital. That social capital can in turn compensate for initial low levels of power in negotiations.

Context

Although the Andean (highland) sector of Cotacachi covers only 20% of the canton, it is home to > 60% of its population. The highlands are the ancestral home of the indigenous population of the canton. The traditional *haciendas* are also located here. Cotacachi was largely untouched by the land reform of the 1960s, but disputes are principally over scarce water rather than land *per se*. Because of their smallholdings, indigenous peasants practise 'circular migration' – young people and male heads of household work in other parts of Ecuador, but return home for holidays (see Flora, Chapter 18, this volume). Women generally tend the small plots of land. The economy of this highland microregion is based on three main activities: agriculture, artisanry (particularly leather goods) and tourism. In the past decade, agroindustrial firms that specialize in non-traditional export products – flowers, asparagus and fruit – have come to provide significant local employment.

The tourist and hotel trade emerged in the 1970s. The Cotacachi-Cayapas Ecological Reserve was established in 1968, and includes Cotacachi mountain and the crater Lake Cuicocha, an important tourist point. The city of Cotacachi is only a few miles off the Pan-American highway. Its nearness to the famous Otavalo market undoubtedly leads to some spillover tourism.

In Cotacachi, there is a thick organizational network (bonding social capital), particularly in the rural part of the Andean zone. The community (*comuna*) is the traditional organizational form of the indigenous population of the highlands. The rural population of the semi-tropical zone, consisting mostly of *mestizos*, is also organized into cooperatives, agricultural and livestock associations, and an environmental organization that has spearheaded the opposition to mining.

The most important peasant organization in the highlands is UNORCAC. Since UNORCAC's founding in 1978, it has focused on cultural and political issues. It has fostered a strong bilingual education movement and, over the years, has brought political pressure to bear for government services in rural highland areas. It has been rather effective in building links with the outside (bridging social capital) and has succeeded in obtaining grants from national and international foundations and NGOs (Báez *et al.*, 1999, pp. 64–65).

Methodology

Issues chosen for analysis involved some local mobilization, cross-cut sectors and were policy relevant. We analysed documents produced by each key organization involved in the chosen issues to determine publicly expressed and collectively desired futures and mental causal models. Then key organizational leaders were interviewed to understand how the issue has unfolded, the role of their organization and others in that process, and to elicit names of other institutional actors. Interviews were conducted in snowball fashion as relevant organizations were identified, starting with interviews with core organizations in each likely coalition. Desired futures and mental causal models were then mapped based on both documents and the interviews. The interviews supplied basic information to form focus groups consisting of local organizations that are or have the potential to become an advocacy coalition. Preliminary assessments regarding desired futures and composition of advocacy coalitions were tested against the interpretations of relevant actors in the focus groups.

The advocacy coalition diagrams presented below were drawn based on information garnered principally from the in-depth key informant interviews with a representative of the main organizations in the coalition. We entered all documents and transcriptions of each interview and focus group into a database, which we analysed using N-Vivo software. Analysis of desired future states and mental causal models allowed us to determine the degree to which desired future states are homogeneous within a coalition and heterogeneous between coalitions focused on the same issue.

Tourism and Management of the Cotacachi Cayapas Ecological Reserve

In the context of the discourse regarding tourism as an alternative 'development pole', the Mayor's office contracted a consultant to recommend how tourism initiatives could be put into practice. He proposed that a mixed public–private firm be established to manage the tourist 'circuit' that extends from the city of Cotacachi to Lake Cuicocha. Bolstered by the legal structure for decentralizing management of natural resources, the Mayor's office initiated a petition to the Minister of Environment to concede the administration of the concessions around Lake Cuicocha to a mixed tourism management company (referred to hereafter as the Mixed Company). The firm was immediately organized with private capital, largely from urban *mestizo* stockholders, and with funds from the Municipality.

Shortly thereafter, UNORCAC asked the Ministry of the Environment to transfer the management of various tourist points within the reserve to UNORCAC in order to maintain the management integrity of the natural resources in the entire highland portion of the reserve.

In response to these two requests, the Ministry asked the Mayor's office to develop an integrated proposal to be based on agreement among all parties interested in managing the resources of the reserve: the Municipality, UNORCAC and Incamaki (an indigenous artisan and tourism organization that currently manages the boating service at Lake Cuicocha). The Ministry suggested that the three principal interested parties form a 'management board' to administer resources of the reserve, but suggested no mechanisms for reaching a mutually acceptable compromise. The Municipality (jointly with the Mixed Company), UNORCAC and the Inkamaki Association each presented the Ministry with distinct proposals for management of different parts of the reserve. The Ministry does not have the organizational capacity – or the will – to broker an agreement. Its most recent move was to invest the Municipality, because it forms part of the Ecuadorian government hierarchy, with the mandate to organize the administration of the bioreserve. This has given the green light to the Mixed Company to take charge of certain concessions, but has not resulted in a decision regarding a management plan for the entire bioreserve. This may be because the Municipality also does not have the resources (either organizational or financial) or the political will to broker a management plan for the reserve. The national government has transferred no resources to the Municipality for administering the reserve, although if a plan were adopted, presumably the Municipality would collect the entrance fees to the park and assign them to whatever entities that would be given responsibility for protecting the park's natural resources and biodiversity. Thus, it appears that a *de facto* privatization of the most lucrative activities within the reserve has occurred, thereby reducing the degrees of freedom available for designing a decentralized administration of the entire bioreserve. The Ministry of Environment continues to enforce environmental laws within the reserve.

What are the diverse interests behind the various proposals and can they be reconciled?

The Mayor's ethnic and organizational background is central in understanding his potential and limitations in representing the general interest. His legitimacy with the mestizo population is based squarely on his ability to respond 'even handedly' (as perceived by the mestizos) to the interests of the two main social groups in the canton: mestizos and indígenas. In this case, the Mayor must overcome the image held by mestizos that any indigenous person is lacking in the knowledge of how to carry out public functions. Arguably he did that when he won re-election in April 2000 with 80% of the vote, including the support of a majority of mestizos and virtually all indigenous voters. He has emphasized institutional modernization of local government and efficiency in managing natural resources, which, in the case of management of the bioreserve, involves pursuing the dual goals of conservation and commercial tourism development. The Mixed Company is a concrete manifestation of that vision. Also, his training as an economist (at the University of Havana) may have contributed to his favouring a more technocratic and less participatory approach to administration of the reserve.

Identification of advocacy coalitions around governance of the reserve

The discourse that cements the coalition centred on the Municipality is that of entrepreneurial development, which resulted in greater support for the Mayor from the sector of the mestizo population that is involved in tourism and retail business (Fig. 19.1).

UNORCAC has built another advocacy coalition with itself at the centre. This coalition consists of the leaders and technical people (staff and contracted) of the SLO (Secondary Level Organization) itself its constituent communities that are located inside the reserve and within its buffer zone; the Incamaki Association, whose interests are compatible with the UNORCAC proposal although it has presented its own tightly focused proposal; and a Dutch NGO, called AGRITERRA, that works closely with UNORCAC.

The desired futures of the two coalitions are not so different. Both support: (i) administrative decentralization of natural resource management that benefits the locality; and (ii) management of natural resources as natural capital to be invested for both present and future generations. For the group that has formed the Mixed Company, the most appropriate rationale is an entrepreneurial one that would generate resources for local self-management and for individual stockholders. The coalition around UNORCAC, without discarding entrepreneurship, believes it is essential to build a future in which not only are the conditions for the reproduction of the peasant communities within maintained, but their quality of life is raised. Many in the entrepreneurial coalition believe that UNORCAC lacks management capacity (even though it has been shown in other areas) and would therefore put at risk the tourism project that

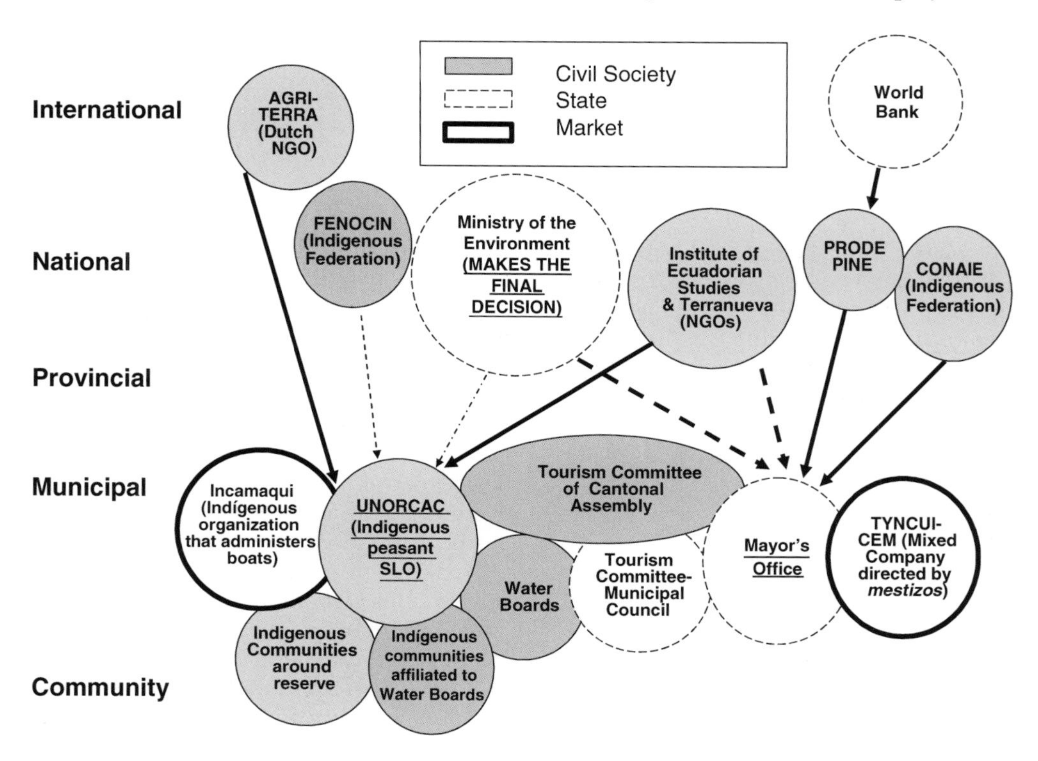

Fig. 19.1. Advocacy coalitions around issue of governance of the Cotacachi Cayapas Biological Reserve, Cotacachi, Imbabura, Ecuador.

is a central axis of development of the canton. For the indigenous-led coalition, the entrepreneurial tourism proposal would benefit a social group that already has better economic conditions, and therefore would exacerbate the substantial inequality of access to resources that already exists.

Inspection of Fig. 19.1 shows that both coalitions consist of entities from all three sectors and at various levels from local to international. Of considerable importance is the fact that there are a number of institutions that form a bridge between the two coalitions. At the time of this writing, none of these 'bridging' entities has stepped forward to broker a solution.

The Mining Controversy

The second example is that of the controversy over mining in Intag, the semi-tropical part of the canton. The opposition coalition is centred on a local environmental NGO, DECOIN (*Defensa y Conservación Ecológica de Intag*; Defense and Ecological Conservation of Intag) which includes middle-class environmentalists who have chosen to live in the area and longer-term residents of various backgrounds as its members. The other coalition centres on the Ministry of Energy and Mines. Their view, backed up by advisors from the World Bank, is that the western slope of the Andes is not apt for agriculture and tourism, and that the need for increasing foreign exchange is a central national concern during this period of great financial stress.

The opposition coalition felt it needed information on the environmental impacts of open pit mining, so, with financial help from the national and international environmental groups in the coalition, it sent a delegation to visit open pit mines in Peru. Their visit included the Oroya mine in the central Peruvian highlands. This coalition's rules of evidence are experientially, not scientifically, based. They visited people and places where what they opposed had already occurred. It could be argued the Oroya mine, in a desolate part of the highlands of Peru, which continues to pollute air and water, is an inappropriate comparison for a mine in the semi-tropical mountain area of Ecuador. However, the evidence of impact is stark, and the community members who went to Peru reported back to Ecuador that even the music from Oroya is sad – surely a result of what has happened to the land.

The opposition group, DECOIN, also organized a trip to Japan and asked the Mayor of Cotocachi to lead the delegation. They had previous contacts with environmental groups in that country, who organized their trip within Japan. The opposition coalition also attempted to engage the Ecuadorian Ministry of Energy and Mines on the residents' own turf. The Ministry, in a counter offer, invited them to Quito. The standoff was broken by DECOIN's occupation of the prospecting camp that had been built by Mitsubishi Metals. They then invited the Assistant Secretary of Mines to come and meet with them. When they had received no response by the third day, they carefully inventoried all the machinery, carted it off the premises and left it in the safekeeping of the Mayor. Then they burned the camp down. While this discouraged Mitsubishi Metals from opening what would most probably have been an open pit copper mine, the Ministry of Energy and Mines remains determined to regroup.

Local people are split on the issue. Some are strongly in opposition and others believe that the lack of jobs and income in this physically rugged area and the primitive nature of access roads all point to mining as an effective engine of growth and modernization for the area. The opposition argues just as vehemently that a way of life would be destroyed. They recognize that it is not enough to show that mining is harmful to the environment and to residents, and have proposed and implemented some economic alternatives. For instance, they organized some 200 farmers into an organic coffee cooperative and, using contacts they made in Japan, have made an initial shipment of coffee at a premium price to that country. The Mining section of the World Bank is strongly supportive of the mining alternative but, unlike the Ecuadorian Ministry they are advising, World Bank personnel insist that local people must participate in the decision-making process, although they are not sure how to carry that out in view of presumed illegitimate tactics used by those opposing the mines. They were concerned about the burning of the prospecting camp, although the opposition group was careful not to damage the equipment.

Figure 19.2 shows the two advocacy coalitions of the mining controversy. Two aspects are striking: the lack of civil society organizations in the pro-mining coalition, and a shortage of entities that bridge the two coalitions and which might serve to mediate between the two factions. The World Bank is committed to local involvement, but the Ministry of Energy and Mines is not. Neither knows how to mobilize the local population directly, nor would such a top-down approach be likely to be successful. Local supporters of mining, who formed the backbone of the opposition to the Mayor in the recent election, are hesitant to even meet together publicly for fear of intimidation by the Mayor's supporters, who coincided with the opposition to mining.

Only two entities have potential to bridge the two coalitions. One is the Ministry of the Environment, which has little clout vis-à-vis the Ministry of Energy and Mines, and the Mayor of Cotacachi, who was mentioned favourably by both sides. His strong win in the recent elections and the apparent imperious behaviour of his local supporters (many of whom oppose the mining project) in victory may have dimmed his star as a possible mediator. The inherent winner-take-all nature of a decision about mining (strip mining will either happen or not happen and mitigation of its impact can only go so far) also makes it difficult to find individuals or institutions willing to broker a solution.

The local anti-mining coalition was effective in stopping the mining company from coming into the area. That coalition then filed a complaint with the World Bank

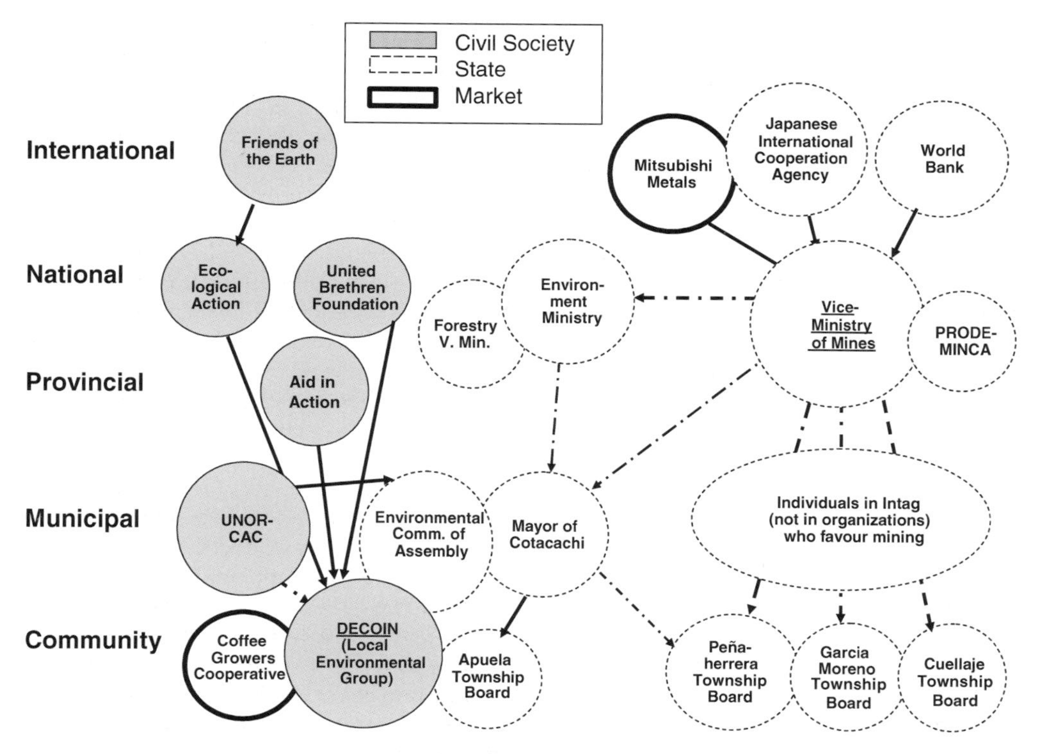

Fig. 19.2. Advocacy coalitions around issue of mining in Intag, Cotacachi, Imbabura, Ecuador.

concerning the management of a larger project on the geology of Ecuador, which they contended would inflict environmental damage because, by publishing the results of a study of mineral formations, mining interests would know exactly where to mine to extract the minerals and despoil the land. They further claimed that the study was carried out without due regard for environmental damage in the bioreserve area.

The World Bank appointed an Investigative Committee, which found that basically the rules of the World Bank were followed and reaffirmed that no mining would take place in biological reserves. This was viewed as a victory by the local advocacy coalition. With support of the Work Bank and with Ecuadorian governmental funds, the Ministry of Energy and Mining, in collaboration with the Ministry of the Environment, is putting together a handbook on community use of the geological information for local planning purposes. They have also hired two Ecuadorian environmental NGOs – Fundación Pronatura and CEDA (Centro Ecuatoriana para la Defensa del Ambiente) – to form 'puntos focales' (monitoring groups) with local organizations and institutions around bioreserves to report activity that threatens it. They have also set up a web page, where individuals can post suspected violations of the laws against exploitation of the bioreserves. The World Bank is promoting these two innovations in other countries. It is doubtful that all this would have occurred without the locally based advocacy coalition with its multiple ties and concerted action.

Ironically, the mental causal model and desired future conditions of the original nucleus of the anti-mining advocacy coalition has hardened to the point that they turned down the opportunity to form the community-based punto focal for the Cotacachi Cayapas Reserve. Thus another group, adjacent to a different part of the reserve, was recruited to provide the information to protect the reserve. The olive branch offered by the Ministry of Mines to DECOIN was rejected. Nine such organizations, inspired in part by the experience in Cotacachi, are now functioning across Ecuador, and more are being established. There was policy impact at the regional, national and international level.

Conclusion

The advocacy coalition approach is a useful way of understanding how issues develop over time to influence policies broader than their immediate concerns. In the case of governance of the bioreserve, there is considerable room for compromise and reconciliation. Yet stalemate has resulted, in spite of the fact that desired futures of different institutional actors are not very different. Why? We believe that the focus on activities and short-term outputs (in this case, the rush to develop plans for tourism and governing parts of the reserve), rather than long-term outcomes, has obscured common values and shared desired futures. Once alliances were initially made, the groups focused on building bonding social capital and reinforcing their positions, rather than looking for shared interests. The Ministry of Environment may be at fault in calling for proposals without first bringing interested parties together to discuss goals for the reserve in order to articulate shared interests from the beginning.

The mining issue presents much starker choices. Few institutions bridge the two entrenched advocacy coalitions. The World Bank became the brokering institution after the initial blockage of the mining effort in order to increase local participation in decisions surrounding the exploitation of natural resources. New systems of education and vigilance have been instituted by the Ecuadorian government with local organizations in order to protect bioreserves from exploitation and increase the options of local entities – market, state and civil society – in appropriately using geological resources in their area.

The ACF helps in description and analysis of grassroots-identified policy resolution. It helps pinpoint key areas where the power of negotiation can be equalized among groups. However, it also reveals areas where contested power shifts the conflict from how to achieve ends to who controls the means of getting there. The shift back to specific positions from shared interests reduces the size of the advocacy coalitions, and decreases attention to processes and relationships, to the potential detriment of the community (Daniels and Walker, 2001).

When the advocacy coalitions fall apart, it is not only because of the lack of a bridging institutional actor. When members of the coalition focus too much on bonding social capital, and fail to maintain bridging social capital with the diverse groups who share pieces of their desired future conditions and agreement on at least some of the ways to achieve them (mental causal models), the coalitions fail. In such cases, those at the core distrust anyone who does not wholely share their desired future conditions and their blueprint of how to get there. They have moved beyond negotiation and back to the less effective tactic of making demands. Now they have fewer allies to stand behind those demands. Thus internal as well as external forces can destroy advocacy coalitions, even when they are successful in changing national and international policies.

Note

[1] See, for instance, Granovetter's (1973) concept of *strong ties*; Freudenberg's (1986) concept of *density of acquaintanceship*, and Coleman's (1988) concept of *closure.*

References

Báez, S., García, M., Guerrero, F. and Larrea, A.M. (1999) *Cotacachi: Capitales Comunitarios y Propuestas de Desarrollo Local.* Ediciones Abya Yala, Quito, Ecuador.

Bourdieu, P. and Wacquant, L.J.D. (1992) *An Invitation To Reflexive Sociology.* University of Chicago Press, Chicago, Illinois.

Clarkson Centre for Business Ethics (1999) *Principles of Stakeholder Management.* CCBE, University of Toronto, Toronto. Available at: http://www.mgmt.utoronto.cal-stake/principles.htm

Coleman, J.S. (1988) Social capital in the creation of human capital. *American Journal of Sociology* 94 (Supplement S95–S120), 95–119.

Daniels, S.F. and Walker, G.B. (2001) *Working Through Environmental Conflict: the Collaborative Learning Approach.* Praeger Publishing, Westport, Connecticut.

Flora, G. (1998) Circular Migration and Community Attachment in the Highland Indigenous Communities of Cotacachi, Ecuador. MA thesis. University of Georgia, Athens, Georgia.

Freeman, R.E. (1984) *Strategic Management: a Stakeholder Approach.* Pitman Publishing, Boston, Massachusetts.

Freudenberg, W.R. (1986) The density of acquaintanceship: an overlooked variable in community research? *American Journal of Sociology* 92, 27–63.

Granovetter, M.S. (1973) The strength of weak ties. *American Journal of Sociology* 78, 1360–1380.

Grimble, R. and Chan, M. (1995) Stakeholder analysis for natural resource management in developing countries. *Natural Resources Forum* 19, 113–124.

Mason, R.O. and Mitroff, I.I. (1981) *Challenging Strategic Planning Assumptions.* John Wiley and Sons, New York.

Münch, R. with Lahusen, C., Kurth, M., Borgards, C., Stark, C. and Jaub, C. (2000) *Democracy at Work – A Comparative Sociology of Environmental Regulation in the United Kingdom, France, Germany, and the United States.* Greenwood Publishing Group, Westport, Connecticut.

Narayan, D. (1999) *Bonds and Bridges – Social Capital and Poverty.* Poverty Research Working Paper No. 2167. World Bank, Washington, DC. Available at: http://www.worldbank.org/poverty/scapital/library/narayan.pdf

Sabatier, P.A. and Jenkins-Smith, H.C. (eds) (1993) *Policy Change and Learning: an Advocacy Coalition Approach.* Westview Press, Boulder, Colorado.

Sharp, J.S., Flora, J.L. and Killacky, J. (2003) The growth machine and voluntary sector: analysis of business elite involvement in civic organizations of a nonmetropolitan city. *Journal of the Community Development Society* 34, 36–56.

20 Future Visioning for the Cotacachi Andes: Scientific Models and Local Perspectives on Land Use Change

Robert E. Rhoades[1] and Xavier Zapata Ríos[2]

[1]*University of Georgia, Department of Anthropology, 250 Baldwin Hall, Athens, GA 30605, USA*; [2]*SANREM–Andes Project, PO Box 17-12-85, Quito, Ecuador*

Introduction

After more than a decade of sustainability research involving multiple stakeholders operating at multiple scales with conflicting objectives, a number of operational problems still face researchers and development practitioners. One of these is the necessity to create understandable and culturally relevant platforms for community-based discussion and debate about the future of the landscape or watershed in question. The projections of scientists and the components they have chosen to highlight in analysis may not coincide with the values of local people who actually live in the watershed. This chapter, therefore, discusses the SANREM experience with developing a future visioning methodology which attempts to combine and contrast scientific land use (LU) and local perceptions. First, we will discuss why the future time dimension is critical for sustainability science and development. Second, we will present an argument as to why conventional participatory community planning without scientific information has limitations. Third, we will present a methodology for future visioning and scenario building. Finally, we will present the results of our land use modelling projections and scenarios compared with Cotacacheños interpretations of the same land use change (LUC).

Why the Future in Sustainability?

After a decade-long struggle to define, measure, operationalize and deconstruct the concept of 'sustainability', most theorists and practitioners have abandoned the search, leaving the concept as little more than a catchall 'Motherhood' statement (Rhoades, 2001). Part of the frustration is that much of what passes for sustainable development is merely old wine in new bottles. Many projects are often no more than outdated and re-named Farming Systems Research, conventional agriculture with 'green' window dressing, or restoration ecology as narrowly practised over the years. The pressure from funders and other impatient overseers of development work has led to the scramble for immediate results which has allowed precious little room for experimentation with innovative or radical approaches. 'Business as usual' meant thinking about sustainability in limited time frames, involving little more than mapping past LU or analysing present LU patterns.

A central problem for sustainability scientists and applied practitioners is the inability to deal effectively with uncertainty and the future. As many early writers on sustainability argued, the concept is about creating conditions at least a generation in the future (20 years), not about outcomes of an annual planning or planting cycle (World Commission on Environment and Development, 1987, p. 8). Sustainability has in some cases became a synonym for monitoring present distribution of plants, animals and other biological components. Although such information can be helpful to people who live in a particular landscape (or at any other scale), the true essence of sustainability is about societal values and what society, in all its negotiated complexity, desires for the future (or 'desired future conditions'). Any programme that addresses sustainability must come up with a methodology that deals with the time dimension and the uncertainty of what lies ahead.

Why Participation is Insufficient for Future Visioning

In the early stages of participatory sustainability research and development, it was thought that all we needed to do was get local people, non-governmental organizatons (NGOs), scientists – the 'stakeholders' – around a debate table to discuss problems, resolve differences and come to some sort of a consensus plan for action. By bringing together diverse people from different walks of life, advised by scientists and planners, and arguing out consequences and trade-offs of decision, a common understanding of the problem could hypothetically be achieved. This was fine in theory, but all too often in such 'participatory' community meetings there was little that people could grab on to which allowed them a clear, empirical image or vision of the future. Since the future is unknown, ambiguous and fuzzy, people have a hard time thinking beyond vague generalities guided by their own ill-informed biases. In these participatory encounters, no matter how well intended, there was precious little empirical reality around which precise boundaries could be drawn so as to keep the visioning dialogue on target. Feel-good consensus-building exercises do not yield understandable, empirical or visual information capable of dealing with the future or the consequences of different decisions. As a result, the consensus-building process tends to break down since it is not easy to understand the consequences of different decisions by different groups or alliances. All too often, scientists dazzle and confuse non-scientists with statistics, equations, charts and models that tend to be static. In all such meetings, there are invariably power plays, by groups and individuals who dominate, and those who have the ability to distort the dialogue, especially scientists. This lack of a methodology for envisaging the future in a clear and empirical way has generated much of the dissatisfaction with sustainability as a useful concept for development.

The Future Visioning Methodology

The future visioning methodology compares scientifically generated LUC scenarios based on robust, predictive models with visions of the same landscape created by local people. There is no assumption of superiority or correctness of either, as each represents its own understanding of the landscape and what is important. By presenting past, present and future (in locally understandable images) and as various stakeholder groups discuss actions and consequences of decisions based on different scenarios, it should be clear that each 'view of the landscape' has its own assumptions and explanatory power. Ultimately, however, scientists must recognize that local people will live by the consequences of planning and outside scientists will not. We will expect there to be disagreement over these assumptions and viewpoints. Indeed, highly rational and 'scientific' external models of the landscape resources may prove to be meaningless or irrelevant to local people. The purpose of the contrasting future scenarios is to present alternatives to the future from both scientific and local perspectives and establish a clear platform for debate and planning.

Our 'sustainable futures methodology' was first developed and tested on four communities in Nanegal Parish, Quito Province, located in northwest Ecuador (Rhoades, 2001, 2003; Stewart, 2001). The Cotacachi application presented here is a further modification and additional test of the methodology. The methodology has a four-step procedure.

1. Develop scientific scenarios of LUC. The past and present LUC were analysed and explained between 1963 and 2000 (Zapata Ríos *et al.*, Chapter 4, this volume). The LUCs were then linked with human drivers such as markets, economic activities and population growth. Then, a LUC model for the year 2030 was elaborated based on a linear regression model for each LUC category. The dominant LUC categories were identified and a LUC map for the year 2030 was projected using geographic information systems (GIS).
2. Create culturally relevant images of the scientific scenarios. A panoramic photograph of the landscape (west view of Cotacachi mountain and landscape) was taken from a well known and culturally significant point for the year 2000. The photo was then manipulated according to the results of the LU maps in 1963 and 2030 to reproduce past and future landscape scenarios for these years.
3. Generate local people's own future visioning. In a future visioning workshop, a contour line-only panorama was given to the local people in which they can draw their own 'desired future' of the landscape in any way they want. They are asked to respond to the question: 'what would you like Cotacachi to look like in 30 years?'
4. Compare these images in a community-based dialogue for future planning. In the same workshop, local and scientific scenarios along with other sets of information are presented and discussed for the purposes of debate and action planning. In the participatory analysis, changes will be linked to past and ongoing behaviour, and 'if–then' outcomes are hypothesized. The alternatives (e.g. do nothing, ecological development, industrial development, etc.) will be presented. Trade-offs will be made clear in a way understandable to the very people whose lives are to be impacted by the planning process.

Developing Scientific LUC Scenarios for Cotacachi: 1963–2030

The basis for the future visioning exercise is the elaboration of LU maps from a point in the past to the present. From aerial photographs, a series of LU maps were developed for the years 1963, 1978, 1993 and 2000 (Zapata Ríos *et al.*, Chapter 4, this volume). Twelve LU categories were identified and each of their surfaces quantified (Table 4.1, Chapter 4, this volume). The significant LUCs in the Cotacachi region over the last 40 years are: (i) an expansion of the urban areas; (ii) an intensification of the agricultural land use; (iii) a fragmentation of the agricultural lands; (iv) a strong increase of introduced forest and greenhouses; and (v) a reduction of primary and secondary native forests and bushlands (Zapata Ríos *et al.*, Chapter 4, this volume).

Based on archival research, study of government statistics and key informant interviews, we were able to identify some human drivers of these LUCs. According to data from the Instituto Nacional de Estadística y Censos (2001), the population in the urban areas in the Cotacachi region has increased from 17,338 to 25,223 people between 1974 and 2001. The growth of urban areas has gone hand in hand with population growth. A corresponding development of paved roads and new transportation arteries has occurred. According to interviews with local people, the migration of young people to the city and livestock theft are reducing the pressure on and the interest in exploiting the páramo–brush land which was traditionally used to graze animals. The lumber export has brought an increase in areas devoted to plantation forest with introduced species, such as eucalyptus. Agrarian reforms, population growth pressure and inheritances have led to a loss of hacienda lands, which have been converted into holdings of < 5 ha. The greenhouse industry has grown in conjunction with the paving of the Pan-American highway and development of international markets.

Plate 3 shows these dominant tendencies among the LU categories in Cotacachi based on linear projection from 1963 through 2000 and 2030. Each figure separated below illustrates the significant changes, either to increase or decrease, in LU categories.

1. Páramo–brush land (will increase from 959 ha in 1963 to 1834 ha in 2030, or a 91% increase).
2. Introduced forest (will increase from 106 ha in 1963 to 1428 ha in 2030 or a 1247% increase).
3. Urban areas (will increase from 83 ha in 1963 to 266 ha in 2030 or a 220% increase).
4. Croplands with extensions < 3 ha (will increase from 3961 ha in 1963 to 6155 ha in 2030, or a 55% increase).
5. Croplands with extensions between 5 and 5 ha (will increase from 146 ha in 1963 to 1700 ha in 2030, or a 1064% increase).

The predominant categories which will have declined include:

6. Brush land (a decline from 4677 ha in 1963 to 2684 ha in 2030, or a 43% decline).
7. Croplands > 5 ha (a decline from 5523 ha in 1963 to 1496 ha in 2030, or a 73% decline).

Based on these straightforward linear regression analyses, we then derived LUC rules which, in turn, were phrased in terms understandable to local people. These change rules are:

- Community interest and need to exploit the páramo is declining.
- Brush land will continue decreasing and the páramo–brush land increasing.
- Urban areas will continue to grow.
- Haciendas will continue to be subdivided and converted into croplands of < 5 ha.
- Market conditions and demand for eucalyptus forest will continue growing and the intensive agricultural production under greenhouses increasing.

By considering these assumptions, the linear regression for each category was prepared (Fig. 20.1). The area derived for each category from the interpolation for the year 2030 can be found in Table 20.1.

The LU map for the year 2030 was built based on the calculated areas for each category and by maintaining the LUC tendency in the last 40 years through the use of GIS (Plate 4). The next step involved taking a panoramic photo from a well-known point in the city of Cotacachi. Based on the LU maps for the years 1963 and 2030, the current photo was manipulated with photo editing software to reproduce the LU conditions and the landscape (Fig. 20.2). These manipulated scenario photographs of 1963 and 2030 were then blown up to the size of 30 cm × 150 cm. In addition, a panoramic photograph of the actual 2000 landscape was used for comparison.

Developing Local People's LUC Scenarios

To understand how local people view LUC in the Cotacachi landscape, a workshop was held with approximately 100 community members (Fig. 20.3). The advantage of using the visioning methodology is that more alternatives and issues are discovered through community meetings and forums than through hierarchical decision-making processes. Visioning generates discussion among alternatives, so that the act of choosing becomes a powerful communal event. It also insists upon everyone at the table having specific input, which helps a community to better understand the values of its members and identify different forces and trends. During the workshop, the participants had the task to look into the future and draw on a blank, contour-only landscape, the elements, resources and spatial conditions that they thought the future should have. Visual reference materials were

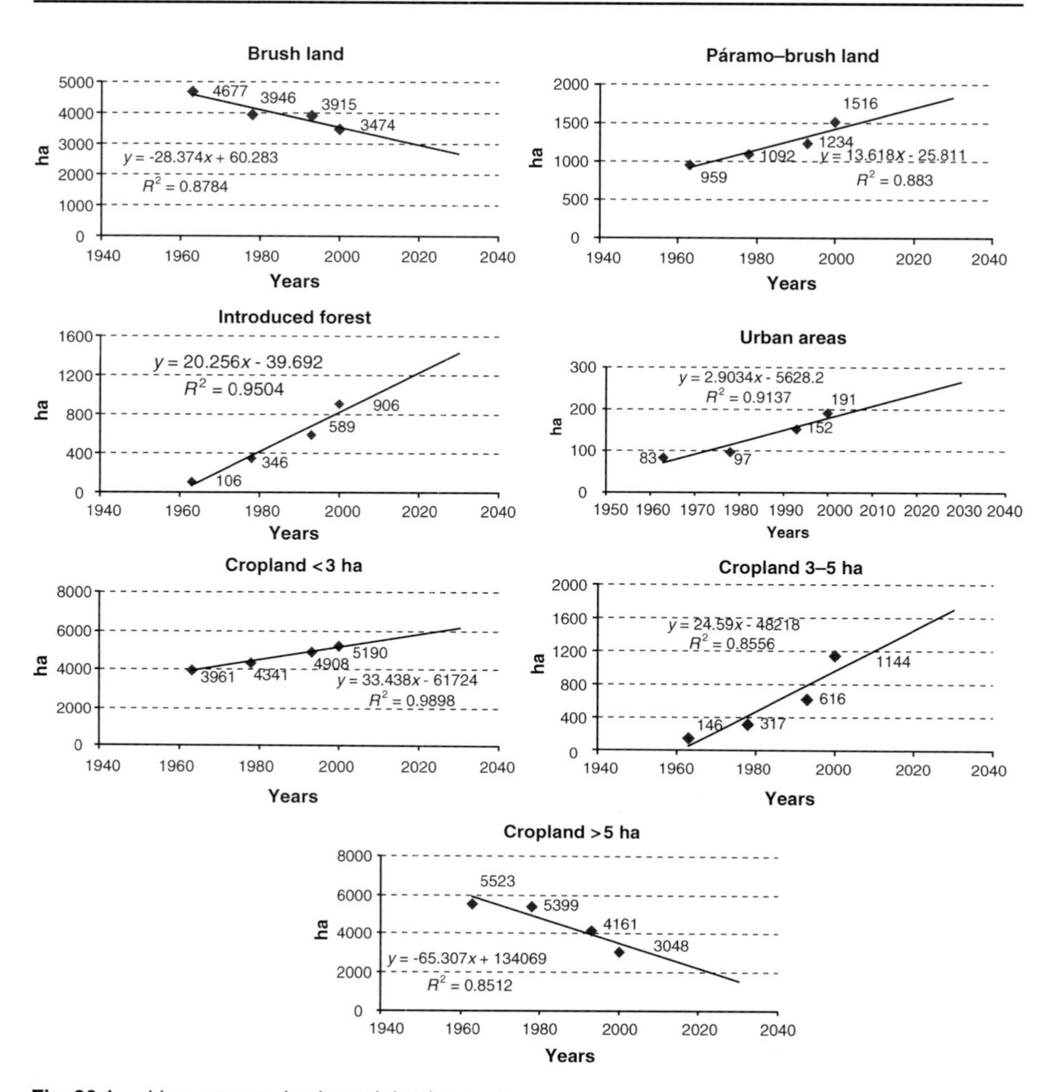

Fig. 20.1. Linear regression in each land use category.

used to stimulate the interaction and discussion among stakeholders. The final step was to compare the people's visions with our scientific scenarios, especially for the year 2030. The comparison allows differential understanding of values, perceptions, time and spatial horizons.

In the workshop, participants were animated by the images of the future we had created from our LUC maps. We made it clear that ours was a vision from a scientific point of view and that they should correct our vision based on their knowledge and values. In order to stimulate exchange, we divided the participants into small groups based on gender and age. This also helped us identify intra-group differences in landscape perception and future visioning. The groups were: (i) elders (> 50 years old); (ii) family men and women (25–49 years old); (iii) young adults (15–25 years old); and (iv) children below 15 years old. Each group was given a sheet of paper exactly the same size as the panoramic photo containing only perspective contour lines. They were given colour pens and asked to draw the landscape of their community 30 years in the future. We asked 'how would you

Table 20.1. Area of the land use change categories in the year 2030.

Code	Categories	1963 (ha)	2030 (ha)	% increase or decrease
1	Páramo	4096	4117	1
2	Brush land	4677	2684	–43
3	Native forest	137	61	–55
4	Introduced forest	106	1428	1247
5	Pastures	190	117	–38
7	Urban areas	83	266	220
8	Mountain	530	530	0
9	Water	401	401	0
10	Greenhouses	0	92	0
11	Others	1093	1021	–7
21	Páramo–brush land	959	1834	91
61	Cropland < 3 ha	3961	6155	55
62	Cropland 3–5 ha	146	1700	1064
63	Cropland > 5 ha	5523	1496	–73
Total		21,902	21,902	

Fig. 20.2. Landscape visioning plates. Years 1963, 2003 and 2030 (Photos: Robert E. Rhoades and Xavier Zapata Ríos).

like your community and surroundings to look 30 years form now?' Each group was allocated 1 h to complete their vision of the future and prepare a short presentation before the workshop participants (Figs 20.4 and 20.5).

Characteristics common to all groups and their desirable future for Cotacachi were a land with more basic infrastructure, such as schools, public services, sport facilities, transportation, and drinking and irrigation water systems. With the exception of elders, all groups drew abundant rain and wide rivers in their desirable future. Although water constitutes a key element for the development of the region, during the last decades the availability of water has decreased considerably due primarily to the loss of glaciers in the Cotacachi mountain.

Fig. 20.3. Indigenous leaders Alfonso Morales (right) and Rafael Guitarra (left) lead discussion comparing scientific scenarios with the people's visions for the future (Photo: Robert E. Rhoades).

Reforestation is also considered as an essential component for the communities.

When shown the scientific scenarios (1963, 2000 and 2030), each group had its own perspective. The elders examined and corrected the 1963 image by drawing more snow on the mountain, more forest in some places and fewer houses in others. Young people added sports facilities and public transportation to the 2030 image, reflecting not only their interests in travelling to work in Quito or elsewhere but their youthful social activitives. Similarly, family men and women (25–49 years of age) emphasized civic and social infrastructure, with men paying more attention to water supply and women to social services such as church, market, women's house and Municipality.

In creating their own future scenarios, the Cotachacheños projected a strong cultural overlay to their landscape and a deep attachment to ethnic ties within communities. Reflecting their historical sense of place, they named their communities, drew Andean symbols of Pachamama (Mother Earth) and her husband Inti (Sun God), identified community houses and noted sacred places in the landscape. There was a clear effort to connect their future with the indigenous cultural and political movement and activism now sweeping the Andes. The future scenarios also contained written statements pointing to issues of contestation over access to land and the Cotacachi Cayapas Ecological Reserve. They specified which modern aspects of the landscape should be

Fig. 20.4. Children's and elders' groups creating their visions for the future (Photos: Robert E. Rhoades).

Fig. 20.5. Future visioning drawings of the men and women's groups of Cotacachi.

removed in the future ('no high radio/TV antennas' 'no floriculture greenhouses', 'no haciendas', 'no urban development'). In viewing their future scenarios, it is clear that their social values of strong cultural identity and attachment to community were more salient than vegetation cover and change. Any effort to address sustainable LU without addressing these cultural values will probably be ignored by Cotacacheños who proudly claim they want 'development with identity', not just development.

Conclusion

The scientific LUC maps and their panoramic derivates generated animated debate among local people. While SANREM scientists focused on vegetation and forest patterns, it is clear that Cotacacheños of all groups were interested in development of infrastructure, services, water and education. While this varied by age and gender, the need for a more modern lifestyle was emphasized. In line with the slogan of 'Development with Identity,' however, they also emphasized their own cultural values and the need for development to respect indigenous identity.

The purpose of the future visioning methodology is to provide local decision makers with insights about the impact of their decisions and actions. This stands in clear contrast to the conventional cost–benefit analysis, conventional public meetings aimed at consensus building, or other tools of the trade in natural resource management (Gregory, 2000). The difference also lies in the fact that there is an upfront admission by scientists that we need to understand the perspectives of all stakeholders, including those of scientists. The notion that 'objective' science has the answers for local people is drawn into question. Science is merely one of the perspectives on the table. The approach attempts to pinpoint the consequences or impacts of alternative actions related to the objectives and goals/values of the community. It means there is a very close tie between analysis and deliberation of issues wherein stakeholders' judgements are informed not only by science but also by their own worldviews (Twery *et al.*, 1998).

We realize that this approach is very different from conventional approaches in that it requires scientists and policymakers to accept that local values and objectives are primary. Scientists can play a role by demonstrating the trade-offs or impacts of different values and proposed alternatives, but they cannot determine the values themselves.

References

Gregory, R. (2000) Using stakeholder values to make smarter environmental decisions. *Environment* 42, 34–44.

Instituto Nacional de Estadística y Censos (2001) VI Censo de Población y V de Vivienda: estadísticas del Cantón Cotacachi. Instituto Nacional de Estadística y Censos. Quito, Ecuador.

Rhoades, R.E. (ed.) (2001) *Bridging Human and Ecological Landcapes: Participatory Research and Sustainable Development in an Andean Agriculture Frontier.* Kendall/Hunt Publishers, Dubque, Iowa.

Rhoades, R.E. (2003) Modelos científicos y visiones locales de la sustentabilidad en el futuro: un nuevo método para la planificación comunitaria del uso de recursos naturales. In: *La Conservación y el Desarrollo Integrado. Lecciones Aprendidas al Vincular Pueblos, Proyectos y Políticas en América Tropical.* Abya-yala, Quito, Ecuador, pp. 15–21.

Stewart, D.J. (2001) Creating land use change scenarios: past patterns and future trajectories. In: Rhoades, R.E. (ed.) *Bridging Human and Ecological Landscapes: Participatory Research and Sustainable Developments in an Andean Agriculture Frontier.* Kendall/Hunt Publishers, Dubuque, Iowa, pp. 179–189.

Twery, M.J., Stout, S. and Loftis, D. (1998) Using desired future conditions to integrate multiple resource prescriptions: the northeast decision model. In: El-Swaify, S.A. and Yakowitz, D.S. (eds.) *Multiple Objective Decision Making for Land, Water, and Environmental Management.* Lewis Publishers, Boca Raton, Florida, pp. 197–203.

World Commission on Environment and Development (1987) *Our Common Future.* Oxford University Press, Oxford, UK.

21 Sustainability Science in Indigenous Communities: Reconciling Local and Global Agendas

Robert E. Rhoades
University of Georgia, Department of Anthropology, 250 Baldwin Hall, Athens, GA 30602, USA

Introduction

Over the past decade, indigenous peoples in Latin America have started to demand that scientific research conducted in their communities leave behind some tangible benefit (see Tituaña, Foreword this volume). The age has passed when outsiders, regardless of nationality and affiliation, freely enter indigenous territory or homes without prior consent and take information, biological samples, botanical specimens or cultural materials. Although the colonial mentality of superiority and authoritarian dominance over indigenous people still remains in many universities and government agencies, most experienced sustainability scientists consider it highly unethical to pursue only 'extractive' research aimed at clients outside the local context (Fairhead and Leach, 2003). Indigenous people are not opposed to scientific research, as the case of Cotacachi illustrates, but they are also looking for research which is 'enriching' for them (Waters-Bayer, 1994). This basic principle of human rights and ethics of research has been respected by the SANREM–Andes research team throughout 7 years of activities in Cotacachi (see Rhoades, Chapter 1, this volume). We were aware that if our research benefited only ourselves as scientists or our donors in Quito or Washington, DC, our Cotacachi hosts would have little interest in giving time, resources and their knowledge to something which returns nothing of value.

The problem we faced, however, was how to carry out technical scientific activities on global research questions which on the surface did not have an obvious or immediate local benefit. Our interdisciplinary team was composed of several academic disciplines with specific data and methodological needs. As funded researchers and members of the academic community, scientists also faced pressure from fellow scientists, donors and technical reviewers to meet high standards of rigorous methods and data gathering. Field activities, such as collecting soil or water samples, setting up rain simulators or climate monitoring stations, analysing people's ablution activities, administering long questionnaires or measuring farm sizes, were not locally beneficial in the minds of local people. We always tried to explain why an activity was required; for example, digging a soil profile trench would help us to improve 'their' agriculture. Such explanations typically fell on deaf ears. With a history of > 500 years of exploitation, more plausible explanations to indigenous people were that we were seeking gold buried by the Inca or robbing antiquities to take back to the USA. In the context of the global news system and activist education by non-governmental organizations (NGOs) and the indigenous movement, politically charged words such as 'biopirate' or 'imperialist' are now common currency in Cotacachi. For scientists willing to listen to indigenous people, however, the rewards of modifying research to make it relevant far outweigh the tensions of facing local accountability. After all, accountability upward to donors and administrators is a constant and time-consuming necessity; why not the same accountability to local people?

The purpose of this chapter is to describe and analyse the trade-offs and experiences of the SANREM–Andes team as they carried out basic sustainability research in the social and political contexts of indigenous communities in Cotacachi, Ecuador. In presenting this case, we will look at how the encounter between outside scientists (and their social and knowledge systems) with indigenous people (and their social and knowledge system) unfolds through time. Honest assessments and descriptions of 'how it really works' as opposed to 'how it should work' are rarely written up in participatory manuals and guidebooks on working with local communities. Although this chapter is from my view as an outside scientist, and ideally we need another chapter from the indigenous view, I will attempt to move from sugar coating the participatory research process in order to discuss in realistic terms what is needed for the global–local encounter of sustainable development to succeed.

To describe our experiences with what worked and what did not, I will first frame the conditions under which our interdisciplinary team agreed to work in Cotacachi and what it meant for the scientific method. This will be followed by a discussion of the required negotiations, trade-offs and compromises reached by scientists and local people. Thirdly, I will present six case studies of applied research which benefited communities in exchange for doing more basic research. Finally, an analysis of pluses and minuses of 'enriching research,' especially the sharing and use of data generated in the project, will be made.

Sustainability Research in Indigenous Cotacachi: Conditions and Implications

Indigenous Cotacacheños do not easily understand what scientists do or why. If research means anything at all to villagers, it carries negative connotations based on past exploitative experiences. Investigators came to their villages before, asked a lot of questions, carried away information or items and never returned anything to the community. Some NGOs working the area add to the mistrust by giving lectures on the neo-colonial evils of researchers, especially foreign ones. Debate in anthropology has tended to pit 'advocacy research' against 'basic research' in which advocacy supporters say that researchers should only work on problems of interest to local people and basic researchers argue that this compromises the integrity of the research findings (Gross and Plattner, 2002). This polarized debate was never a factor in Cotacachi simply because indigenous communities require relevancy as an a priori condition to other activities such as research.

The rules governing the conduct of sustainability research in Cotacachi were agreed upon in 1997 between myself (acting on behalf of SANREM) and officials of the canton (Mayor's office) and UNORCAC. In the Foreword of this book by the indigenous Cotacachi Mayor Auki Tituana Males and in my acknowledgements, the first contact between our project and the people of Cotacachi was described as a somewhat serendipitous event. After almost 5 years of conducting research in the mestizo Nanegal region of Pinchincha Province, our team was interested in transferring activities to a high Andean location. During several temporary visits by myself and other researchers, we formally approached Cotacachi's leadership to explore areas of mutual interest and to seek research permission. Cotacachi is not a location in which outside researchers, development practitioners or tourists can do whatever they wish. Indigenous communities demand directly or through the agency of UNORCAC that research or development take place on their terms and on topics they see of benefit to the community. Permission to do research, while not always a formal bureaucratic procedure, requires approval by a range of local authorities, including government, UNORCAC, *comunas* or individual landlords. Within the Cotacachi-Cayapas Ecological Reserve, for example, this means that the Ministry of the Environment must issue a research permit and a pass to enter the reserve. To enter the large hacienda, El Hospital, one needs written permission from the Ministry of Environment which is then presented to the *mayordomo* (foreman) who has absolute power to grant passage across the reserve part of the hacienda. Within the communities outside the reserve, one will be met with cold stares and even aggressive rejection unless the activities are explained and approved by at least one indigenous body such as the *comuna* President, directorate or the assembly of UNORCAC. Project researchers, individually and jointly, have presented their research proposals to numerous levels of indigenous bodies. As the programme manager, I presented the overall SANREM project to the Assembly of UNORCAC, composed of 43 communities, and later the same proposal to the assemblies of several communities where specific activities would take place. In cases of projects by individual scientists, permission from the President or the Committee of Directors of the affected community would be requested by the scientist in visits to the *casa comunal* (communal house). If the scientist was working on water issues, then the permission from the *junta de agua* (water committee) of the water system being studied was required. Regardless of the local approval route, it was always necessary that local people know who we were, that the research was explained and approved upfront, and that someone (preferably a Quichua speaker) representing UNORCAC accompanied the researchers. Our scientists always respected this permission protocol and, as a result, received excellent support over the 7 years of the research activity. Finally, it is worth noting that unlike many NGOs and bilateral projects in the area, SANREM did not as a policy utilize logos, banners, posters, marked cars or public advertisement, preferring instead to maintain a quiet presence in the landscape.

The dilemma arising from this local approval process is that scientists frequently had basic research priorities and data gathering requirements that were not understandable or obviously relevant to local people. Our university-based researchers were interested in basic research using the conventional methodology of testing hypotheses related to basic scientific questions. The project was truly interdisciplinary, with scientists representing soils, aquatic resources, hydrology, biology and botany, economics, rural sociology and anthropology. Each discipline has its own themes and methods of inquiry. Within the team, a great deal of time was spent trying to communicate between these various disciplines characterized by differences in scale, subject matter, research design and methodology. Some researchers were graduate students who needed to complete a Master's thesis or PhD dissertation with theoretical as well as empirical results. Researchers also arrived at different times during the research programme and a constant new set of permissions was required. SANREM's annual work plans had to be approved by a US-based technical committee which had little to no knowledge of working conditions in the field. Juggling between these various interests and levels of understanding was always a frustrating and uncertain process for the researcher. Rarely was it immediately obvious to local people how such research would benefit them.

Three possible scenarios of differences between local people and scientists were: (i) scientists wanted to do research not understood by and of no interest to local people; (ii) scientists had needs that required local resources (land, time, water, plants and knowledge) which local people controlled; and (iii) local people had interest in seeking solutions to problems of no interest to scientists. Only in rare cases was there an automatic overlap of scientists' and local people's interests. Such differences in needs and understanding are commonplace but are rarely discussed in methodologies of sustainability science or development. A great deal of ink is spilled in the development literature over how different disciplines should work together, but little is said about how scientific interest can be reconciled with community needs. In general, NGOs do not have the same problem since they rarely conduct research and tend to work on applied problems of concern to local people. Science, however, with its global questions and principles to test, operates under a different set of commands.

Negotiations, trade-offs and compromises: give and take of participatory research

An operational challenge arises as to how the interests of scientists and local people can be addressed so that both are reasonably satisfied. While it is clear why SANREM researchers wanted to work in Cotacachi, it is not so obvious why Cotacacheños wanted to have researchers in their presence. Ours was not a donor programme providing large sums of funds for local development; rather our operating budget was small and earmarked only for research. For diverse reasons, however, Cotacacheños believed we possessed valuable resources for them, beyond money. The leadership of UNORCAC and the canton expressed genuine interest in having data and any information which they could use for obtaining grants, justifying projects to the government, leveraging in negotiations with donors, or political reasons. For example, in the mayoral election of 2000, political interest peaked in whether or not drinking water in Cotacachi's urban area was safe. The opposition party to the incumbent Mayor publicly declared it was unsafe, but the Mayor retaliated with evidence to the contrary from the SANREM water monitoring project (see Ruiz-Córdova *et al.*, Chapter 16, this volume). Our water data were published in the local papers and, according to the Mayor at the victory celebration, our water data played a key role in his party's victory. Although we brought little money to communities, we hired local indigenous assistants and provided minor equipment support such as computers, recorders, supplies and the promise to leave our project vehicle behind. SANREM also gave small funds for workshops, tickets for international travel, and networking with foreigners who were sympathetic with their indigenous goals. Cotacachi, unlike nearby Otavalo, does not yet have a global network of connections nor an economic niche such as textiles and Andean music to sell on the global market (Meisch, 2002). Cotacachi is in the early stages of this globalization process. SANREM's small capital investment could not compensate for the time and resources required by local people to support our activities. Fortunately, throughout our stay in Cotacachi, specific interests based on real problems were expressed by Cotacacheños on ways that we could help them achieve their goal of 'development with identity.'

The process of arriving at a common consensus of exchange of research for local benefits did not occur in a straightforward or systematic way. A great deal of negotiation and 'give and take' on the part of researchers and local people took place on many levels: the leadership of UNORCAC; with individual communities and their leaders; with farmers; with schools and professors; and with water associations. Agreements were often reached through informal conversations or gatherings in which local people would express certain needs which we could fulfill. By addressing these special requests, SANREM was allowed to stay and continue to do research of interest only to scientists. In the process of this negotiation, friendships were formed, co-madres and co-padres declared and alliances sealed. Over time, trust and confidence between researchers and local people put a very human face on our project and its interactions.

Case Studies or Examples of Research Addressing Requests of Indigenous People

By accepting the responsibility of directing some resources, time and energy to priority research topics identified by local people, a social credit was created for us to use in pursuit of other research questions not prioritized by local people. This sometimes meant in the course of daily interaction we would be informed of a local interest or request and it was up to us how to make it fit with research. Six examples will illustrate this process.

Memory banking and the scholarship fund (investigators: Virginia Nazarea and Maricel Piniero)

The leadership of UNORCAC and many families with whom we collaborated always stressed the critical need for education of indigenous children. The illiteracy rate is very high and few indigenous people can fill roles of leadership in their own communities, organizations or schools. Many key positions for managing UNORCAC are held by mestizos or educated indigenous personnel from Otavalo. Through our main collaborator in UNORCAC, Magdelena Fueres, we agreed to provide funding to select indigenous children to attend primary, secondary and even college if they were willing to conduct memory banking research with their elders on the topic of landraces. Each child was allocated around US$20 (sufficient for entering school each term) if they would interview their parents, grandparents or other elders using the memory banking method.

The purpose of the memory banking project was to document and preserve local knowledge associated with agricultural crops so that neither seeds nor knowledge would be lost (Nazarea, 1998). While old varieties are of keen interest to local people, it is unlikely they would give their own time and energy to recovering this knowledge on a wider scale than their own household. However, if memory and seed collection are connected with the possibility that a child in their family can attend school, then parents were enthusiastic to participate. From July 1999, until the SANREM project closed in 2004, 15 indigenous children annually received funding for school activities if they documented vanishing agrobiodiversity knowledge. They asked a series of basic questions ('What varieties did your grandparents grow?' 'How did farming change over time?'). In addition to learning how to interview, record information and store it in the project database, they also collected culturally significant plants (leaves, seeds and roots) and prepared them for display and herbarium storage. In addition, the scholarship students (*becarios*) planted and maintained a biodiversity garden on the grounds of Jambi Mascaric. Children also made public exhibitions of the plants and elders' knowledge during special days at the school. Two goals were accomplished: scientists gained information about changing agrobiodiversity and children were able to attend school. The memory banking collaborative project illustrates how both scientists and local people gained in a collaborative effort of dovetailing interests.

Farm of the ancestral futures (investigators: Robert Rhoades and A. Shiloh Moates)

An outgrowth of the memory banking project was the establishment of a participatory farm in the high zone of Cotacachi called the Ancestral Futures's Farm. Two main objectives of the ancestral farm were: (i) to create an *in situ* farm whereby vanishing or culturally significant Andean crops could be grown out and re-distributed to local people; and (ii) to collaborate with the national genetic resources bank (INIAP: Instituto Nacional Autónomo de Investigaciones Agropecurias) by obtaining and growing out disappearing Andean crops with local participation. The local farm would be a mechanism to recuperate Andean crops that have been mostly or completely abandoned and an educational tool for dialogue between indigenous collaborators, scientists (SANREM, NGOs and INIAP), and young people with elders, on the issue of loss of local landraces.

Cotacachi is losing its traditional agrobiodiversity due to multiple factors including climate change, the Green Revolution, declining family labour and minifundization (see Skarbø, Chapter 9, this volume). The Ancestral Futures Farm activity was seen by UNORCAC as an important element in the larger indigenous cultural revitalization movement. In Andean cultures, an inextricable link exists between biology/culture and the past and present. The focus of that ritual nurturance is on the field (*chacra*) and the *ayllu* (disappearing in Cotacachi but still known) or the family. Bringing back the old crops was symbolic of the recovery of their traditions.

The Ancestral Farm was established in Ushugpungo in October 2002, with the families of 13 memory banking students. Shiloh Moates, coordinator of SANREM–Andes, and Magdalena Fueres, leader of Jambi Mascaric, took the lead. The parents provided knowledge and labour in exchange for scholarships for their children. Since the children were mainly in school when the farm was worked, the parents organized weekly *minga* (planting, weeding and harvesting). The farm was divided into halves: one dedicated to local varieties and planted under the direction of the parents; the other half was planted in INIAP varieties using their specifications on spacing and planting depth. The local side was driven as much as possible by indigenous planting rules, their perceptions and desires. In keeping with traditional practices, the farm was entirely organic and managed using local knowledge. The parents led this process by planting 'ally' crops together and keeping the crops that 'don't get along' apart, just as they had learned from their parents. The crop they had least knowledge of was *mashua*, but it was planted similarly to *oca* which it strongly resembles (Moates, 2003). The mothers provided the seed they planted in their part of the farm. They provided two types of quinua, peas, chochos, habas, mellocos, chaucha potato, mashuas and ocas, all from their own farms.

We signed an agreement with Dr Jaime Estrella, Head of Ecuador's national gene bank (INIAP), for the donation of clean seeds. The contract specified that 25 varieties of *oxa* and *mashua* as well as four varieties of *achira* be donated in exchange for rights to visit the plot and the return of well-documented data on the production of their varieties. Through our later focus group survey with the parents, we learned that maize and beans would have been more relevant crops for repatriation or returning from the gene bank due to their strong ritual significance. At the time of harvest, we conducted structured interviews with the participants to see how they regarded the experience and the varieties. There was a kitchen follow-up on the food preparation, cooking and consumption of the crops. The women provided detailed insights on the cooking quality, taste and overall performance of different varieties.

The Ancestral Futures Farm turned out to be positive for both scientists and local people. Local people received new varieties, exchanged them among themselves, took home some food and received support for their children to go to school. SANREM, in turn, received a great deal of information about local varieties: why some had disappeared; what still existed; and what was desired through repatriation. The Ancestral Futures Farm continued operation after the SANREM programme officially closed in May, 2005.

The participatory 3D 'maqueta' model (investigators: Robert Rhoades and A. Shiloh Moates)

UNORCAC is a second degree indigenous organization which depends almost entirely on external funds for its activities. UNORCAC's leadership needs natural resource information and decision support tools which can be used in negotiation with donors and bilateral projects. They also request information which helps them interact with member communities and to help establish planning priorities. One example of how SANREM contributed to UNORCAC's decision support was the construction of a three-dimensional physical model of the Cotacachi Andean landscape. Called the P3DM (participatory 3-dimensional modelling), or 'maqueta' in Spanish, this tool recently has been made popular by IAPD

(Integrated Approaches to Participatory Development) in the Philippines (see http://www.iapad.org/participatory_p3dm.htm). However, use of the *maqueta* in the Andes goes back to Inca times when physical models were used to plan towns, agricultural fields and irrigation systems (Hyslop, 1990). We built and used the model at a scale of 1:10,000 as part of a participatory process whereby spatial information is combined with people's knowledge for advocacy, awareness raising, community planning, conflict resolution, and participatory monitoring and evaluation.

The relief model of the landscape is built by placing layers of cardboard on top of one another, each layer having been cut out to represent the contour lines of a topographical map. Because of an emphasis on the vertical dimension, this tool is particularly suited to application in mountain environments. Verticality is a fundamental feature of Andean landscapes and livelihoods, which rely on exchanges of goods and services among production zones and social spaces at different altitudinal levels. In Cotacachi, this layered system structures agricultural practices, migratory patterns, community relationships and cognitive perspectives. In participatory mapping exercises, local people tend to draw their communities in terms of vertical arrangements. A relief model is therefore better able to represent key ecosystem linkages as well as to be more consistent with local understanding of the environment than two-dimensional maps or images. Using the *maqueta* is inherently dynamic and never completed, since information is constantly revised and updated as new stakeholders or processes intervene and affect the landscape.

After the basic relief model was laid out, the specific features and properties of the landscape were filled in by local people through a participatory workshop. This exercise offered a unique educational opportunity that enabled local people clearly to visualize and understand issues of human–environment interactions. Among stakeholders, it also provided a platform for dialogue concerning watershed management and mediation of conflicts surrounding natural resource use. For instance, it served as a visual aid to indigenous communities and their neighbours in the facilitation of agreements surrounding grazing rights and resource use in the Cotacachi-Cayapas Ecological Reserve. The model allowed concrete discussion of territorial boundaries, often the subject of tension because of contradictions between ancestral rights, formal deeds and land reform provisions. In the effort to promote rural ecotourism and generate revenues in the region, UNORCAC uses the *maqueta* to highlight locations of cultural and natural interest, such as sacred places (i.e. where ritual baths or cleansings are held) and hiking trails leading to Cuicocha lake and to the summit of Mount Cotacachi.

The *maqueta* was also an important tool for researchers as it was useful for integration and analysis of natural resource data. In interaction with local people, we were able to pinpoint on the *maqueta* local geographical knowledge on the landscape and analyse it in the broader context of regional economy/ecology and natural and social system interactions (see Fig. 21.1). For example, water quality data can be used to pinpoint contamination points and examine how those may be affected by production strategies, settlement patterns and household interaction. Likewise, soil data can be overlain on the model in a workshop to illustrate issues of erosion and fertility in relation to land use and climate at various altitudes. The *maqueta* was used in a climate change workshop to talk about the retreat of the glacier on Mount Cotacachi (see Rhoades and Zapata Ríos, Chapter 5, this volume). The *maqueta* was used by an NGO project studying upstream–downstream interactions surrounding the Pitzambitze river, a tributary of the Ambi river. By plotting their data on the model, they showed how upstream processes of contamination and deforestation affect downstream communities. For instance, the common practice of burning grasslands at high altitudes results in erosion problems and water table disturbances downstream.

Fig. 21.1. Researchers Juana Camacho and Xavier Zapata Ríos listen to Cotacacheños explaining climate change using the participatory 3D model (*maqueta*) (Photo: Jenny Aragundy).

The *maqueta*, therefore, was a product from SANREM research that was seen by the indigenous leadership of UNORCAC as an extremely valuable tool. It is a powerful instrument for information exchange and analytical reflection on landscape change, conflict resolution and watershed management. The *maqueta* is located today in the training conference room of UNORCAC and is used almost daily for training activities.

Studies of Cotacachi folklore: legends, customs, ancestral sayings and food recipes (investigators: Virginia Nazarea, Maricel Piniero and Juana Camacho)

Another direct request of the indigenous community, primarily from Magdalena Fueres of Jambi Mascaric (UNORCAC), was to collect and publish the folklore of Cotacachi, especially folk tales (see Nazarea *et al.*, Chapter 6, this volume), ancestral sayings, dreams and beliefs, customs and rituals, traditional food and medicinal recipes. This research was not originally a part of any project, but Dr Virginia Nazarea of the Ethnoecology team accepted this request and through her field assistants conducted a study of Cotacachi's folklore (see Fig. 21.2). UNORCAC's expressed interest was specifically to rescue the oral traditions before they disappeared as a strategy of cultural revitalization. Thus, they wanted the memory banking concept applied to this cultural recovery and to publish booklets in Spanish, English and Quichua for schools and the emerging tourist trade in the area. The book *Stories of Creation and Resistance* presents the legends and myths (Nazarea and Guitarra, 2004). A similar volume, *Recipes for Life: Counsel, Customs, and Cuisine from the Andean Hearth* by Virginia Nazarea and her team was published by Abya-Yala Press in 2005 (Nazarea *et al.*, 2005).

The collection of folklore also brought joy and pleasantness to the relationship between researchers and local people. Circles of elders were invited to talk not only about their old varieties but also about the legends and myths. The elders enjoyed

Fig. 21.2. Rosita Ramos, SANREM assistant, tape records a folktale for the memory banking project (Photo: Natalia Parra).

immensely talking about the old ways. Although this project activity was criticized by our US technical committee as 'documenting the demise', it not only brought good will but gave us insights into how local knowledge is adaptive in the landscape. The recipes and sayings are time-tested insights into the Cotacachi landscape and offers practical remedies that are certainly worth saving and might even have practical value in the future. As in any culture's folklore, many stories and sayings are for entertainment or education in best manners. One saying goes 'When children stick their tongue out, it is said they will look like lizards' or 'One must not sing while eating because one can bite one's tongue'. There are also omens and secrets, such as 'when the left ear burns, it is because the enemies are talking' or 'when the dogs howl, one must not go out because the evil spirits are passing by'. In a similar vein, there are many local recipes for everyday living, for hunger, famine and illness. This activity bought us a great deal of social capital and, when we asked to do research in the community on a topic they did not understand or maybe even annoyed them (administer a questionnaire), they were open to our request.

Diagnosis of water systems in Cotacachi: university extension project for the benefit of local communities (investigator: Olga H. Mayorca)

Water is the most critical natural resource in Cotacachi because of an essential role in human survival and its increasing scarcity as a clean and ample resource. The local water systems have multiple problems: poor service, inability to collect user fees, high costs, limited supply, sanitation, conflict over access and control of water between communities, individuals and government agencies. They also lack essential information about the water systems. Local water associations (*juntas*) have organized over the past decades in response to new infrastructure and changes in needs of users in the landscape. Given limited government support and subsidies for water, local communities are often left to their own resources and ingenuity.

Due to water's central role in the lives of Cotacacheños, we were regularly called upon to help with some aspects of the water systems: providing training on citizen water monitoring, generating data on quality and quantity of water, GIS mapping of water system, or lending time of our research assistants to help water associations in collecting funds. The SANREM office, in fact, was also the water association office for two of the largest water association, Chamuvi and Cambugan. Interested in improving their service, the *juntas* of these two association requested that SANREM help gather specific data on quality, coverage, trends regarding numbers of users, and supply capacity.

Olga H. Mayorga and her students of the Geography School at the Catholic University-Quito agreed to conduct the study for the water associations (Mayorga, 2004). The Cambugan water system serves six communities (1177 users) and Chumavi serves seven (1504 users). Typically, the administration of a local water system is done through water *juntas*: a central one and others in each one of the communities which are integrated by a president, a vice president, a secretary, a treasurer and an operator. The funds come from the collection of water service fees paid by the communities, although there is little support. There is one operator per community in charge of repairs and installation of new pipelines. There is a central operator who receives a US$100 monthly payment and each community operator receives US$13 per month.

In carrying out the research, Professor Mayorca and her students conducted office work and field work. They surveyed and digitized each of the water systems, designed surveys for users ($n = 490$) and administrators, and held workshops. The information from the points collected with GPS (geographical positioning system) equipment and the information tabulated in Excel were integrated with Arc View. The final report to each water system association was presented in digital format and in hard copy. They also developed maps of the pipelines. Biodegradable garbage was revealed to be disposed of in the fields or thrown in nearby streams which are mainly dry. Plastics were typically incinerated. Garbage was often blown by the wind and scattered throughout the agricultural fields. Sources of water have also been channelled into illegal pipes to provide water to different communities other than those in the water association.

Water from potable water systems has multiple uses beyond drinking: washing clothes, cooking, irrigation and watering farm animals. The majority of women wash their clothes in their homes, but if water is scarce they wash in a stream or irrigation channel near their house. Users of the Cambugan water system distrust the water so much that many users rely on other water systems. People also reported numerous diseases related to water: diarrhoea in children, colds, head and stomach aches, fever and parasites. Treatment is largely by traditional methods of curing, often through a shaman. The Catholic University team submitted reports containing detailed infrastructure information to the two water system authorities, along with suggestions on how to improve the systems. Recommendations are now being implemented by the water associations with outside funding. The study, although requested by local water authorities, also provided excellent information for our own scientific work on water in Cotacachi.

Creating local databases and leaving them behind: SANREM data files, Cotacachi atlas (investigator: Monserrath Mejía S.)

Ironically, the most desired product from our research was data. Indigenous leaders of UNORCAC as well as NGO personnel sought information on varied topics for securing funding or grants. Since little systematic information on Cotacachi existed, we were constantly called upon for data on water, soils, climate, agricultural production or secondary government statistics which were also available to anyone but had not been gathered systematically. The information was put into the SANREM computers or placed in hard copy in two open access filing cabinets. We placed no restriction on these data and left all control in the hands of the UNORCAC's leadership. We only asked that all hard copy information be returned to the filing cabinets after their use.

In addition to these archives, a data node was established in the Catholic University-Quito where all available secondary and primary data on Cotacachi at the cantonal level were processed. 'The Digital Atlas of Cotacachi Canton,' headed by Monserrath Mejia of the School of Geography has produced a rich integrated GIS database which brings together biophysical, socioeconomic and demographic information for the characterization of different parishes in the canton. This digital atlas is geared toward professionals, technical personnel, planners, policymakers and institutions involved in natural resource and rural development. Cotacachi as a canton is in the process of formulating a Cantonal Territorial Ordinance (*Ordenanza Territorial*) and this atlas serves to identify priorities through geographic analysis of vital statistics and biophysical data. The atlas contains an ordered collection of maps which represent spatial distribution and temporal variation of biophysical aspects and socioeconomic indicators of the parishes.

Prior to developing the atlas, Monserrath Mejia and her team met in Cotacachi with individuals from different institutions, including UNORCAC and the Planning Direction of the Cotacachi Municipality. Through this participatory dialogue, it was decided to structure the atlas around four themes: (i) *natural space* with information about biophysical aspects (cartographic base, relief, satellite images, soils, climate, erosion, geomorphology, etc.); (ii) *cultural space* which delineates parish divisions, political–administrative units, evolution of population, tourist activities, etc.; (iii) *space and infrastructure* which shows roads, health facilities, housing, schools, and services such as water, electricity, sewage and garbage collection; and (iv) *synthesis maps* of sociocultural and natural interest sites, natural potentialities, natural and anthrogenic threats and natural limits to agricultural production. These colour maps visualize the spatial characteristics and problems of canton Cotacachi. The final atlas was produced in two forms: a hard copy atlas and a digital version. The digital atlas can be managed from various locations, including from the Catholic University or from Cotacachi itself. Periodic updates with new data and thematic maps are possible, although those making changes would need to be trained in the use of Arc View in order to manage and operate the atlas. The atlas has been incorporated in the canton's initiative on 'transparency through democracy' and will be available at the municipal library for public access.

Conclusion

The purpose of this chapter has been to present our experiences with reconciling differences between the agendas of sustainability scientists and those of local people. Examples other than the six highlighted here could have been mentioned. We also researched and published histories of 15 communities for distribution in the communities. A CD version of all SANREM data was created and distributed in the 'tool book' format with different levels of analysis, including embedded raw data. These examples illustrate creative ways to address the needs of local people while simultaneously pursuing basic research. However, some caveats must be mentioned. Scientists who freely leave project data with local people as a form of exchange should be prepared that the data will be distributed or used in ways not always in accordance with intellectual property ethics of researchers.

Virtually all of these data were quantitative, descriptive or geographic, with no confidential material about individuals or families. Any data of a personal issue remained confidential human subjects matter between the individual researcher and the individual or group who provided them. The project data sources were used not only by the local people but also by unaffiliated researchers who passed through, NGOs looking for information for their own reports or proposals, as well as local officials. Due to the continual rummaging through these files, a state of disorganization prevailed. The archived materials also became a source of leverage and currency for the indigenous people who would exchange the information for new projects or funds. NGOs were especially notorious for extracting the raw data and marketing them as their own. Publications in the form of booklets, pamphlets and flyers based on SANREM research would appear in workshops or annual reports, typically without any credit to the original researchers.

While such trafficking in data generated so arduously by a researcher may seem disconcerting, we have concluded that this form of plagiarism or data lifting is a minor issue and should be overlooked for a number of reasons. First, raw data typically cannot be understood by anyone except the person who generated it in the first place. The odds that such data will end up in a significant publication are remote. Second, the information was generated thanks to the good graces of the local people and it is within reason that such information should be left in the communities, even in raw form. This step renders the need to repatriate the data back to the community unnecessary; although there is the risk the data will be lost or misunderstood. Third, if local people can use the data as currency and gain advantage, we should welcome the opportunity to help in exchange for research support.

Sustainability science projects justify themselves and receive funding to create decision support tools, research findings and policy advisement that will improve natural resource management. Rarely, however, do such projects succeed in the short term in generating impacts that local people understand and appreciate. Sustainability science, as an academic discipline, and sustainable development, as the applied outcome, are still in their infancy. We do not yet fully understand the principles of sustainability or how it is to be achieved, even in the long term. This will require decades of systematic interdisciplinary investigation on the complexities of nature–society interactions. In the meantime, we need the full cooperation of local people who should not be expected to sacrifice for the benefit of researchers. The case studies in this chapter illustrate how to create a win–win situation for both scientists and residents of a watershed.

References

Fairhead, J. and Leach, M. (2003) Does globalised science work for the poor? *Forest Perspectives.* IDS Research Direct 1, 1–4.

Gross, D. and Plattner, S. (2002) Anthropology as social work: collaborative models of anthropological research. *Anthropology News* November, 2002, p. 4.

Hyslop, J. (1990) *Inca Settlement Planning.* University of Texas Press, Austin, Texas.

Mayorga, O.H. (2004) *Piped Water Systems of Cambugan and Cumavi, Canton Cotacachi: A University Extension Project for the Benefit of Local Communities.* SANREM–Andes, Quito. Ecuador.

Moates, S. (2003) *Reduced Biodiversity in Highland Ecuador.* SANREM–Andes, Athens, Georgia.

Meisch, L.A. (2002) Andean Entrepreneurs: Otavalo Merchants and Musicians in the Global Arena. University of Texas Press, Austin, Texas.

Nazarea, V. (1998) *Cultural Memory and Biodiversity.* University of Arizona Press, Tucson, Arizona.

Waters-Bayer, A. (1994) The ethics of documenting rural people's knowledge: investigating milk marketing among Fulani women in Nigeria. In: Scoones, I. and Thompson, J. (eds) *Beyond Farmer First: Rural People's Knowledge, Agricultural Research, and Extension Practice.* Intermediate Technology Publications, London, pp. 144–150.

Index

Numbers in *italic* denote figures, those in **bold** denote tables.